Mathematical Programming
State of the Art 1994

Edited by

John R. Birge and Katta G. Murty
University of Michigan

Mathematical Programming: State of the Art 1994
John R. Birge and Katta G. Murty, Editors

ISBN 0-9642433-0-X

Typeset in Universe and Bodoni by the editors. Printed and bound by
Braun-Brumfield, Inc. in the United States of America.

Contents

Foreword

Mathematical programming has grown dramatically since its inception following World War II and the first symposium in 1949. Access to mathematical programming has spread from a handful of people to virtually anyone with a personal computer. The 15th International Symposium on Mathematical Programming demonstrates this success through a wide range of research presentations, student-oriented sessions and tutorial surveys.

This book presents sixteen of the tutorial lectures given at the meeting. The topics reflect the field's diversity with contributions in varying areas of application, theoretical advances, and computational studies. These articles convey the vitality of mathematical programming. They provide an assessment of the remarkable progress of research in the subject, and indicate some of the problems that will be the focus of future research as we approach the beginning of the next century.

Many people made the symposium and this collection possible. In particular, we are indebted to Samer Takriti for his careful typesetting and formatting of the papers and to Ruby Sowards for her excellent organization of the papers and database. We also thank Penny Tully for her overall supervision of the symposium, Tongnyoul Yi for his assistance in the typesetting, Elizabeth Titus for her production assistance, Carolyn Barritt for her cover and layout design, and Aaron King for his logo design. We further extend our gratitude to all the symposium organizers, topic coordinators, and committee members.

John R. Birge, Katta G. Murty
Ann Arbor, Michigan.

Balanced Matrices

Michele Conforti
Dipartimento di Matematica
Pura ed Applicata
Università di Padova
Padova, Italy

Gérard Cornuéjols
Ajai Kapoor
Kristina Vušković
Carnegie Mellon University
Pittsburgh, Pennsylvania

Mendu R. Rao
Indian Institute of Management
Bangalore, India

1 Introduction

A $0, 1$ matrix is *balanced* if it does not contain a square submatrix of odd order with two ones per row and per column. This notion was introduced by Berge [4]. A $0, \pm 1$ matrix A is *balanced* if, in every submatrix with two nonzero entries per row and per column, the sum of the entries is a multiple of four. This definition is due to Truemper [63]. The class of balanced $0, \pm 1$ matrices includes balanced $0, 1$ matrices and totally unimodular $0, \pm 1$ matrices. (A matrix is *totally unimodular* if every square submatrix has determinant equal to $0, \pm 1$. The fact that total unimodularity implies balancedness follows, for example, from Camion's theorem [13] which states that a $0, \pm 1$ matrix is totally unimodular if and only if, in every square submatrix with an *even number* of nonzero entries per row and per column, the sum of the entries is a multiple of four.)

In Section 2, we characterize balanced $0, \pm 1$ matrices in terms of "bicoloring". This extends the notion of graph bipartition to $0, \pm 1$ matrices. We then discuss integral polytopes associated with "generalized" set packing, partitioning and covering

1

problems. These results extend the integrality of set packing, partitioning and covering polytopes when the constraint matrix is balanced. We then discuss classes of $0, \pm 1$ matrices with related polyhedral properties, such as perfect and ideal $0, \pm 1$ matrices. Finally we introduce the connection with propositional logic and nonlinear $0, 1$ optimization.

In Section 3, we show how to sign a $0, 1$ matrix into a $0, \pm 1$ balanced matrix (when such a signing exists, the matrix is called *balanceable*). It follows that, in order to understand the structure of balanced $0, \pm 1$ matrices, it is equivalent to study $0, 1$ matrices that are balanceable. We then introduce a decomposition theorem for these matrices. The decomposition theorem can be used to obtain a polynomial algorithm to test membership in the class of balanced $0, \pm 1$ matrices. This is discussed in Section 4. Section 5 surveys special classes of balanced matrices while Section 6 states a coloring theorem for graphs whose clique-node matrix is balanced. Finally, in Section 7, we propose some conjectures and indicate some directions for further research.

2 Bicoloring, Logic and Integer Polyhedra

2.1 Bicoloring

Berge [4] introduced the following notion. A $0, 1$ matrix is *bicolorable* if its columns can be partitioned into blue and red columns in such a way that every row with two or more $1's$ contains a 1 in a blue column and a 1 in a red column. This notion provides the following characterization of balanced $0, 1$ matrices.

Theorem 2.1 (Berge [4]) *A $0, 1$ matrix A is balanced if and only if every submatrix of A is bicolorable.*

A $0, 1$ matrix A can be represented by a hypergraph (the columns of A represent nodes and the rows represent edges). Then the definition of balancedness for $0, 1$ matrices is a natural extension of the property of not containing odd cycles for graphs, and the notion of bicoloring is a natural extension of bipartition in graphs. Berge's theorem can be viewed as an extension to hypergraphs of the fact that a graph is bipartite if and only if it contains no odd cycle. In fact, this is the motivation that led Berge to introduce the notion of balancedness. Several results on bipartite graphs generalize to balanced hypergraphs, such as König's bipartite matching theorem, as stated in the next theorem. In a hypergraph, a *matching* is a set of pairwise non-intersecting edges and a *transversal* is a node set intersecting all the edges.

Theorem 2.2 (Berge, Las Vergnas [9]) *In a balanced hypergraph, the maximum cardinality of a matching equals the minimum cardinality of a transversal.*

The next result generalizes a theorem of Gupta [43] on bipartite multigraphs.

Theorem 2.3 (Berge [6]) *In a balanced hypergraph, the minimum number of nodes in an edge equals the maximum cardinality of a family of disjoint transversals.*

Ghouila-Houri [40] introduced the notion of *equitable bicoloring* for a $0, \pm1$ matrix A as follows. The columns of A are partitioned into blue columns and red columns in such a way that, for every row of A, the sum of the entries in the blue columns differs from the sum of the entries in the red columns by at most one.

Theorem 2.4 (Ghouila-Houri [40]) *A $0, \pm1$ matrix A is totally unimodular if and only if every submatrix of A has an equitable bicoloring.*

A $0, \pm1$ matrix A is *bicolorable* if its columns can be partitioned into blue columns and red columns in such a way that every row with two or more nonzero entries either contains two entries of opposite sign in columns of the same color, or contains two entries of the same sign in columns of different colors. For a $0, 1$ matrix, this definition coincides with Berge's notion of bicoloring. Clearly, if a $0, \pm1$ matrix has an equitable bicoloring as defined by Ghouila-Houri, then it is bicolorable.

Theorem 2.5 (Heller, Tompkins [45]) *Let A be a $0, \pm1$ matrix with at most two nonzero entries per row. A is totally unimodular if and only if A is bicolorable.*

A consequence of Camion's theorem is that a $0, \pm1$ matrix with at most two nonzero entries per row is balanced if and only if it is totally unimodular. So Theorem 2.5 shows that a $0, \pm1$ matrix with at most two nonzero entries per row is balanced if and only if it is bicolorable. The following theorem extends Theorem 2.1 to $0, \pm1$ matrices and Theorem 2.5 to matrices with more than two nonzero entries per row.

Theorem 2.6 (Conforti, Cornuéjols [20]) *A $0, \pm1$ matrix A is balanced if and only if every submatrix of A is bicolorable.*

Cameron and Edmonds [12] observed that the following simple algorithm finds a valid bicoloring of a balanced matrix. They described their algorithm for $0, 1$ matrices, but it also works for $0, \pm1$ matrices.

Algorithm

Input: A $0, \pm1$ matrix A.
Output: A bicoloring of A or a proof that the matrix A is not balanced.
    Stop if all columns are colored or if some row is improperly colored. Otherwise, color a new column red or blue as follows.
    If no row of A forces the color of a column, arbitrarily color one of the uncolored columns.
    If some row of A forces the color of a column, color this column accordingly.
    When the algorithm fails to find a bicoloring, the sequence of forcings that resulted in an improperly colored row identifies a submatrix with two nonzeros per row and column which violates balancedness. However, this algorithm cannot be used as a recognition of balancedness for the following reason: When the matrix A is not balanced, the algorithm may still find a bicoloring if one exists.

2.2 Integrality of Packing, Partitioning and Covering Polytopes

A polytope is *integral* if all its extreme points have only integer-valued components. Given a $0,1$ matrix A, the *set packing polytope* is

$$P(A) = \{x :\ Ax \leq 1,\ 0 \leq x \leq 1\}.$$

The integrality of the set packing polytope is related to the notion of perfect graph. A graph G is *perfect* if, for every node induced subgraph H of G, the chromatic number of H equals the size of its largest clique. The fundamental connection between the theory of perfect graphs and integer programming was established by Fulkerson [37], Lovász [53] and Chvátal [17]. The *clique-node matrix* C_G of a graph G is a $0,1$ matrix whose columns are indexed by the nodes of G and whose rows are the incidence vectors of the maximal cliques of G.

Theorem 2.7 (Lovász [53], Fulkerson [37], Chvátal [17]) *A graph G is perfect if and only if the set packing polytope $P(C_G)$ is integral.*

A $0,1$ matrix is *perfect* if $P(A)$ is integral. It follows from Theorem 2.7 that a $0,1$ matrix is perfect if and only if its rows of maximal support form the clique-node matrix of a perfect graph. Berge [5] showed that every balanced $0,1$ matrix is perfect. In fact, the next theorem characterizes a balanced $0,1$ matrix A in terms of the set packing polytope $P(A)$ as well as the set covering polytope $Q(A)$ and the set partitioning polytope $R(A)$:

$$Q(A) = \{x :\ Ax \geq 1,\ 0 \leq x \leq 1\},$$
$$R(A) = \{x :\ Ax = 1,\ 0 \leq x \leq 1\}.$$

Theorem 2.8 (Berge [5], Fulkerson, Hoffman, Oppenheim [38]) *Let M be a $0,1$ matrix. Then the following statements are equivalent:*

 (i) *M is balanced.*

 (ii) *For each submatrix A of M, the set covering polytope $Q(A)$ is integral.*

 (iii) *For each submatrix A of M, the set packing polytope $P(A)$ is integral.*

 (iv) *For each submatrix A of M, the set partitioning polytope $R(A)$ is integral.*

Conforti and Cornuéjols [20] generalize this result to $0,\pm1$ matrices. Given a $0,\pm1$ matrix A, let $n(A)$ denote the column vector whose i^{th} component is the number of -1's in the ith row of matrix A.

Theorem 2.9 (Conforti, Cornuéjols [20]) *Let M be a $0,\pm1$ matrix. Then the following statements are equivalent:*

 (i) *M is balanced.*

 (ii) *For each submatrix A of M, the generalized set covering polytope $\{x :\ Ax \geq 1 - n(A),\ 0 \leq x \leq 1\}$ is integral.*

(iii) *For each submatrix A of M, the generalized set packing polytope* $\{x : Ax \leq 1 - n(A), 0 \leq x \leq 1\}$ *is integral.*

(iv) *For each submatrix A of M, the generalized set partitioning polytope* $\{x : Ax = 1 - n(A), 0 \leq x \leq 1\}$ *is integral.*

A system of linear constraints is *totally dual integral* (TDI) if, for each integral objective function vector c, the dual linear program has an integral optimal solution (if an optimal solution exists). Edmonds and Giles [35] proved that, if a linear system $Ax \leq b$ is TDI and b is integral, then $\{x : Ax \leq b\}$ is an integral polyhedron.

Theorem 2.10 (Fulkerson, Hoffman, Oppenheim [38]) *Let* $A = \begin{pmatrix} A_1 \\ A_2 \\ A_3 \end{pmatrix}$ *be a balanced $0, 1$ matrix. Then the linear system*

$$\left\{ \begin{array}{rcl} A_1 x & \geq & 1 \\ A_2 x & \leq & 1 \\ A_3 x & = & 1 \\ x & \geq & 0 \end{array} \right.$$

is TDI.

So Theorem 2.10 and the Edmonds-Giles theorem imply Theorem 2.8. Note that the total dual integrality of the set packing problem when the constraint matrix is a balanced $0, 1$ matrix also follows from the perfect graph theorem of Lovász [53].

Theorem 2.11 (Conforti, Cornuéjols [20]) *Let* $A = \begin{pmatrix} A_1 \\ A_2 \\ A_3 \end{pmatrix}$ *be a balanced $0, \pm 1$ matrix. Then the linear system*

$$\left\{ \begin{array}{rcl} A_1 x & \geq & 1 - n(A_1) \\ A_2 x & \leq & 1 - n(A_2) \\ A_3 x & = & 1 - n(A_3) \\ 0 & \leq & x \leq 1 \end{array} \right.$$

is TDI.

It may be worth noting that this theorem does not hold when the upper bound $x \leq 1$ is dropped from the linear system, see [20]. In fact, the resulting polyhedron may not be integral. For comparison, we state a result that follows from the Hoffman-Kruskal theorem [46].

Theorem 2.12 (Hoffman, Kruskal [46]) *Let* $A = \begin{pmatrix} A_1 \\ A_2 \\ A_3 \end{pmatrix}$ *be a totally unimodular matrix and* $b = \begin{pmatrix} b_1 \\ b_2 \\ b_3 \end{pmatrix}$ *an integral vector of appropriate dimensions. Then the linear* *system*

$$
\begin{cases}
A_1 x & \geq & b_1 \\
A_2 x & \leq & b_2 \\
A_3 x & = & b_3 \\
x & \geq & 0
\end{cases}
$$

is TDI.

2.3 Related Classes of $0, \pm 1$ Matrices

In this section, we first introduce a family of integral polytopes obtained by spanning the spectrum from totally unimodular to balanced $0, \pm 1$ matrices. In the second part of the section we consider two natural extensions of the concept of balancedness, namely perfection and idealness.

The matrix A is *minimally non-totally unimodular* if it is not totally unimodular, but every proper submatrix has that property.

Theorem 2.13 (Camion [15] and Gomory (cited in [15])) *Let A be a $0, \pm 1$ minimally non-totally unimodular matrix. Then A is square, $det(A) = \pm 2$ and A^{-1} has only $\pm \frac{1}{2}$ entries. Furthermore, each row and column of A has an even number of nonzeros.*

Let $\mathcal{H}$ be the class of minimally non-totally unimodular matrices. Recent results of Truemper [65] (see also [66]), give a simple construction and several characterizations of all matrices in $\mathcal{H}$. For a $0, \pm 1$ matrix A, denote by $t(A)$ the column vector whose i^{th} component is the number of nonzeroes in row i. Finally, let $\mathcal{J}$ be the family of matrices that can be obtained from the identity matrix by changing some $+1's$ into $-1's$.

Theorem 2.14 (Conforti, Cornuéjols, Truemper [24]) *The following two statements are equivalent for a $0, \pm 1$ matrix A and a nonnegative integral vector c.*

(i) *A does not contain a submatrix $A' \in \mathcal{H}$ such that $t(A') \leq 2c'$, where c' is the subvector of c corresponding to the rows of A'.*

(ii) *The polytope $\{(x, s) : Bx + Js = b - n(B), 0 \leq x \leq 1, s \geq 0\}$ is integral for all column submatrices B of A, all $J \in \mathcal{J}$ and all integral vectors b such that $0 \leq b \leq c$.*

We make the following remarks about this theorem.

- When $2c \geq t(A)$, Theorem 2.14 gives a characterization of totally unimodular matrices which can be deduced from the Hoffman-Kruskal theorem (Theorem 2.12).

- It is easy to see that A is balanced if and only if A does not contain a submatrix $A' \in \mathcal{H}$ with $t(A') \leq 2$. So, when $c = 1$ in Theorem 2.14, we get a variation of Theorem 2.9.

- When A is a $0, 1$ matrix, Theorem 2.14 reduces to a result of Truemper and Chandrasekaran [67].

Now we consider two extensions of the concept of balanced $0, \pm 1$ matrix. A $0, \pm 1$ matrix A is *ideal* if the generalized set covering polytope $Q(A) = \{x : Ax \geq 1 - n(A), \ 0 \leq x \leq 1\}$ is integral. A generalized set covering inequality $ax \geq 1 - n(a)$ is *dominated* by $bx \geq 1 - n(b)$, if $\{k : b_k = 1\} \subseteq \{k : a_k = 1\}$ and $\{k : b_k = -1\} \subseteq \{k : a_k = -1\}$. A *prime implication* of $Q(A)$ is a generalized set covering inequality $ax \geq 1 - n(a)$ which is satisfied by all the $0, 1$ vectors in $Q(A)$ but is not dominated by any other generalized set covering inequality valid for $Q(A)$. A *row monotonization* of A is any $0, 1$ matrix obtained from a row submatrix of A by multiplying some of its columns by -1. A row monotonization of A is *maximal* if it is not a proper submatrix of any row monotonization of A. Little is known about ideal $0, \pm 1$ matrices but ideal $0, 1$ matrices have been studied [51], [52], [59], [56], [30].

Theorem 2.15 (Hooker [49]) *Let A be a $0, \pm 1$ matrix such that the generalized set covering polytope $Q(A)$ contains all of its prime implications. Then A is ideal if and only if all the maximal row monotonizations of A are ideal $0, 1$ matrices.*

A $0, \pm 1$ matrix A is *perfect* if the generalized set packing polytope $P(A) = \{x : Ax \leq 1 - n(A), \ 0 \leq x \leq 1\}$ is integral. For $0, 1$ matrices, the concept of perfection is well studied (through Theorem 2.7 and the extensive literature on perfect graphs), but very little is known about perfect $0, \pm 1$ matrices. Therefore it seems natural to relate the notion of perfection for $0, \pm 1$ matrices to that for $0, 1$ matrices.

We say that a polytope Q contained in the unit hypercube $[0, 1]^n$ is *irreducible* if, for each j, both polytopes $Q \cap \{x_j = 0\}$ and $Q \cap \{x_j = 1\}$ are nonempty. A generalized set packing inequality $ax \leq 1 - n(a)$ is *dominated* by $bx \leq 1 - n(b)$, if $\{k : a_k = 1\} \subseteq \{k : b_k = 1\}$ and $\{k : a_k = -1\} \subseteq \{k : b_k = -1\}$. Given a $0, \pm 1$ matrix A, the *completion* of A is the matrix A^* obtained by adding to A all row vectors a that induce a generalized set packing inequality $ax \leq 1 - n(a)$ which is valid for $P(A)$ and not dominated by any other inequality in A^*. A $0, 1$ matrix B obtained from A^* by multiplying through some columns by -1 and replacing all negative entries of the resulting matrix by 0 is called a *monotone completion* of A.

Theorem 2.16 (Conforti, Cornuéjols, De Francesco [21]) *Let A be a $0, \pm 1$ matrix such that the generalized set packing polytope $P(A)$ is irreducible. Then A is perfect if and only if all the monotone completions of A are perfect $0, 1$ matrices.*

2.4 Propositional Logic

In propositional logic, *atomic propositions* $x_1, \ldots, x_j, \ldots, x_n$ can be either *true* or *false*. A *truth assignment* is an assignment of "true" or "false" to every atomic proposition. A *literal* is an atomic proposition x_j or its negation $\neg x_j$. A *clause* is a disjunction of literals and is *satisfied* by a given truth assignment if at least one of its literals is true.

A survey of the connections between propositional logic and integer programming can be found in [48]. The following formulation appears in Dantzig [34]. A truth assignment satisfies the set S of clauses

$$\bigvee_{j \in P_i} x_j \vee (\bigvee_{j \in N_i} \neg x_j) \text{ for all } i \in S$$

if and only if the corresponding $0, 1$ vector satisfies the system of inequalities

$$\sum_{j \in P_i} x_j - \sum_{j \in N_i} x_j \geq 1 - |N_i| \text{ for all } i \in S.$$

The above system of inequalities is of the form

$$Ax \geq 1 - n(A). \tag{1}$$

We consider three classical problems in logic. Given a set S of clauses, the *satisfiability problem* (SAT) consists of finding a truth assignment that satisfies all the clauses in S or showing that none exists. Equivalently, SAT consists of finding a $0, 1$ solution x to (1) or showing that none exists.

Given a set S of clauses and a weight vector w whose components are indexed by the clauses in S, the *weighted maximum satisfiability problem* (MAXSAT) consists of finding a truth assignment that maximizes the total weight of the satisfied clauses. MAXSAT can be formulated as the integer program

$$\begin{aligned} \min \quad & \sum_{i=1}^{m} w_i s_i \\ & Ax + s \geq 1 - n(A) \\ & x \in \{0,1\}^n, s \in \{0,1\}^m. \end{aligned}$$

Given a set S of clauses (the premises) and a clause C (the conclusion), *logical inference* in propositional logic consists of deciding whether every truth assignment that satisfies all the clauses in S also satisfies the conclusion C.

To the clause C, using transformation (1), we associate an inequality

$$cx \geq 1 - n(c),$$

where c is a $0, \pm 1$ vector. Therefore C cannot be deduced from S if and only if the integer program

$$\min \{cx : Ax \geq 1 - n(A), \ x \in \{0,1\}^n\} \tag{2}$$

has a solution with value $-n(c)$.

These three problems are NP-hard in general but SAT and logical inference can be solved efficiently for Horn clauses, clauses with at most two literals and several related classes [11], [16], [64]. MAXSAT remains NP-hard for Horn clauses with at most two literals [39]. A set S of clauses is *balanced* if the corresponding $0, \pm 1$ matrix A defined in (1) is balanced. Similarly, a set of clauses is *ideal* if A is ideal. If S is ideal, SAT, MAXSAT and logical inference can be solved by linear programming. The following theorem is an immediate consequence of Theorem 2.9.

Theorem 2.17 *Let S be a balanced set of clauses. Then SAT, MAXSAT and logical inference can be solved in polynomial time by linear programming.*

This has consequences for probabilistic logic as defined by Nilsson [55]. Being able to solve MAXSAT in polynomial time provides a polynomial time separation algorithm for probabilistic logic via the ellipsoid method, as observed by Georgakopoulos, Kavvadias and Papadimitriou [39]. Hence probabilistic logic is solvable in polynomial time for ideal sets of clauses.

Remark 2.18 *Let S be an ideal set of clauses. If every clause of S contains more than one literal then, for every atomic proposition x_j, there exist at least two truth assignments satisfying S, one in which x_j is true and one in which x_j is false.*

Proof   Since the point $x_j = 1/2$, $j = 1, \ldots, n$ belongs to the polytope $Q(A) = \{x : Ax \geq 1 - n(A), \ 0 \leq x \leq 1\}$ and $Q(A)$ is an integral polytope, then the above point can be expressed as a convex combination of $0, 1$ vectors in $Q(A)$. Clearly, for every index j, there exists in the convex combination a $0, 1$ vector with $x_j = 0$ and another with $x_j = 1$.   ∎

A consequence of Remark 2.18 is that, for an ideal set of clauses, SAT can be solved more efficiently than by general linear programming.

Theorem 2.19 (Conforti, Cornuéjols [19]) *Let S be an ideal set of clauses. Then S is satisfiable if and only if a recursive application of the following procedure stops with an empty set of clauses.*
Recursive Step
   If $S = \emptyset$ then S is satisfiable.
   If S contains a clause C with a single literal (unit clause), *set the corresponding atomic proposition x_j so that C is satisfied. Eliminate from S all clauses that become satisfied and remove x_j from all the other clauses. If a clause becomes empty, then S is not satisfiable* (unit resolution).
   If every clause in S contains at least two literals, choose any atomic proposition x_j appearing in a clause of S and add to S an arbitrary clause x_j or $\neg x_j$.

The above algorithm for SAT can also be used to solve the logical inference problem when S is an ideal set of clauses, see [19]. For balanced (or ideal) sets of clauses, it is an open problem to solve MAXSAT in polynomial time by a direct method, without appealing to polynomial time algorithms for general linear programming.

2.5   Nonlinear 0, 1 Optimization
Consider the nonlinear $0, 1$ maximization problem

$$\max_{x \in \{0,1\}^n} \sum_k a_k \prod_{j \in T_k} x_j \prod_{j \in R_k} (1 - x_j),$$

where, w.l.o.g., all ordered pairs (T_k, R_k) are distinct and $T_k \cap R_k = \emptyset$. This is an NP-hard problem. A standard linearization of this problem was proposed by Fortet [36]:

$$\max \sum a_k y_k$$

$$y_k - x_j \leq 0 \qquad \text{for all } k \text{ s.t. } a_k > 0, \text{ for all } j \in T_k$$

$$y_k + x_j \leq 1 \qquad \text{for all } k \text{ s.t. } a_k > 0, \text{ for all } j \in R_k$$

$$y_k - \sum_{j \in T_k} x_j + \sum_{j \in R_k} x_j \geq 1 - |T_k| \qquad \text{for all } k \text{ s.t. } a_k < 0$$

$$y_k, \ x_j \in \{0, 1\} \qquad \text{for all } k \text{ and } j.$$

When the constraint matrix is balanced, this integer program can be solved as a linear program, as a consequence of Theorem 2.11. Therefore, in this case, the non-linear $0, 1$ maximization problem can be solved in polynomial time. The relevance of balancedness in this context was pointed out by Crama [31].

3 Decomposition Theorems

In this section, we state a decomposition theorem for balanced $0, \pm 1$ matrices due to Conforti, Cornuéjols and Rao [23] and Conforti, Cornuéjols, Kapoor and Vušković [22]. Section 3.1 discusses the problem of changing the sign of some of the entries of a $0, 1$ matrix so that the resulting $0, \pm 1$ matrix becomes balanced. In Section 3.2, we present the main theorem. Section 3.3 contains an outline of its proof.

3.1 Signing 0, 1 Matrices to Be Balanced

In this section, we consider the following question: given a $0, 1$ matrix A, is it possible to turn some of the 1's into -1's in order to obtain a balanced $0, \pm 1$ matrix? It turns out, as we will see shortly, that if such a signing exists, it is unique up to multiplication of rows and columns by -1. Furthermore, there is a very simple algorithm to perform the signing. As a consequence, in order to understand the structure of balanced $0, \pm 1$ matrices, it will be sufficient to concentrate on the zero-nonzero pattern.

Given a $0, \pm 1$ matrix A, the *signed bipartite graph representation of* A is a bipartite graph G together with an assignment of weights $+1$ or -1 to the edges of G, defined as follows. The nodes of G correspond to the rows and columns of A, ij is an edge of G if and only if the entry a_{ij} of A is nonzero and the weight of edge ij is the value of a_{ij}. We say that G is balanced if A is. A *hole* in a graph is a chordless cycle of length four or greater. Thus, a signed bipartite graph G is balanced if and only if, for every hole H of G, the sum of the weights of the edges in H is a multiple of four. (Beware that the material presented in this paper is unrelated to the notion of balanced signed graphs introduced in [44], [1] in connection with a problem in attitudinal psychology. There, a signed graph is balanced if every cycle contains an even number of edges with weight -1.)

A bipartite graph G is *balanceable* if there exists a signing of its edges so that the resulting signed graph is balanced. Equivalently, a $0, 1$ matrix is *balanceable* if it is possible to turn some of the 1's into -1's and obtain a balanced $0, \pm 1$ matrix.

Since a cut and a cycle in a graph have even intersection, it follows that if a signed bipartite graph G is balanced, then the signed bipartite graph G' obtained by switching signs on the edges of a cut, is also balanced. For every edge uv of a spanning tree there is a cut containing uv and no other edge of the tree. These cuts are

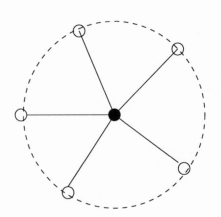 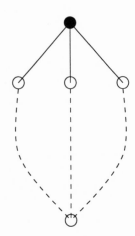

Figure 1: Odd wheel and 3-path configuration

known as *fundamental cuts* and every cut is the symmetric difference of fundamental cuts. Thus, if G is a balanceable bipartite graph, its signing into a balanced signed bipartite graph is unique up to the (arbitrary) signing of a spanning tree of G. This was already observed by Camion [14] in the context of $0, 1$ matrices that can be signed to be totally unimodular. It follows that a bipartite graph G is balanceable if and only if the following signing algorithm produces a balanced signed bipartite graph:

Signing Algorithm

Input: A bipartite graph G.
Output: A signing of G which is balanced if and only if G is balanceable.

Choose a spanning tree of G and sign its edges arbitrarily. Then recursively choose an unsigned edge uv which closes a hole H of G with previously signed edges, and sign uv so that the sum of the weights of the edges in H is a multiple of four.

Note that the recursive step of the signing algorithm can be performed efficiently. Indeed, the unsigned edge uv can be chosen to close the smallest length hole with signed edges. Such a hole H is also a hole in G, else a chord of H in G contradicts the choice of uv.

It follows from this signing algorithm, and the uniqueness of the signing (up to the signing of a spanning tree), that the problem of recognizing whether a bipartite graph is balanceable is equivalent to the problem of recognizing whether a signed bipartite graph is balanced.

Figure 1 shows two classes of bipartite graphs which are important in this study. In all figures, solid lines represent edges and dashed lines represent chordless paths of length at least one. The black and white nodes are on opposite sides of the bipartition.

Let G be a bipartite graph. Let u, v be two nonadjacent nodes in opposite sides

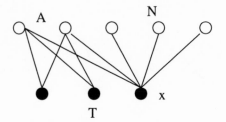

Figure 2: Extended star

of the bipartition. A *3-path configuration connecting u and v* is defined by three chordless paths P_1, P_2, P_3 connecting u and v, having no common intermediate nodes, such that the subgraph induced by the nodes of these three paths contains no edge other than those of the paths. See Figure 1. Since paths P_1, P_2, P_3 of a 3-path configuration are of length one or three modulo four, the sum of the weights of the edges in each path is also one or three modulo four. It follows that two of the three paths induce a hole of weight two modulo four. So a 3-path configuration is not balanceable.

A *wheel* is defined by a hole H and a node $x \notin V(H)$ having at least three neighbors in H, say $x_1, x_2, \ldots, x_n$. If n is even, the wheel is an *even wheel*, otherwise it is an *odd wheel*, see Figure 1. An edge xx_i is a *spoke*. A subpath of H connecting x_i and x_j is called a *sector* if it contains no intermediate node x_l, $1 \le l \le n$. Consider a balanceable wheel. By the signing algorithm, all spokes of the wheel can be assumed to be signed positive. This implies that the sum of the weights of the edges in each sector is two modulo four. Hence the wheel must be an even wheel.

So, bipartite graphs that are balanceable contain neither odd wheels nor 3-path configurations as node induced subgraphs. The following important theorem of Truemper states that the converse is also true.

Theorem 3.1 (Truemper [63]) *A bipartite graph is balanceable if and only if it does not contain an odd wheel or a 3-path configuration as a node induced subgraph.*

3.2 Decomposition Theorem
In this section we give the main decomposition theorem for balanceable bipartite graphs. The theorem states that if a balanceable bipartite graph does not belong to a restricted class, called basic, then it has one of three cutsets.

Cutsets

A set S of nodes (edges) of a connected graph G is a *node (edge) cutset* if the subgraph of G obtained by removing the nodes (edges) in S, is disconnected.

A *biclique* is a complete bipartite graph with at least one node from each side of the bipartition and it is denoted by K_{BD} where B and D are the node sets in each side of the bipartition.

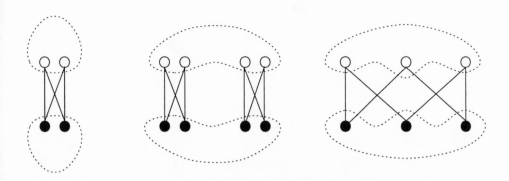

Figure 3: A 1-join, a 2-join and a 6-join

For a node x, let $N(x)$ denote the set of all neighbors of x. In a bipartite graph, an *extended star* is defined by disjoint subsets T, A, N of $V(G)$ and a node $x \in T$ such that

(i) $A \cup N \subseteq N(x)$,

(ii) node set $T \cup A$ induces a biclique (with T on one side of the bipartition and A on the other),

(iii) if $|T| \geq 2$, then $|A| \geq 2$.

This concept was introduced in [23] and is illustrated in Figure 2. An *extended star cutset* is one where $T \cup A \cup N$ is a node cutset. Since the nodes in $T \cup A$ induce a biclique, an extended star cutset with $N = \emptyset$ is called a *biclique cutset*. An extended star cutset having $T = \{x\}$ is called a *star cutset*. Note that a star cutset is a special case of a biclique cutset.

Let K_{BD} be a biclique with the property that its edge set $E(K_{BD})$ is an edge cutset of the connected bipartite graph G and no connected component of $G \setminus E(K_{BD})$ contains both a node of B and a node of D. Let G_B be the union of the components of $G \setminus E(K_{BD})$ containing a node of B. Similarly, let G_D be the union of the components of $G \setminus E(K_{BD})$ containing a node of D. The set $E(K_{BD})$ is a *1-join* if the graphs G_B and G_D each contains at least two nodes. This concept was introduced by Cunningham and Edmonds [33].

Let K_{BD} and K_{EF} be two bicliques of a connected bipartite graph G, where B, D, E, F are disjoint node sets with the property that $E(K_{BD}) \cup E(K_{EF})$ is an edge cutset and every component of $G \setminus E(K_{BD}) \cup E(K_{EF})$ either contains a node of B and a node of E but no node of $D \cup F$, or contains a node of D and a node of F but no node of $B \cup E$. Let G_{BE} be the union of the components of $G \setminus E(K_{BD}) \cup E(K_{EF})$ containing a node of B and a node of E. Similarly, let G_{DF} be the union of the

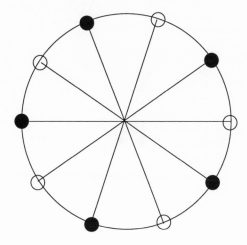

Figure 4: R_{10}

components in $G \setminus E(K_{BD}) \cup E(K_{EF})$ containing a node of D and a node of F. The set $E(K_{BD}) \cup E(K_{EF})$ is a *2-join* if neither of the graphs G_{BE} and G_{DF} is a chordless path with no intermediate nodes in $B \cup D \cup E \cup F$. This concept was introduced by Cornuéjols and Cunningham [29].

In a connected bipartite graph G, let A_i, $i = 1, \ldots, 6$, be disjoint nonempty node sets such that, for each i, every node in A_i is adjacent to every node in $A_{i-1} \cup A_{i+1}$ (indices are taken modulo 6), and these are the only edges in the subgraph A induced by the node set $\cup_{i=1}^{6} A_i$. Assume that $E(A)$ is an edge cutset but that no subset of its edges forms a 1-join or a 2-join. Furthermore assume that no connected component of $G \setminus E(A)$ contains a node in $A_1 \cup A_3 \cup A_5$ and a node in $A_2 \cup A_4 \cup A_6$. Let G_{135} be the union of the components of $G \setminus E(A)$ containing a node in $A_1 \cup A_3 \cup A_5$ and G_{246} be the union of components containing a node in $A_2 \cup A_4 \cup A_6$. The set $E(A)$ constitutes a *6-join* if the graphs G_{135} and G_{246} contain at least four nodes each. This concept was introduced in [22].

Two Basic Classes

A bipartite graph is *restricted balanceable* if its edges can be signed so that the sum of the weights in each *cycle* is a multiple of four. This class of bipartite graphs is well studied in the literature, see [18], [60], [68], [26]. We discuss it in a later section. R_{10} is the bipartite graph on ten nodes defined by the cycle $x_1, \ldots, x_{10}, x_1$ of length ten with chords $x_i x_{i+5}$, $1 \leq i \leq 5$, see Figure 4.

We can now state the decomposition theorem for balanceable bipartite graphs:

Theorem 3.2 *A balanceable bipartite graph that is not restricted balanceable is either R_{10} or contains a 2-join, a 6-join or an extended star cutset.*

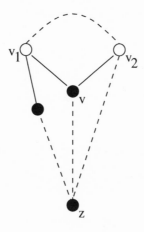

Figure 5: Parachute

3.3 Outline of the Proof

The key idea in the proof of Theorem 3.2 is that if a balanceable bipartite graph G is not basic, then G contains one of several node induced subgraphs, which force a decomposition of G with one of the cutsets described in Section 3.2.

Parachutes

A *parachute* is defined by four chordless paths of positive lengths, $T = v_1, \ldots, v_2$; $P_1 = v_1, \ldots, z$; $P_2 = v_2, \ldots, z$; $M = v, \ldots, z$, where v_1, v_2, v, z are distinct nodes, and two edges vv_1 and vv_2. No other edges exist in the parachute, except the ones mentioned above. Furthermore $|E(P_1)| + |E(P_2)| \geq 3$. See Figure 5.

Note that if G is balanceable then nodes v, z belong to the same side of the bipartition, else the parachute contains a 3-path configuration connecting v and z or an odd wheel (H, v) with three spokes.

Connected Squares and Goggles

Connected squares are defined by four chordless paths $P_1 = a, \ldots, b$; $P_2 = c, \ldots, d$; $P_3 = e, \ldots, f$; $P_4 = g, \ldots, h$, wh ere nodes a and c are adjacent to both e and g and b and d are adjacent to both f and h, as in Figure 6. No other adjacency exists in the connected squares. Note that nodes a and b belong to the same side of the bipartition, else the connected squares contain a 3-path configuration connecting a and b or, if $|E(P_1)| = 1$, an odd wheel with center a. Therefore the nodes a, b, c, d are in one side of the bipartition and e, f, g, h are in the other.

Goggles are defined by a cycle $C = h, P, x, a, Q, t, R, b, u, S, h$, with two chords ua and xb, and chordless paths P, Q, R, S of length greater that one, and a chordless path $T = h, \ldots, t$ of length at least one, such that no intermediate node of T belongs to C. No other edge exists, connecting nodes of the goggles, see Figure 6.

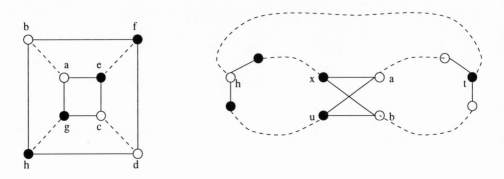

Figure 6: Connected squares and goggles

Connected 6-Holes

A *triad* consists of three internally node-disjoint paths $t, \ldots, u$; $t, \ldots, v$ and $t, \ldots, w$, where t, u, v, w are distinct nodes and u, v, w belong to the same side of the bipartition. Furthermore, the graph induced by the nodes of the triad contains no other edges than those of the three paths. Nodes u, v and w are called the *attachments* of the triad.

A *fan* consists of a chordless path $x, \ldots, y$ together with a node z adjacent to at least one node of the path, where x, y and z are distinct nodes all belonging to the same side of the bipartition. Nodes x, y and z are called the *attachments* of the fan.

A *connected 6-hole* Σ is a graph induced by two disjoint node sets $T(\Sigma)$ and $B(\Sigma)$ such that each induces either a triad or a fan, the attachments of $T(\Sigma)$ and $B(\Sigma)$ induce a 6-hole and there are no other adjacencies between the nodes of $T(\Sigma)$ and $B(\Sigma)$. Figure 7 depicts the four types of connected 6-holes.

The following theorem proved by Conforti, Cornuéjols and Rao [23] concerns the class of balanceable bipartite graphs that do not contain a connected 6-hole or R_{10} as a node induced subgraph.

Theorem 3.3 *A balanceable bipartite graph not containing R_{10} or a connected 6-hole as a node induced subgraph either is restricted balanceable or contains a 2-join or an extended star cutset.*

We now discuss the proof of this theorem. A bipartite graph is *strongly balanceable* if its edges can be signed so that each cycle whose sum of weights is congruent to $2 \bmod 4$ has at least two chords. It follows from the definition that every restricted balanceable bipartite graph is strongly balanceable. Conforti and Rao [26] prove the following:

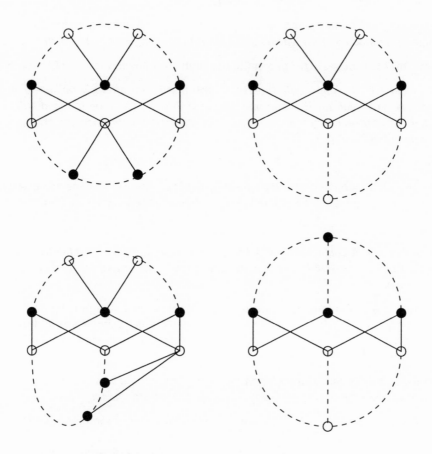

Figure 7: The four types of connected 6-holes

Theorem 3.4 *A strongly balanceable bipartite graph either is restricted balanceable or contains a 1-join.*

Let K_{BD} and K_{EF} be two bicliques on disjoint node sets such that the node set $B \cup E$ induces another biclique K_{BE}. If K_{BE} is a biclique articulation of G and the removal of the edges $E(K_{BD}) \cup E(K_{EF})$ disconnects G, then $E(K_{BD}) \cup E(K_{EF})$ is a *strong 2-join*. Part II of [23] proves the following:

Theorem 3.5 *A balanceable bipartite graph that is not strongly balanceable either contains a parachute or a wheel as a node induced subgraph or has a strong 2-join.*

Part III of [23] disposes of the parachutes through the following variant of Theorem 3.5:

Theorem 3.6 *A balanceable bipartite graph that is not strongly balanceable, that contains no wheel, no R_{10} and no connected 6-hole as a node induced subgraph, either contains an extended star cutset or contains connected squares or goggles as a node induced subgraph.*

Part IV contains a decomposition result for connected squares:

Theorem 3.7 *A balanceable bipartite graph that contains connected squares but no wheel as a node induced subgraph, has a biclique articulation or a 2-join.*

Part V decomposes goggles:

Theorem 3.8 *A balanceable bipartite graph that contains goggles but no wheel, no R_{10} and no connected 6-hole as a node induced subgraph, has an extended star cutset or a 2-join.*

So the theorems contained in Parts II-V give a decomposition theorem for balanceable bipartite graphs that do not contain R_{10}, connected 6-holes or wheels as node induced subgraphs. Part VI gives a decomposition when wheels are present as node induced subgraphs:

Theorem 3.9 *A balanceable bipartite graph containing a wheel but no connected 6-hole as a node induced subgraph, has an extended star cutset.*

Since a graph that has a 1-join has a biclique articulation, Theorems 3.4 and 3.6-3.9 prove Theorem 3.3.

So it remains to find a decomposition of balanceable bipartite graphs that contain R_{10} or connected 6-holes as node induced subgraphs. This is accomplished by Conforti, Cornuéjols, Kapoor and Vušković in [22].

Theorem 3.10 *A balanceable bipartite graph containing R_{10} as a proper node induced subgraph has a biclique articulation.*

Theorem 3.11 *A balanceable bipartite graph that contains a connected 6-hole as a node induced subgraph, has an extended star cutset or a 6-join.*

Now the proof of Theorem 3.2 follows from Theorems 3.3, 3.10 and 3.11.

4 Recognition Algorithm

Conforti, Cornuéjols, Kapoor and Vušković [22] give a polynomial time algorithm to check whether a $0, \pm1$ matrix is balanced. It generalizes the algorithm of Conforti, Cornuéjols and Rao [23] to check whether a $0, 1$ matrix is balanced. Note that, together with the signing algorithm described in Section 3.1, the algorithm to check whether a $0, \pm1$ matrix is balanced tests whether a $0, 1$ matrix is balanceable. We describe the recognition algorithm using the bipartite representation introduced in Section 3.

4.1 Balancedness Preserving Decompositions

Let G be a connected signed bipartite graph. The removal of a node or edge cutset disconnects G into two or more connected components. From these components we construct *blocks* by adding some new nodes and signed edges. We say that a decomposition is *balancedness preserving* when it has the following property: all the blocks are balanced if and only if G itself is balanced. The central idea in the algorithm is to decompose G using balancedness preserving decompositions into a polynomial number of basic blocks that can be checked for balancedness in polynomial time.

For the 2-join and 6-join, the blocks can be defined so that the decompositions are balancedness preserving. For the extended star cutset this is not immediately possible.

2-Join Decomposition

Let $E(K_{BD}) \cup E(K_{EF})$ be a 2-join of G and let G_{BE} and G_{DF} be the graphs defined in Section 3.2. We construct the block G_1 from G_{BE} as follows:

- Add two nodes d and f, connected respectively to all nodes in B and to all nodes in E.

- Let P be a shortest path in G_{DF} connecting a node in D to a node in F. If the weight of P is 0 or 2 mod 4, nodes d and f are connected by a path of length 2 in G_1. If the weight of P is 0 mod 4, one edge of Q is signed $+1$ and the other -1, and if the weight of P is 2 mod 4, both edges of Q are signed $+1$. Similarly if the weight of P is 1 or 3 mod 4, nodes d and f are connected by a path of length 3 with edges signed so that Q and P have the same weight modulo 4. Let d' and f' be the endnodes of P in D and F respectively. Sign the edges between node d and the nodes in B exactly the same as the corresponding edges between d' and the nodes of B in G. Similarly, sign the edges between f and E exactly the same as the corresponding edges between f' and the nodes in E.

The block G_2 is defined similarly.

Theorem 4.1 *Let G be a signed bipartite graph with a 2-join $E(K_{BD}) \cup E(K_{EF})$ where K_{BD} and K_{EF} are balanced and neither $D \cup F$ nor $B \cup E$ induces a biclique. Then G is balanced if and only if both blocks G_1 and G_2 are balanced.*

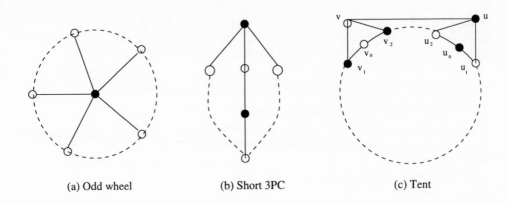

(a) Odd wheel   (b) Short 3PC   (c) Tent

Figure 8: Odd wheel, short 3-path configuration and tent

6-Join Decomposition

Let G be a signed bipartite graph and let $A_1, \ldots, A_6$ be disjoint nonempty node sets such that the edges of the graph A induced by $\cup_{i=1}^6 A_i$ form a 6-join. Let G_{135} and G_{246} be the graphs defined in Section 3.2. We construct the block G_1 from G_{135} as follows:

- Add a node a_2 adjacent to all the nodes in A_1 and A_3, a node a_4 adjacent to all the nodes in A_3 and A_5 and a node a_6 adjacent to all the nodes in A_5 and A_1.

- Pick any three nodes $a_2' \in A_2$, $a_4' \in A_4$ and $a_6' \in A_6$ and, in G_1, sign the edges incident with a_2, a_4 and a_6 according to the signs of the corresponding edges of G incident with a_2', a_4' and a_6'.

The block G_2 is defined similarly.

Theorem 4.2 *Let G be a signed bipartite graph with a 6-join $E(A)$ such that A is balanced. Then G is balanced if and only if both blocks G_1 and G_2 are balanced.*

4.2 Extended Star Cutset Decomposition

Consider the following way of defining the blocks for the extended star decomposition of a connected signed bipartite graph G. Let S be an extended star cutset of G and $G_1', \ldots, G_k'$ the connected components of $G \setminus S$. Define the blocks to be $G_1, \ldots, G_k$ where G_i is the subgraph of G induced by $V(G_i') \cup S$ with all edges keeping the same weight as in G.

The extended star decomposition defined in this way is not balancedness preserving. Consider, for example, a signed odd wheel (H, x) where H is an *unquad hole* (a hole of weight congruent to 2 *mod* 4). If we decompose (H, x) by the extended star cutset $\{x\} \cup N(x)$, then it is possible that all of the blocks are balanced, whereas (H, x) itself is not since H is an unquad hole. Two other classes of bipartite graphs that can present a similar problem when decomposing with an extended star cutset are tents and short 3-path configurations, see Figure 8. A *tent*, denoted by $\tau(H, u, v)$, is a bipartite graph induced by a hole H and two adjacent nodes $u, v \notin V(H)$ each having two neighbors on H, say u_1, u_2 and v_1, v_2 respectively, with the property that u_1, u_2, v_2, v_1 appear in this order on H. A *short 3-path configuration* is a 3-path configuration in which one of the paths contains three edges.

To overcome the fact that our extended star decomposition is not balancedness preserving, we proceed in the following way. We transform the input graph G into a graph G' that contains a polynomial number of connected components, each of which is a node induced subgraph of G, and which has the property that if G is not balanced, then G' contains an unquad hole that will either never be broken by any of the decompositions we use, or else be detected while performing the decomposition. We call this process a *cleaning procedure*. To do this, we have to study the structure of signed bipartite graphs that are not balanced, in particular the structure of a smallest (in the number of edges) unquad hole. For such a hole we prove the following theorem.

Theorem 4.3 *In a non-balanced signed bipartite graph, a smallest unquad hole H^* contains two edges x_1x_2 and y_1y_2 such that:*

- *The set $N(x_1) \cup N(x_2) \cup N(y_1) \cup N(y_2)$ contains all nodes with an odd number (greater than 1) of neighbors in H.*

- *For every tent $\tau(H^*, u, v)$, u or v is contained in $N(x_1) \cup N(x_2) \cup N(y_1) \cup N(y_2)$.*

Let x_0, x_1, x_2, x_3 and y_0, y_1, y_2, y_3 be subpaths of H^*. The above theorem shows that if we remove from G the nodes $N(x_1) \cup N(x_2) \cup N(y_1) \cup N(y_2) \setminus \{x_0, x_1, x_2, x_3, y_0, y_2, y_3, y_4\}$, then H^* will be *clean* (i.e. it will not be contained in any odd wheel or tent). If H^* is contained in a short 3-path configuration, this can be detected during the decomposition (before it is broken). It turns out that, by this process, all the problems are eliminated. So the cleaning procedure consists of enumerating all possible pairs of chordless paths of length 3, and in each case, generating the subgraph of G as described above. The number of subgraphs thus generated is polynomial and, if G is not balanced, then at least one of these subgraphs contains a clean unquad hole.

4.3 Algorithm Outline
The recognition algorithm takes a signed bipartite graph as input and recognizes whether or not it is balanced. The algorithm consists of four phases:

- **Preprocessing:** The cleaning procedure is applied to the input graph.

- **Extended Stars:** Extended star decompositions are performed, until no block contains an extended star cutset.

- **6-joins:** 6-join decompositions are performed until no block contains a 6-join.

- **2-joins:** Finally, 2-join decompositions are performed until no block contains a 2-join.

The 2-join and 6-join decompositions cannot create any new extended star cutset, except in one case which can be dealt with easily. Also a 2-join decomposition does not create any new 6-joins. So, when the algorithm terminates, none of the blocks have an extended star cutset, a 2-join or a 6-join. By the decomposition theorem (Theorem 3.2), if the original signed bipartite graph is balanced, the blocks must be copies of R_{10} or restricted balanced (i.e. the weight of every cycle is a multiple of four). R_{10} is a graph with only ten nodes and so it can be checked in constant time. Checking whether a signed bipartite graph is restricted balanced can be done using the following algorithm of Conforti and Rao [26]:

Construct a spanning forest in the bipartite graph and check if there exists a cycle of weight 2 mod 4 which is either fundamental or is the symmetric difference of fundamental cycles. If no such cycle exists, the signed bipartite graph is restricted balanced.

A different algorithm for this recognition problem, due to Yannakakis [68], has linear time complexity and will be mentioned in Section 5.1.

The preprocessing phase and the decomposition phases using 2-joins and 6-joins are easily shown to be polynomial. For the extended star decomposition phase, it is shown that each bipartite graph which is decomposed has a path of length three which is not present in any of the blocks. This bounds the number of such decompositions by a polynomial in the size of the graph. Thus the entire algorithm is polynomial. See [22] for details.

4.4 Two Related Recognition Problems

The algorithm presented in the previous section recognizes in a polynomial time whether a signed bipartite graph contains an unquad hole. Interestingly Kapoor [50] has shown that it is NP-complete to recognize whether a signed bipartite graph contains an unquad hole going through a prespecified node.

Theorem 4.4 (Kapoor [50]) *Given a bipartite graph G and a node v of G, it is NP-complete to check if G has an unquad hole which contains node v.*

One can also ask the following question: given a signed bipartite graph, does it contain a *quad* hole (i.e. a hole of weight 0 *mod* 4)? A linear algorithm for this recognition problem is given by Conforti, Cornuéjols and Vušković [25].

A signed bipartite graph is *unbalanced* if it does not contain a quad hole. Bipartite graphs which can be signed to be unbalanced are called *unbalanceable*. If a bipartite graph is unbalanceable, there is a simple algorithm to perform the signing (similar to the signing algorithm of Section 3.1). The class of unbalanced signed bipartite graphs is structurally much simpler than the class of balanced signed bipartite graphs, one of the reasons being the following property: in a signed bipartite graph G, all holes of G are unquad if and only if all cycles of G are unquad. The recognition algorithm in [25] is based on the following decomposition theorem.

Theorem 4.5 (Conforti, Cornuéjols, Vušković [25]) *An unbalanceable bipartite graph is either a hole or it contains a one node or a two node cutset.*

5   Classes of Balanceable Matrices

In this section we survey decomposition theorems for several classes of balanceable matrices. We relate these decompositions to Theorem 3.2.

5.1   Seymour's Decomposition of Totally Unimodular Matrices

Seymour [58] gave an important decomposition theorem for $0,1$ matrices that can be signed to be totally unimodular. The decompositions involved in his theorem are 1-separations, 2-separations and 3-separations. A $0,1$ matrix B has a k-*separation* if its rows and columns can be partitioned so that

$$B = \left(\begin{array}{cc} A^1 & D^2 \\ D^1 & A^2 \end{array} \right)$$

where $r(D^1) + r(D^2) = k - 1$ and the number of rows plus number of columns of A^i is at least k, for $i = 1, 2$. ($r(M)$ denotes the $GF(2)$-rank of the $0,1$ matrix M).

The basic matrices used in Seymour's decomposition theorem are

$$R_{10} = \left(\begin{array}{ccccc} 1 & 1 & 0 & 0 & 1 \\ 1 & 1 & 1 & 0 & 0 \\ 0 & 1 & 1 & 1 & 0 \\ 0 & 0 & 1 & 1 & 1 \\ 1 & 0 & 0 & 1 & 1 \end{array} \right), \quad R'_{10} = \left(\begin{array}{ccccc} 1 & 1 & 1 & 1 & 1 \\ 1 & 1 & 1 & 0 & 0 \\ 1 & 0 & 1 & 1 & 0 \\ 1 & 0 & 0 & 1 & 1 \\ 1 & 1 & 0 & 0 & 1 \end{array} \right),$$

graphic matrices and their transpose. A $0,1$ matrix M is *graphic* if there exists a tree T such that the rows of M are indexed by the edges of T and the columns of M are incidence vectors of paths of T. The transpose of a graphic matrix is said to be *cographic*.

Theorem 5.1 (Seymour [58]) *A $0,1$ matrix that can be signed to be totally unimodular is either R_{10}, R'_{10}, graphic, cographic, or it contains a 1-, 2- or 3-separation.*

For a 1-separation $r(D^1) + r(D^2) = 0$. Thus both D^1 and D^2 are matrices all of whose entries are 0. The bipartite graph corresponding to the matrix B is disconnected.

For the 2-separation $r(D^1) + r(D^2) = 1$, thus w.l.o.g. D^2 has rank zero and is identically zero. Since $r(D^1) = 1$, after permutation of rows and columns, $D^1 = \left(\begin{array}{cc} 0 & 1 \\ 0 & 0 \end{array} \right)$, where 1 denotes a matrix all of whose entries are 1 and 0 is a matrix all of whose entries are 0. The 2-separation in the bipartite graph representation of B corresponds to a 1-join.

For the 3-separation $r(D^1) + r(D^2) = 2$. If both D^1 and D^2 have rank 1 then, after permutation of rows and columns,

$$D^1 = \left(\begin{array}{cc} 0 & 1 \\ 0 & 0 \end{array} \right), \quad D^2 = \left(\begin{array}{cc} 0 & 0 \\ 1 & 0 \end{array} \right).$$

This 3-separation in the bipartite graph representation of B corresponds to a 2-join. Now w.l.o.g. we assume $r(D^1) = 2$ and $r(D^2) = 0$. Up to permutation of rows and columns D^1 is of the form

$$D^1 = \begin{pmatrix} P & N \\ M & Q \end{pmatrix}$$

where N is a 2×2 nonsingular matrix (over $GF(2)$). Again, up to permutation of rows and columns, there are exactly two possible cases for N:

$$\begin{pmatrix} 0 & 1 \\ 1 & 0 \end{pmatrix}, \quad \begin{pmatrix} 1 & 0 \\ 1 & 1 \end{pmatrix}.$$

We first examine the structure of D^1 when N is of the first kind. Given N, P and Q, the matrix M is completely determined by the formula $M = QN^{-1}P$, because $r(D^1) = 2$. So, the bipartite graph representation of D^1 has node sets C_1, C_2 and C_3 corresponding to columns of $(P \; N)$ of the type $\begin{pmatrix} 1 \\ 1 \end{pmatrix}$, $\begin{pmatrix} 1 \\ 0 \end{pmatrix}$ and $\begin{pmatrix} 0 \\ 1 \end{pmatrix}$ respectively, and the node sets R_1, R_2 and R_3 corresponding to rows of $\begin{pmatrix} N \\ Q \end{pmatrix}$ of the type $(1\ 1)$, $(1\ 0)$ and $(0\ 1)$ respectively. Either or both node sets C_1 and R_1 may be empty. When all the node sets are nonempty the 3-separation is a 6-join. When one of C_1 or R_1 is empty it is called a *4-join* and when both are empty it is a 2-join.

When N is of the second type, node sets C_2 and R_3 may be empty. When neither is empty, we get a 6-join in the bipartite graph representation. When one is empty, we get a 4-join and when both are empty, a 3-join. Note that, when a bipartite graph contains a 1-join, a 3-join or a 4-join, it also contains an extended star cutset. So the only 1-, 2-, and 3-separations which do not induce an extended star cutset are the 2-join and the 6-join. By noting that R'_{10} contains an extended star cutset, Seymour's theorem 5.1 implies a result resembling Theorem 3.2:

Corollary 5.2 *A $0, 1$ matrix that can be signed to be totally unimodular is either R_{10}, graphic, cographic, or its bipartite representation contains an extended star cutset, a 2-join or a 6-join.*

5.2 More Decomposition Theorems

A signed bipartite graph is *strongly balanced* if every cycle of weight $2 \bmod 4$ has at least two chords. Strongly balanced $0, \pm 1$ matrices are defined accordingly. It follows from the definition that restricted balanced $0, \pm 1$ matrices are strongly balanced, and it can be shown that strongly balanced $0, \pm 1$ matrices are totally unimodular, see [26]. Strongly balanceable $0, 1$ matrices can be signed to be strongly balanced with the signing algorithm described in Section 3.1. Conforti and Rao [26] have shown that a strongly balanceable $0, 1$ matrix that is not restricted balanceable has a 2-separation (the bipartite graph representation has a 1-join).

Theorem 5.3 (Conforti, Rao [26])*A strongly balanceable bipartite graph either is restricted balanceable or contains a 1-join.*

Crama, Hammer and Ibaraki [32] say that a $0, \pm 1$ matrix A is *strongly unimodular* if every basis of (A, I) can be put in triangular form by permutation of rows and columns.

Theorem 5.4 (Crama, Hammer, Ibaraki [32])*A $0, \pm 1$ matrix is strongly unimodular if and only if it is strongly balanced.*

Yannakakis [68] has shown that a restricted balanceable $0, 1$ matrix having both a row and a column with more than two nonzero entries has a very special 3-separation: the bipartite graph representation has a 2-join consisting of two single edges. A bipartite graph is *2-bipartite* if all the nodes in one side of the bipartition have degree at most 2.

Theorem 5.5 (Yannakakis [68]) *A restricted balanceable bipartite graph either is 2-bipartite or contains a cutnode or contains a 2-join consisting of two edges.*

Based on this theorem, Yannakakis designed a linear time algorithm for checking whether a $0, \pm 1$ matrix is restricted balanced. A different algorithm for this recognition problem was stated in an earlier section of this survey.

A bipartite graph is *linear* if it does not contain a cycle of length 4. Note that an extended star cutset in a linear bipartite graph is always a star cutset, due to Condition (iii) in the definition of extended star cutsets. Conforti and Rao [27] proved the following theorem for linear balanced bipartite graphs:

Theorem 5.6 (Conforti, Rao [27])*A linear balanced bipartite graph either is restricted balanced or contains a star cutset.*

5.3 Totally Balanced 0, 1 Matrices

A bipartite graph is *totally balanced* if every hole has length 4. Totally balanced bipartite graphs arise in location theory and were the first balanced graphs to be the object of an extensive study. Several authors (Golumbic and Goss [42], Anstee and Farber [2] and Hoffman, Kolen and Sakarovitch [47] among others) have given properties of these graphs.

An edge uv is *bisimplicial* if either u or v has degree 1 or the node set $N(u) \cup N(v)$ induces a biclique. Note that if uv is a bisimplicial edge and nodes u and v have degree at least 2, then G has a strong 2-join formed by the edges adjacent to exactly one node in the set $\{u, v\}$. The 2-join is strong since $N(u) \cup N(v) \setminus \{u, v\}$ induces a biclique. The following theorem of Golumbic and Goss [42] characterizes totally balanced bipartite graphs.

Theorem 5.7 (Golumbic, Goss, [42])*A totally balanced bipartite graph has a bisimplicial edge.*

Since wheels and parachutes contain holes of length greater than 4, neither of these two graphs can occur as a node induced subgraph of a totally balanced bipartite graph. Furthermore if uv is a bisimplicial edge such that the degree of both u and v is greater that 2, then nodes u, v together with their neighbors induce a strong 2-join. So the above theorem is related to Theorem 3.5.

A $0, 1$ matrix A is in *standard greedy form* if it contains no 2×2 submatrix of the form $\begin{pmatrix} 1 & 1 \\ 1 & 0 \end{pmatrix}$, where the order of the rows and columns in the submatrix is the same as in the matrix A. This name comes from the fact that the linear program

$$\max \sum y_i$$
$$yA \leq c \qquad (3)$$
$$0 \leq y \leq p$$

can be solved by a greedy algorithm. Namely, given $y_1, \ldots, y_{k-1}$ such that $\sum_{i=1}^{k-1} a_{ij} y_i \leq c_j, j = 1, \ldots, n$ and $0 \leq y_i \leq p_i$, $i = 1, \ldots, k-1$, set y_k to the largest value such that $\sum_{i=1}^{k} a_{ij} y_i \leq c_j$, $j = 1, \ldots, n$ and $0 \leq y_k \leq p_k$. The resulting greedy solution is an optimum solution to this linear program. What does this have to do with totally balanced matrices? The answer is in the next theorem.

Theorem 5.8 (Hoffman, Kolen, Sakarovitch [47]) *A $0, 1$ matrix is totally balanced if and only if its rows and columns can be permuted in standard greedy form.*

This transformation can be performed in time $O(nm^2)$ [47].

Totally balanced $0, 1$ matrices come up in various ways in the context of facility location problems on trees. For example, the *covering problem*

$$\min \sum_{1}^{n} c_j x_j + \sum_{1}^{m} p_i z_i$$
$$\sum_{j} a_{ij} x_j + z_i \geq 1, \quad i = 1, \ldots, m \qquad (4)$$
$$x_j, z_i \in \{0, 1\}$$

can be interpreted as follows: c_j is the set up cost of establishing a facility at site j, p_i is the penalty if client i is not served by any facility, and $a_{ij} = 1$ if a facility at site j can serve client i, 0 otherwise.

When the underlying network is a tree and the facilities and clients are located at nodes of the tree, it is customary to assume that a facility at site j can serve all the clients in a *neighborhood subtree* of j, namely, all the clients within distance r_j from node j.

An *intersection matrix* of the set $\{S_1, \ldots, S_m\}$ versus $\{R_1, \ldots, R_n\}$, where S_i, $i = 1, \ldots, m$, and R_j, $j = 1, \ldots, n$, are subsets of a given set, is defined to be the $m \times n$ $0, 1$ matrix $A = (a_{ij})$ where $a_{ij} = 1$ if and only if $S_i \cap R_j \neq \emptyset$.

Theorem 5.9 (Giles [41]) *The intersection matrix of neighborhood subtrees versus nodes of a tree is totally balanced.*

It follows that the above location problem on trees (4) can be solved as a linear program (by Theorem 2.8 and the fact that totally balanced matrices are balanced). In fact, by using the standard greedy form of the neighborhood subtrees versus nodes matrix, and by noting that (4) is the dual of (3), the greedy solution described earlier for (3) can be used, in conjunction with complementary slackness, to obtain an elegant solution of the covering problem. The above theorem of Giles has been generalized as follows.

Theorem 5.10 (Tamir [61]) *The intersection matrix of neighborhood subtrees versus neighborhood subtrees of a tree is totally balanced.*

Other classes of totally balanced $0, 1$ matrices arising from location problems on trees can be found in [62].

6 A Coloring Theorem on Graphs

Let G be a graph and q a positive integer no greater than its chromatic number $\chi(G)$. A *partial q-coloring* of G is a family of q pairwise disjoint stable sets, say $S_1, \ldots, S_q$. If $x \in S_i$, node x is said to have color i. Not all nodes of G need have a color. The partial q-coloring is *optimal* if the number of colored nodes is as large as possible.

A family of cliques $C_1, \ldots, C_r$ is said to be *associate* of a partial q-coloring $S_1, \ldots,$ S_q if these cliques are pairwise disjoint, each clique C_j intersects each S_i, and every node of G is either colored or belongs to one of the cliques (or both).

Theorem 6.1 (Berge [8]) *Let G be a graph and $q \leq \chi(G)$ a positive integer. If the clique-node matrix of G is balanced, then every optimal partial q-coloring has an associate family of cliques.*

In the special case where $q = 1$, an optimal partial q-coloring is a maximum stable set. So the existence of an associate family also follows from Lovász's perfect graph theorem [53]. Indeed, let $\alpha(G)$ denote the stability number of G. When G is perfect, a minimum cardinality clique partition of G, say $C_1, \ldots, C_{\alpha(G)}$, is an associate family of cliques for any stable set S_1 of cardinality $\alpha(G)$.

7 Some Conjectures and Open Questions
7.1 Eliminating Edges
The following conjecture extends a conjecture of Conforti and Rao [27] to $0, \pm 1$ matrices.

Conjecture 7.1 *In a balanced signed bipartite graph G, either every edge belongs to some R_{10}, or some edge can be removed from G so that the resulting signed bipartite graph is still balanced.*

The condition on R_{10} is necessary since removing any edge from R_{10} yields a wheel with three spokes or a 3-path configuration as a node induced subgraph.

The truth of the above conjecture would imply that given a $0, \pm 1$ balanced matrix we can sequentially turn the nonzero entries to zero in some specific order until every nonzero belongs to some R_{10} matrix, while maintaining balanced $0, \pm 1$ matrices in the intermediate steps.

For $0, 1$ matrices, the above conjecture reduces to the following:

Conjecture 7.2 (Conforti, Rao [27]) *Every balanced bipartite graph contains an edge which is not the unique chord of a cycle.*

It follows from the definition that restricted balanced signed bipartite graphs are exactly the ones such that the removal of *any* subset of edges leaves a restricted balanced signed bipartite graph.

Conjecture 7.1 holds for signed bipartite graphs that are strongly balanced since, by definition, the removal of any edge leaves a chord in every unquad cycle.

Theorem 5.7 shows that the graph obtained by eliminating a bisimplicial edge in a totally balanced bipartite graph is totally balanced. Hence Conjecture 7.2 holds for totally balanced bipartite graphs.

7.2   Strengthening the Decomposition Theorems

The extended star decomposition is not balancedness preserving. This heavily affects the running time of the recognition algorithm for balancedness. Therefore it is important to find strengthenings of Theorem 3.2 so that only operations that preserve balancedness are used. We have been unable to obtain these results even for linear balanced bipartite graphs [28].

Let H be a hole where nodes $u_1, \ldots, u_p, v_1, \ldots, v_q, w_1, \ldots, w_p, x_1, \ldots, x_q$ appear in this order when traversing H, but are not necessarily adjacent. Let $Y = \{y_1, \ldots, y_p\}$ and $Z = \{z_1, \ldots, z_q\}$ be two node sets having empty intersection with $V(H)$ and inducing a biclique K_{YZ}. Node y_i is connected with u_i and w_i for $1 \le i \le p$. Node z_i is connected with v_i and x_i for $1 \le i \le q$. Such a graph, denoted with W_{pq} is balanceable and contains no 2-join, no 6-join and no biclique cutset. But this is the only balanceable bipartite graph with this property that we know. This suggests the following conjecture:

Conjecture 7.3 *Every balanceable bipartite graph that is not W_{pq}, R_{10} or restricted balanceable, has a 2-join, a 6-join or a biclique cutset.*

Another direction in which the main theorem might be strengthened is as follows.

Conjecture 7.4 *Every balanceable bipartite graph which is not signable to be totally unimodular has an extended star cutset.*

In [23], it was shown that the bipartite representation of every balanced $0, 1$ matrix which is not totally unimodular, has an extended star cutset.

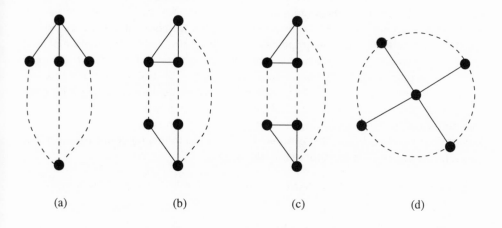

(a)   (b)   (c)   (d)

Figure 9: 3-path configurations and wheel

7.3 Holes in Graphs

α-Balanced Graphs

Let G be a signed graph (not necessarily bipartite) and let α be a vector whose components are in one-to-one correspondence with the chordless cycles of G and take values in $\{0, 1, 2, 3\}$. G is said to be α-balanced if the sum of the weights on each chordless cycle H of G is congruent to $\alpha_H \bmod 4$. In the special case where G is bipartite and $\alpha = 0$, this definition coincides with the notion of balanced signed bipartite graph, introduced earlier in this survey.

A graph is α-balanceable if there is a signing of its edges such that the resulting signed graph is α-balanced. A *3-path configuration* is one of the three graphs represented in Figure 9 (a), (b) or (c). A *wheel* consists of a chordless cycle H and a node $v \notin V(H)$ with at least three neighbors on H, see Figure 9 (d).

Theorem 7.5 (Truemper [63])*A graph G is α-balanceable if and only if*

- $\alpha_H \equiv |H| \bmod 2$ *for every chordless cycle H of G,*

- *every 3-path configuration and wheel of G is α-balanceable.*

Theorem 3.1 is the special case of this theorem where G is bipartite and $\alpha = 0$, while Theorem 4.5 provides an independent proof of Theorem 7.5 in the special case where G is bipartite and $\alpha = 2$. A difficult open problem is to extend the decomposition theorem 3.2 to α-balanceable graphs.

Odd Holes

A long standing open problem in graph theory is that of testing whether a graph contains an odd hole. No polynomial time algorithm is known for this problem. It was shown by Bienstock [10] that it is NP-complete to decide whether a graph has an odd (even respectively) hole containing a given node. One might be lead to believe that testing whether a graph has an odd hole is also NP-complete. However, recall that testing whether a bipartite graph has a hole of length 2 *mod* 4 is polynomial time (Section 4.3) whereas testing whether a bipartite graph has a hole of length 2 *mod* 4 containing a given node is NP-complete (Theorem 4.4). This encourages us to believe that deciding whether a graph has an odd hole can also be done in polynomial time.

A related open question is that of testing whether a graph or its complement has an odd hole. Berge [3] suggested that this question is in fact nothing but the recognition problem for perfect graphs. He made the following conjecture in 1961, when he introduced the concept of perfect graphs.

Conjecture 7.6 (Berge [3])*A graph is perfect if and only if neither it nor its complement has an odd hole.*

This conjecture has been shown to hold for several classes of graphs and, for some of these classes, a polynomial time recognition algorithm is known as well. Such algorithms often rely on a decomposition theorem. So, a general algorithm for the recognition of perfect graphs may well require a decomposition with the flavor of Theorem 3.2. In particular, 2-join decompositions seem relevant for perfect graphs [29].

We know of two important classes of perfect $0, \pm 1$ matrices:

- the matrices obtained from perfect $0, 1$ matrices by switching signs in a subset of columns, and

- balanced $0, \pm 1$ matrices.

It is an open problem to construct all perfect $0, \pm 1$ matrices, starting from these two basic classes.

Even Holes

Another open problem is testing in polynomial time whether a graph contains an even hole. Even holes are related to β-perfect graphs introduced by Markossian, Gasparian and Reed [54]. A graph G is *β-perfect* if, for every node induced subgraph H of G, the chromatic number of H equals max $\{\delta(F)+1 \ : \ F$ is a node induced subgraph of $H\}$, where $\delta(F)$ denotes the smallest node degree in F. No β-perfect graph contains an even hole, but the converse is not true. Also, the complement of a β-perfect graph need not β-perfect.

Theorem 7.7 (Markossian, Gasparian, Reed [54])*A graph G and its complement $\bar{G}$ are both β-perfect if neither G nor $\bar{G}$ contains an even hole.*

Theorem 7.8 (Markossian, Gasparian, Reed [54])*If G contains no even hole and no even cycle with precisely one chord, where this chord forms a triangle with two edges of the cycle, then G is β-perfect.*

A linear time algorithm to determine if either G or its complement $\bar{G}$ contains an even hole follows from their structural characterization of such graphs. It still remains open to refine Theorem 7.8 in order to determine exactly which graphs are β-perfect. Also open is the complexity of deciding if a given graph is β-perfect.

Acknowledgment

We thank Klaus Truemper for his suggestions regarding this survey.

References

[1] J. Akiyama, D. Avis, V. Chvátal, and H. Era, "Balancing signed graphs," *Discrete Applied Mathematics*, **3** (1981), 227-233.

[2] R. Anstee and M. Farber, "Characterizations of totally balanced matrices," *Journal of Algorithms*, **5** (1984), 215-230.

[3] C. Berge, "Färbung von Graphen deren sämtliche bzw. deren ungerade Kreise starr sind (Zusammen-fassung)," *Wissenschaftliche Zeitschrift*, Martin Luther Universität Halle-Wittenberg, Mathematisch-Naturwissenschaftliche Reihe (1961) 114-115.

[4] C. Berge, "Sur certains hypergraphes généralisant les graphes bipartites," *Combinatorial Theory and its Applications I* (P. Erdös, A. Rényi and V. Sós eds.), *Colloq. Math. Soc. János Bolyai*, **4**, North Holland, Amsterdam (1970) 119-133.

[5] C. Berge, "Balanced matrices," *Mathematical Programming*, **2** (1972), 19-31.

[6] C. Berge, "Balanced matrices and the property G," *Mathematical Programming Study*, **12** (1980), 163-175.

[7] C. Berge, "Minimax theorems for normal and balanced hypergraphs. A survey," *Topics on perfect graphs*, C. Berge and V. Chvátal (editors), *Annals of Discrete Mathematics*, **21** (1984), 3-21.

[8] C. Berge, "Minimax relations for the partial q-coloring of a graph," *Discrete Mathematics*, **74** (1989), 3-14.

[9] C. Berge and M. Las Vergnas, "Sur un théorème du type König pour hypergraphes," *International Conference on Combinatorial Mathematics, Annals of the New York Academy of Sciences*, **175** (1970), 32-40.

[10] D. Bienstock, "On the complexity of testing for odd holes and induced odd paths," *Discrete Mathematics*, **90** (1991), 85-92.

[11] E. Boros, Y. Crama, and P.L. Hammer, "Polynomial-time inference of all valid implications for Horn and related formulae," *Annals of Mathematics and Artificial Intelligence*, **1** (1990), 21-32.

[12] K. Cameron and J. Edmonds, "Existentially Polynomial Theorems," *DIMACS Series in Discrete Mathematics and Theoretical Computer Science 1*, American Mathematical Society, Providence, R.I., (1990) 83-100.

[13] P. Camion, "Caractérisation des matrices unimodulaires," *Cahiers du Centre d' Études de Recherche Opérationnelle*, **5** (1963), 181-190.

[14] P. Camion, *Matrices Totalement Unimodulaires et Problèmes Combinatoires*, thesis, Université Libre de Bruxelles, Brussels, 1963.

[15] P. Camion, "Characterization of totally unimodular matrices," *Proceedings of the American Mathematical Society*, **16** (1965), 1068-1073.

[16] V. Chandru and J.N. Hooker, "Extended Horn sets in propositional logic," *Journal of the ACM*, **38** (1991), 205-221.

[17] V. Chvátal, "On certain polytopes associated with graphs," *Journal of Combinatorial Theory B*, **18** (1975), 138-154.

[18] F. G. Commoner, "A sufficient condition for a matrix to be totally unimodular," *Networks*, **3** (1973), 351-365.

[19] M. Conforti and G. Cornuéjols, "A class of logical inference problems solvable by linear programming," *FOCS*, **33** (1992), 670-675.

[20] M. Conforti and G. Cornuéjols, "Balanced 0, ± 1 matrices, bicoloring and total dual integrality," preprint, Carnegie Mellon University, 1992.

[21] M. Conforti, G. Cornuéjols, and C. De Francesco, "Perfect 0, ± 1 matrices," preprint, Carnegie Mellon University, 1993.

[22] M. Conforti, G. Cornuéjols, A. Kapoor, and K. Vušković, "Balanced 0, ± 1 matrices," Parts I-II, preprints, Carnegie Mellon University, 1994.

[23] M. Conforti, G. Cornuéjols, and M. R. Rao, "Decomposition of balanced 0,1 matrices," Parts I-VII, preprints, Carnegie Mellon University, 1991.

[24] M. Conforti, G. Cornuéjols, and K. Truemper, "From totally unimodular to balanced 0, ± 1 matrices: A family of integer polytopes," *Mathematics of Operations Research*, **19** (1994), 21-23.

[25] M. Conforti, G. Cornuéjols, and K. Vušković, "Quad cycles and holes in bipartite graphs," preprint, Carnegie Mellon University, 1994.

[26] M. Conforti and M. R. Rao, "Structural properties and recognition of restricted and strongly unimodular matrices," *Mathematical Programming*, **38** (1987), 17-27.

[27] M. Conforti and M. R. Rao, "Structural properties and decomposition of linear balanced matrices," *Mathematical Programming*, **55** (1992), 129-168.

[28] M. Conforti and M. R. Rao, "Testing balancedness and perfection of linear matrices," *Mathematical Programming*, **61** (1993), 1-18.

[29] G. Cornuéjols and W.H. Cunningham, "Compositions for perfect graphs," *Discrete Mathematics*, **55** (1985), 245-254.

[30] G. Cornuéjols and B. Novick, "Ideal 0, 1 matrices," *Journal of Combinatorial Theory*, **60** (1994), 145-157.

[31] Y. Crama, "Concave extensions for nonlinear 0-1 maximization problems," *Mathematical Programming*, **61** (1993), 53-60.

[32] Y. Crama, P.L. Hammer, and T. Ibaraki, "Strong unimodularity for matrices and hypergraphs," *Discrete Applied Mathematics*, **15** (1986), 221-239.

[33] W. H. Cunningham and J. Edmonds, "A combinatorial decomposition theory," *Canadian Journal of Mathematics*, **22** (1980), 734-765.

[34] G.B. Dantzig, *Linear Programming and Extensions*, Princeton University Press, 1963.

[35] J. Edmonds and R. Giles, "A min-max relation for submodular functions on graphs," *Annals of Discrete Mathematics*, **1** (1977), 185-204.

[36] R. Fortet, "Applications de l'algèbre de Boole en recherche opérationelle," *Revue Française de Recherche Opérationelle*, **4** (1976), 251-259.

[37] D. R. Fulkerson, "Anti-blocking polyhedra," *Journal of Combinatorial Theory B*, **12** (1972), 50-71.

[38] D. R. Fulkerson, A. Hoffman, and R. Oppenheim, "On balanced matrices," *Mathematical Programming Study*, **1** (1974), 120-132.

[39] G. Georgakopoulos, D. Kavvadias, and C. H. Papadimitriou, "Probabilistic satisfiability," *Journal of Complexity*, **4** (1988), 1-11.

[40] A. Ghouila-Houri, "Charactérisations des matrices totalement unimodulaires," *C. R. Acad. Sc. Paris*, **254** (1962), 1192-1193.

[41] R. Giles, "A balanced hypergraph defined by subtrees of a tree," *ARS Combinatorica*, **6** (1978), 179-183.

[42] M. C. Golumbic and C. F. Goss, "Perfect elimination and chordal bipartite graphs," *Journal of Graph Theory*, **2** (1978), 155-163.

[43] R. P. Gupta, "An edge-coloration theorem for bipartite graphs of paths in trees," *Discrete Mathematics*, **23** (1978), 229-233.

[44] F. Harary, "On the measurement of structural balance," *Behavioral Science*, **4** (1959), 316-323.

[45] I. Heller and C.B. Tompkins, "An extension of a theorem of Dantzig's," *Linear Inequalities and Related Systems*, H.W. Kuhn and A.W. Tucker (editors), Princeton University Press, 1956, 247-254.

[46] A.J. Hoffman and J.K. Kruskal, "Integral boundary points of convex polyhedra," *Linear Inequalities and Related Systems*, H.W. Kuhn and A.W. Tucker (editors), Princeton University Press, 1956, 223-246.

[47] A. J. Hoffman, A. Kolen, and M. Sakarovitch, "Characterizations of totally balanced and greedy matrices," *SIAM Journal of Algebraic and Discrete Methods*, **6** (1985), 721-730.

[48] J.N. Hooker, "A quantitative approach to logical inference," *Decision Support Systems*, **4** (1988), 45-69.

[49] J.N. Hooker, "Resolution and the integrality of satisfiability polytopes," preprint, Carnegie Mellon University, 1992.

[50] A. Kapoor, "On the complexity of finding holes in bipartite graphs," preprint, Carnegie Mellon University, 1993.

[51] A. Lehman, "On the width-length inequality," mimeographic notes (1965), *Mathematical Programming*, **17** (1979), 403-417.

[52] A. Lehman, "On the width-length inequality and degenerate projective Planes," *Polyhedral Combinatorics*, W. Cook and P.D. Seymour (editors), *DIMACS Series in Discrete Mathematics and Theoretical Computer Science 1*, American Mathematical Society, Providence, R.I., 1990, 101-105.

[53] L. Lovász, "Normal hypergraphs and the perfect graph conjecture", *Discrete Mathematics*, **2** (1972), 253-267.

[54] S.E. Markossian, G.S. Gasparian, and B.A. Reed, "β-perfect graphs," to appear in *Journal of Combinatorial Theory B*.

[55] N. J. Nilsson, "Probabilistic logic," *Artificial Intelligence*, **28** (1986), 71-87.

[56] M.W. Padberg, "Lehman's forbidden minor characterization of ideal 0, 1 matrices," *Discrete Mathematics*, **111** (1993), 409-420.

[57] A. Schrijver, *Theory of Linear and Integer Programming,*, Wiley, New York, 1986.

[58] P. Seymour, "Decomposition of regular matroids," *Journal of Combinatorial Theory B*, **28** (1980), 305-359.

[59] P. Seymour, "On Lehman's width-length characterization," *Polyhedral Combinatorics*, W. Cook and P. D. Seymour (editors), *DIMACS Series in Discrete Mathematics and Theoretical Computer Science 1*, American Mathematical Society, Providence, R.I., 1990, 107-117.

[60] A. Tamir, "On totally unimodular matrices," *Networks*, **6** (1976), 373-382.

[61] A. Tamir, "A class of balanced matrices arising from location problems," *SIAM Journal on Algebraic and Discrete Methods*, **4** (1983), 363-370.

[62] A. Tamir, "Totally balanced and totally unimodular matrices defined by center location problems," *Discrete Applied Mathematics*, **16** (1987), 245-263.

[63] K. Truemper, "Alpha-balanced graphs and matrices and GF(3)-representability of matroids," *Journal of Combinatorial Theory B*, **32** (1982), 112-139.

[64] K. Truemper, "Polynomial theorem proving I. Central matrices," Technical Report UTDCS 34-90, 1990.

[65] K. Truemper, "A decomposition theory for matroids. VII. Analysis of minimal violation matrices," *Journal of Combinatorial Theory B*, **55** (1992), 302-335.

[66] K. Truemper, *Matroid Decomposition,*, Academic Press, Boston, 1992.

[67] K. Truemper and R. Chandrasekaran, "Local unimodularity of matrix-vector pairs," *Linear Algebra and its Applications*, **22** (1978), 65-78.

[68] M. Yannakakis, "On a class of totally unimodular matrices," *Mathematics of Operations Research* , **10** (1985), 280-304.

Connectivity Augmentation Problems in Network Design

András Frank

Eötvös University

Budapest, Hungary

1 Introduction

In network design it is a fundamental problem to construct graphs or subgraphs of a graph of minimum cost satisfying certain connectivity specifications.

Shortest paths between two specified nodes, or minimum cost spanning trees may be viewed as (well-known) special cases of this problem. Very often a starting graph is already available and the goal is to augment the graph. For example, at least how many new edges must be added to a digraph to make it strongly connected?

Having such a broad class of problems (already a special case, the well-known Steiner-tree problem, has a vast literature), it is of no surprise that a large number of connectivity augmentation problems are NP-complete. But there are interesting special cases, as well, for which polynomial time algorithms are available. Investigations and results of this problem may be categorized into three main directions. One is concerned with heuristics, often based on deep theoretical background such as the polyhedral method, that work well in practice. An excellent survey paper by M. Grötschel, C. L. Monma and M. Stoer [30] summarizes this type of results. Another line is to develop approximation algorithms whose running time is polynomial and the output is provably not much worse than the optimum. In a recent Ph.D. thesis, D. P. Williamson [57] provides a rich class of problems of this type along with approximation algorithms for their solution.

The purpose of the present paper is to survey connectivity augmentation problems for which an algorithm is available to find the exact optimum in (strongly) polynomial time. Though polynomially solvable problems are often too restricted to be used directly in practical applications, they may serve well as building blocks in a more complex procedure. In many cases, such as the k-edge-connectivity augmentation problem on digraphs, the proposed algorithm is purely combinatorial and strongly polynomial.

Such algorithms have the feature (as opposed to the ellipsoid method, say) that they can be run in practice and used for large graphs. In some other cases, such as the k-node-connectivity augmentation problem on digraphs, a theoretical background has been developed which, at least, ensures a polynomial-time algorithm via the ellipsoid method. But such methods might never be used in practice. Their existence should be considered merely as a challenge to design purely combinatorial solution algorithms. Also, these algorithms are important from a theoretical point of view in order to explore the borderline (if there is any) between NP-completeness and polynomial solvability.

We will use two basic measurements for connectivity. Given a directed or undirected graph G, $\lambda(x,y) = \lambda(x,y;G)$ (respectively, $\kappa(x,y) = \kappa(x,y;G)$) denotes the maximum number of edge-disjoint (openly disjoint) paths from x to y. K. Menger's classical theorem asserts that in an undirected graph if x and y are not adjacent, then $\kappa(x,y)$ is equal to the minimum number of nodes whose deletion separates x and y. Other versions of Menger's theorem provide min-max formulae for κ and λ in directed and in undirected graphs. For example, in a digraph $\lambda(x,y)$ is equal to the minimum number of edges leaving an $x\bar{y}$-subset of nodes. (A set X is called an $x\bar{y}$-**set** if $x \in X, y \notin X$.) This is an equivalent formulation of the max-flow min-cut theorem. ($\lambda(x,y)$ may be considered as the maximum flow value from x to y provided that the capacity of each edge is 1.)

An undirected graph G is called k-**edge-connected** if every cut has at least k edges. When $k = 1$ we simply say G is connected. A digraph D is called k-**edge-connected** if every (non-empty, proper) subset of nodes has at least k exiting edges. When $k = 1$ we call D **strongly connected**. By Menger's theorem a digraph or an undirected graph is k-edge-connected if and only if there are k edge-disjoint paths from every node to every other one.

The general form of the augmentation problems we investigate is as follows. Given a directed or undirected graph $G = (V, E)$ and a non-negative integer function $r(x,y)$ on the set of ordered pairs of nodes, serving as a **demand function,** add a minimum number of new edges to G (or, more generally, a minimum cost set of new edges, if a cost-function is given on the set of possible new edges) so that

$$\lambda(x,y;G^+) \geq r(x,y) \tag{1.1a}$$

or

$$\kappa(x,y;G^+) \geq r(x,y) \tag{1.1b}$$

holds for every pair of nodes x, y of the resulting graph (digraph) G^+. Accordingly, we may speak about **edge-connectivity augmentation** problem or **node-connectivity augmentation** problem.

Beside these minimization forms, we will consider degree-constrained augmentation problems, as well, where a lower and upper bound is given at every node v for the number of new edges incident to v.

A natural relaxation of the augmentation problem is the **max flow version**. Suppose that $g(u,v)$ is a non-negative capacity function on the pairs of nodes u, v ($u, v \in V$) and let $r(u,v)$ be a demand function. The problem is to increase the existing capacities so that in the resulting network the maximum flow value between u and v is at least $r(u,v)$ for each pair $\{u,v\}$ of nodes and such that the sum of capacity increments is minimum.

If $g(u,v)$ and $r(u,v)$ are integer-valued and the capacity increments are also re-quired to be integer-valued, then the edge-connectivity augmentation problem is equivalent to the max-flow version. Namely, the latter problem can be formulated as a max-flow version by letting $g(u,v) = 1$ when (u,v) is an edge of G and $g(u,v) = 0$ otherwise. Conversely, if g is integer-valued, we can define a graph having $g(u,v)$ parallel edges between each pair of nodes u and v and then a solution to the edge-connectivity problem yields a solution to the integer-valued max-flow problem.

This equivalence, however, does not mean algorithmic equivalence. We are go-ing to exhibit strongly polynomial time algorithms for the more difficult max-flow augmentation problem. (A polynomial time algorithm is called **strongly polynomial** if it uses only basic operations, such as comparing, adding, subtracting, multiplying, and dividing numbers, and the number of these operations is independent of the numbers occurring in the input.)

We will also investigate the question of when the fractional augmentation allows better solution than the integer-valued one. For example, let V be a set of n nodes, $g \equiv 0$ and $r \equiv 1$. If only integers are allowed for the increments, then the value of the best solution is $n - 1$: take any tree on V and increase the capacity of its edges by one. If we may use fractional increments, then the value of the best solution is $n/2$: take any circuit of n nodes, increase the capacity of its edges by 1/2.

On the other hand, we will see problems (concerning mainly undirected graphs) when the optimum of the integer-valued solution is at most one half bigger than that of the fractional solution, and problems (especially when G is directed) when there is an optimal solution to the max-flow augmentation problem that is integer-valued.

Given two elements s,t and a subset X of a ground-set U, we say that X is an $s\bar{t}$-**set** if $s \in X, t \notin X$. X **separates** s from t (or x and t) if $|X \cap \{s,t\}| = 1$. A family $\{X_1, \ldots, X_t\}$ of pairwise disjoint, non-empty subsets of U is called a **sub-partition**.

Let $G = (U,E)$ an undirected graph. $d_G(X,Y)$ denotes the number of undirected edges between $X - Y$ and $Y - X$. $\bar{d}_G(X,Y) := d_G(X, U - Y)(= d_G(U - X, Y))$. $d_G(X)$ stands for $d_G(X, U - X)$. Observe that $\bar{d}_G(X,Y) = \bar{d}_G(U - X, U - Y)$. When it does not cause ambiguity, we leave out the subscript.

For a directed graph $D = (U,A)$ $\varrho_D(X)$ denotes the number of edges entering X, $\delta_D(X) := \varrho_D(U - X)$ and $\beta_D(X) := \min(\varrho_D(X), \delta_D(X))$. Note that $\beta_D(X) = \beta_D(U - X)$. $d_D(X,Y)$ denotes the number of edges with one end in $X - Y$ and one end in $Y - X$. $\bar{d}_D(X,Y) := d_D(X, U - Y)(= d_D(U - X, Y)$. An **arborescence** F is a directed tree in which every node but one has in-degree 1 and the exceptional node, called the **root** , is of in-degree 0. (Equivalently, there is a directed path from the root to every other node of F.)

Let $M = (U, A \cup E)$ be a mixed graph composed as the union of a directed gaph $D = (U,A)$ and an undirected graph $G = (U,E)$. Let $\varrho_M(X) := \varrho_D(X) + d_G(X)$, $\delta_M(X) := \delta_D(X) + d_G(X)$, and $\beta_M(X) := \min(\varrho_M(X), \delta_M(X))$. We say that a node v of a M is **di-Eulerian** if $\varrho_D(v) = \delta_D(v)$. M is called **di-Eulerian** if every node of M is di-Eulerian.

Splitting off a pair of edges $e = us, f = st$ means that we replace e and f by a new edge ut. The resulting mixed graph will be denoted by M^{ef}. This operation is defined only if both e and f are undirected (respectively, directed) and then the newly added edge ut is considered undirected (directed). Accordingly, we speak of undirected or directed splittings.

For a function $m : V \to R$ we use the notation $m(X) := \sum(m(v) : v \in X)$. For a number x, let $x^+ := \max(x, 0)$.

2   Subgraphs versus Supergraphs

To clarify a simple link between optimal subgraphs and optimal supergraphs, we start with a specific problem. We are given a digraph $D = (V, A)$ with two specified nodes s and t. One of the simplest connectivity property one may consider in D is

$$\lambda(s, t) \geq 1, \tag{2.1}$$

that is, there is a path from s to t. It is well-known that (2.1) holds if and only if every $\bar{st}$-set has an exiting edge.

When (2.1) holds, one may consider the shortest path problem, a starting point of combinatorial optimization, that consists of finding a path from s to t of minimum cost with respect to a given cost function c on E. For non-negative c, this may be considered as a SUBGRAPH problem: given a digraph, find a minimum cost subgraph satisfying (2.1). Dijkstra's classical algorithm for finding a shortest paths is of $O(n^2)$ complexity.

If (2.1) does not hold, then a natural task is to augment optimally D so as to satisfy (2.1). Augmentation may be considered as a SUPERGRAPH problem: given a digraph D and a digraph $H = (V, F)$ of possible new edges which is endowed with a cost function c, in $D + H$ construct a minimum cost supergraph of D satisfying (2.1).

This augmentation problem may be solved by a shortest path computation in the digraph $D + H$ where the cost of the edges of H is determined by c and the cost of the original edges is defined to be 0.

This easy principle may be applied to properties other than (2.1). If one is able to solve the minimum cost subgraph problem, one can solve the corresponding supergraph (that is, the augmentation) problem, as well. Below we list some other connectivity properties when the subgraph problem is efficiently solvable and, therefore, so is the augmentation problem. But we already hasten to emphasize that the focus of this paper will be on polynomially solvable augmentation problems where the corresponding subgraph problem is NP-complete.

Properties in a digraph for which the minimum cost subgraph problem (and the minimum cost augmentation problem, as well) is solvable in strongly polynomial time are:

$$\lambda(s, t) \geq k, \tag{2.2}$$

$$\kappa(s, t) \geq k, \tag{2.3}$$

$$\lambda(s, x) \geq k \text{ for every } x \in V, \text{ and} \qquad (2.4)$$

$$\kappa(s, x) \geq k \text{ for every } x \in V. \qquad (2.5)$$

The minimum cost subgraph problem with respect to (2.2) is equivalent to finding min-cost flow of value k [52]. By an easy elementary construction, observed already in [12], (2.3) goes back to (2.2).

Finding a minimum cost subgraph satisfying (2.4) is trickier. First, we may assume that no edge of D enters s. Since the cost function is supposed to be non-negative, it is enough to consider digraphs satisfying (2.4) which are minimal with respect to edge-deletion. The main observation is that such digraphs are precisely those in which (a) the in-degree of every node $v \neq s$ is precisely k and (b) the underlining undirected graph is the union of k disjoint spanning trees. (The equivalence may be proved by Edmonds' [7] arborescence theorem) By this equivalent formulation the problem is to find a minimum cost common basis of two matroids M_1 and M_2 defined on the edge-set of D. Here M_1 is a partition matroid in which a set is independent if it contains at most k edges entering any node $v \neq s$. M_2 is defined to be the sum of k copies of the circuit matroid of the underlining undirected graph (that is, a subset of edges of D is independent in M_2 if it is the union of k forests).

Since there are strongly polynomial algorithms for the weighted matroid intersection problem [9, 15] the minimum cost subgraph problem with respect to (2.4) is also solvable. By exploiting the particular structure of the two matroids in question, H. Gabow [27] developed a more efficient algorithm.

Note that the special case of (2.4) when $k = 1$ is tantamount to finding a minimum cost arborescence of root s, for which D.R. Fulkerson [25] described a particularly elegant algorithm.

No elementary reduction of Property (2.5) to (2.4) is known. A solution to the subgraph minimization problem with respect to (2.5) was described in [22]. It used a tricky reduction to submodular flows [10], a common generalization of network flows and matroid intersection. Here we do not repeat the reduction but provide a min-max theorem concerning the corresponding augmentation problem, which is deducible from the theory of submodular flows but was not explicitly stated in [22].

Let us call a digraph satisfying (2.5) k-**out-connected (from** s). Let $D = (V, E)$ be a digraph with a specified node s and assume our task is to augment D to obtain a k-out-connected digraph. Let $H = (V, F)$ denote the digraph of possible new edges and $c : F \rightarrow R_+$ a cost function. In order to have a solution at all we assume that the union graph $D + H$ is k-out-connected.

Let $\mathcal{F}$ denote the family of pairs (A, B) of two non-empty disjoint subsets of nodes so that $s \in A$. For a pair (A, B) let $\delta(A, B) := \delta_D(A, B)$ denote the number of edges of D from A to B. By a version of Menger's theorem a digraph is k-out-connected from s if and only if $|V - (A \cup B)| + \delta(A, B) \geq k$. Define the **deficiency** $p_{def}(A, B)$ of a pair (A, B) by $k - (|V - (A \cup B)| + \delta(A, B))$ when this number is positive and by zero otherwise. Clearly, adding a subset X of edges of H to D yields a k-out-connected digraph if and only if there are at least $p(A, B)$ edges in X going from A to B for every pair $(A, B) \in \mathcal{F}$.

Theorem 2.1 *The minimum cost of edges of H whose addition to D results in a digraph which is k-out-connected from s is equal to* $\max(\sum y(A, B) p(A, B))$ *where $y \geq 0$ is such that, for every edge $xy \in F$, $\sum(y(A, B) : x \in A, y \in B) \leq c(xy)$. Moreover, if c is integer-valued, y may be chosen integer-valued.*

Actually, this theorem asserts that a certain linear program is totally dual integral. It turns out that the theorem can be stated in a more abstract form. Let $H = (V, F)$ be a directed graph endowed with a cost function $c : F \to R$ and a capacity function $g : F \to Z_+$. Let p be a non-negative integer-valued function on the pairs (A, B) of disjoint subsets of V. We say that p is **intersecting bi-supermodular** if

$$p(X, Y) + p(X', Y') \leq p(X \cap X', Y \cup Y') + p(X \cup X', Y \cap Y') \qquad (2.6)$$

holds whenever $p(X, Y), p(X', Y') > 0, Y \cap Y' \neq \emptyset$.

For a vector x defined on the edge-set F let $\delta_x(A, B) := \sum(x(uv) : uv \in F, u \in A, v \in B)$.

Theorem 2.2 *Let $g : F \to Z_+$ be an integer-valued capacity function so that $\delta_g(A, B) \geq p(A, B)$ for every pair (A, B). Then the linear program*

$$\min(cx : 0 \leq x \leq g, \delta_x(A, B) \geq p(A, B) \text{ for every disjoint } A, B) \qquad (2.7)$$

is totally dual integral. In particular, (2.7) has an integer-valued optimum and if in addition c is integer-valued, the dual linear program also has an integer-valued optimum.

(This theorem may be proved by using the standard uncrossing technique as was done in [13, 16] for the special special case when $p(A, B)$ may be positive only on complementary pairs (i.e. $A \cup B = V$.)

Note that the role of the two variables of p is not symmetric. It becomes symmetric if (2.6) is required only when $p(X, Y), p(X', Y') > 0, X \cap X', Y \cap Y' \neq \emptyset$. In this case p is called **crossing bi-supermodular.** Theorem 2.2 is no more true for crossing bi-supermodular functions. But in Section 4 we will prove that a min-max theorem does hold when H is a complete directed graph (i.e. each of the possible $n(n-1)$ edges belong to H) and $c \equiv 1$. Such a result will allow us to solve the node-connectivity augmentation problem in directed graph when arbitrary edges may be added.

3 Edge-Connectivity Augmentation of Digraphs

In the previous section we have seen that the minimum cost subgraph problem, and therefore the minimum cost augmentation problem, is tractable for properties (2.1) and (2.2).

The next natural property to be investigated is strong connectivity. The minimum cost subgraph problem reads: find a minimum cost strongly connected spanning subgraph of a given digraph. This is NP-complete even if the cost function is identically 1 since if one is able to find a strongly connected subgraph of minimum cardinality, then one is able to decide if a digraph contains a strongly connected subgraph of cardinality n (the number of nodes) and this latter property is equivalent for a digraph to have a Hamiltonian circuit.

The corresponding augmentation problem asks, given a digraph $D = (V, E)$ and another digraph $H = (V, F)$ (endowed with a non-negative cost-function c), for the minimum cost of edges of H whose addition to D creates a strongly connected digraph.

A similar argument shows that the augmentation problem is also NP-complete, even for $c \equiv 1$, if no restriction is made for H. One interesting restriction is when there is a path in D from v to u for each edge uv of H. (For example, when H arises from D by re-orienting each edge.) For the cardinality case, a theorem of Lucchesi and Younger [43] asserts in the present context that *the minimum cardinality of new edges of H to be added to D to provide a strongly connected digraph is equal to the maximum number of H-independent source-sets of D.* (A proper non-empty subset X of V is called a **source-set** if no edge of D enters X and a family of source-sets is H-**independent** if no edge of H enters more than one of them. For later purposes we define **sink-sets** as the complement of source-sets). The theory of submodular functions (established in [10] extends this theorem to the weighted case. In [14] a strongly polynomial time algorithm was developed to find the minimum in question.

The augmentation problem for strong connectivity was solved by K.P. Eswaran and R.E. Tarjan [11] in the case when any possible new edge is allowed to be added and $c \equiv 1$. In a digraph the sink-sets are closed under taking intersection and union. Hence the minimal sink-sets (with respect to containment) are pairwise disjoint. Let p_1 denote their number. Similarly, the minimal source-sets are pairwise disjoint. Let p_2 denote their number. Since in a strongly connected digraph there are no source-sets and sink-sets, at least $\max(p_1, p_2)$ new edges must be added. The next theorem asserts that this bound is achievable. Note that it is not difficult to calculate p_1 (or p_2) since p_1 is the number of sink-nodes (nodes with no leaving edges) of the digraph arising from D by contracting each strong component into one node.

Theorem 3.1 (K.P. Eswaran and R.E. Tarjan [11]) *Given a directed graph $D = (V, E)$ the minimum number of new edges whose addition to D creates a strongly connected digraph is $\max(p_1, p_2)$.*

The proof of Eswaran and Tarjan is constructive and gives rise to a linear-time algorithm.

As we mentioned before, the minimum cost version of the problem is NP-complete. However, the minimum node-cost augmentation problem is solvable as will be shown in a more general context.

In order to generalize the cardinality case of the strong connectivity augmentation problem, suppose that a subset T of nodes is specified in a digraph $D = (V, E)$ and our purpose is to add a minimum number of new edges so that every element of T be reachable from every other element of T. It is not difficult to see [18, 19] that this problem is NP-complete. We will show, however, (even in a more general context) that this problem is solvable in polynomial time if the new edges are required to have both end-nodes in T.

Let us turn to this general case when we require k-edge-connectivity for the augmented digraph, that is the demand function $r(u,v) \equiv k$. The directed k-edge-connectivity augmentation problem was solved by D.R. Fulkerson and L.S. Shapley [26] when the starting digraph $D = (V, \emptyset)$ has no edges at all, by Y. Kajitani and S. Ueno (1986) when the starting digraph is a directed tree and by Frank [18] for an arbitrary starting digraph. The major idea that led to the solution to the minimization problem came from the recognition that degree-prescribed augmentation problems serve as useful intermediate problems.

Let $D = (V, E)$ be a digraph and $m_o, m_i : V \to Z_+$ two integer-valued functions so that $m_o(V) = m_i(V)$.

Theorem 3.2 *A directed graph $D = (V, E)$ can be made k-edge-connected by adding a set F of new edges satisfying*

$$\varrho_F(v) = m_i(v) \text{ and } \delta_F(v) = m_o(v) \qquad (3.1)$$

for every node $v \in V$ if and only if both

$$\varrho(X) + m_i(X) \geq k \text{ and } \delta(X) + m_o(X) \geq k \qquad (3.2)$$

hold for every $X \subseteq V$.

Note that F may contain parallel edges or even loops. It is an important open problem to find characterizations for the existence of an F without loops and parallel edges. To get rid of the loops is at least easy (see, Corollary 3.6).

A crucial observation is that Theorem 3.2 is nothing but a re-formulation of W. Mader's directed splitting off theorem:

Theorem 3.3 (Mader [46]) *Let $D = (V + s, A)$ be a directed graph for which $\varrho(s) = \delta(s)$ and $(*)$ $\lambda(x, y) \geq k$ for every $x, y \in V$. Then the edges entering and leaving s can be partitioned into $\varrho(s)$ pairs so that splitting off all these pairs leaves a k-edge-connected digraph.*

To derive Theorem 3.2, extend D by a new node s and for each $v \in V$ adjoin $m_i(v)$ (respectively, $m_o(v)$) parallel edges from s to v (from v to s). Now by (3.2) the hypotheses of Theorem 3.3 are satisfied and hence we can split off γ pairs of edges to obtain a k-edge-connected digraph. The resulting set of γ new edges (connecting original nodes) satisfies the requirement.

Mader stated his theorem in the form that there is a pair of edges, entering and leaving s, which is splittable in the sense that their splitting off does not destroy $(*)$. Since $\varrho(s) = \delta(s)$, by repeated applications of this one gets Theorem 3.3. An example (in which $\varrho(s) = 1, \delta(s) = 2$) shows that the existence of a splittable pair is not necessarily true without the assumption $\varrho(s) = \delta(s)$. As a slight generalization of Mader's theorem I can prove that there is a splittable pair if $\varrho(s) \leq \delta(s) < 2\varrho(s)$ but I do not know any application of this result.

In [19, 18] the following characterization was derived for the minimization problem.

Theorem 3.4 *A directed graph* $D = (V, E)$ *can be made k-edge-connected by adding at most* γ *new edges if and only if*

$$\sum_i (k - \varrho(X_i)) \leq \gamma \text{ and } \sum_i (k - \delta(X_i)) \leq \gamma \qquad (3.3)$$

holds for every sub-partition $\{X_1, \ldots, X_t\}$ *of* V.

This was proved with the help of Theorem 3.2. (The proof method gives rise to a polynomial time algorithm which is actually strongly polynomial even in the capacitated case). Here we prove an extension of Theorem 3.2 and, using that, derive an extension of Theorem 3.4.

Since (3.3) is a necessary condition for the fractional augmentation, as well, we can conclude that the integer-valued optimum is the same as the fractional optimum.

Let us turn to the generalization of Theorem 3.2 and 3.4.

Theorem 3.5 *Let* T *be a ground-set,* p *a non-negative, integer-valued function defined on subsets of* T *for which* $p(\emptyset) = p(T) = 0$ *and* $p(X) + p(Y) \leq p(X \cap Y) + p(X \cup Y)$ *holds whenever* $p(X) > 0, p(Y) > 0, X \cap Y \neq \emptyset, T - (X \cup Y) \neq \emptyset$. *Let* m_i, m_o *be two modular non-negative integer-valued functions on* T *for which* $m_i(T) = m_o(T) = \gamma$. *There exists a digraph* $H = (T, F)$ *for which*

$$\varrho_H(X) \geq p(X) \text{ for every } X \subseteq T \qquad (3.4)$$

and

$$\varrho_H(v) = m_i(v) \text{ for every } v \in T \qquad (3.5a)$$

$$\delta_H(v) = m_o(v) \text{ for every } v \in T \qquad (3.5b)$$

if and only if

$$m_i(X) \geq p(X) \text{ for every } X \subseteq T \qquad (3.6a)$$

and

$$m_o(T - X) \geq p(X) \text{ for every } X \subseteq T \qquad (3.6b)$$

Proof The necessity of (3.6) is straightforward. To see the sufficiency let $m := m_i + m_o$ and call a set X **in-tight** (resp., **out-tight)** if (3.6a) (resp., (3.6b)) is satisfied with equality. We need 4 easy lemmas.

Lemma 1 *If X, Y are two disjoint out-tight sets, then $m(T - (X \cup Y)) = 0$.*

Proof We have $m_i(X) \geq p(X) = m_o(T - X) = \gamma - m_o(X)$ and $m_i(Y) \geq p(Y) = m_o(T - Y) = \gamma - m_o(Y)$ from which $m_i(X) + m_i(Y) \geq 2\gamma - m_o(X) - m_o(Y)$. Therefore $m(X) + m(Y) \geq 2\gamma = m(T)$ and hence $m(T - (X \cup Y)) = 0$, as required. ∎

Lemma 2 *If X, Y are two in-tight sets for which $T = X \cup Y$, then $m(X \cap Y) = 0$.*

Proof We have $m_o(T - X) + m_o(T - Y) \geq p(X) + p(Y) = m_i(X) + m_i(Y) = 2\gamma - m_i(T - X) - m_i(T - Y)$. Therefore $m(T - X) + m(T - Y) \geq 2\gamma$ and hence $m(X \cap Y) = 0$, as required. ∎

Lemma 3 *If X is out-tight, Y is in-tight and the supermodular inequality holds for $p(X)$ and $p(Y)$ (for example, if $X \subseteq Y$ or $Y \subseteq X$ or X, Y are crossing), then $m(Y - X) = 0$.*

Proof From $m_o(T - X) = p(X)$ and $m_i(Y) = p(Y)$ we have $m_o(T - X) + m_i(Y) = p(X) + p(Y) \leq p(X \cap Y) + p(X \cup Y) \leq m_o(T - (X \cup Y)) + m_i(X \cap Y) = m_o(T - X) - m_o(Y - X) + m_i(Y) - m(Y - X)$ and hence $0 \leq m(Y - X) \leq 0$, as required. ∎

Lemma 4 *The intersection and the union of two crossing in-tight (respectively, out-tight) sets X, Y are in-tight (resp., out-tight).*

Proof We prove the lemma only when X, Y are in-tight. Then $m_i(X) + m_i(Y) = p(X) + p(Y) \leq p(X \cap Y) + p(X \cup Y) \leq m_i(X \cap Y) + m_i(X \cup Y) = m_i(X) + m_i(Y)$ from which equality holds everywhere and the lemma follows. ∎

To prove the theorem let t be a node for which $m_i(t)$ is positive. If there is no in-tight set containing t, define $Z_i := \emptyset$. By Lemma 2 if there are two in-tight sets containing t, then their union Z is not T, and then, by Lemma 4, Z is in-tight. Therefore the union Z_i of all in-tight sets containing t is also in-tight. If there is no out-tight set in $T - t$, define $Z_o = T$. By Lemma 1 if there are two out-tight sets in $T - t$, then their intersection is non-empty, and then, by Lemma 4, the intersection Z_o of all out-tight sets in $T - t$ is out-tight.

It follows from Lemma 3 that the supermodular inequality does not hold for $p(Z_o)$ and $p(Z_i)$. Therefore $Z_o \cap Z_i = \emptyset$ or $Z_o \cup Z_i = T$. We claim that $m_o(Z_o - Z_i) > 0$. For otherwise, if $Z_o \cup Z_i = T$, then $0 = m_o(Z_o - Z_i) = m_o(T - Z_i) \geq p(Z_i) = m_i(Z_i) \geq 0$. Hence $Z_i = \emptyset$ and $m_o(T) = 0$, a contradiction. If $m_o(Z_o - Z_i) = 0$ and $Z_o \cap Z_i = \emptyset$, then $m_i(Z_o) \geq p(Z_o) = m_o(T - Z_o) = \gamma$ and hence $m_i(T - Z_o) = 0$. But then $Z_o = T, Z_i = \emptyset$ and $m_o(T) = 0$, a contradiction.

Choose an element s in $Z_o - Z_i$ for which $m_o(s) > 0$. Define $p'(X) := p(X) - 1$ if X is a $t\bar{s}$-set, $p(X) > 0$ and define $p'(X) := p(X)$ otherwise. Clearly p' satisfies the hypothesis of the theorem. Let $m'_o(s) = m_o(s) - 1, m'_o(v) = m_o(v)$ if $v \neq s$. Let $m'_i(t) = m_i(t) - 1, m'_i(v) = m_i(v)$ if $v \neq t$. Because there is no in-tight ts-set, (3.6a) holds with respect to p', m'_i. Similarly, there is no out-tight set in $T - \{s, t\}$ and therefore (3.6b) holds with respect to p', m'_o.

By induction, there is a digraph $H' = (T, F')$ satisfying the requirements of the theorem with respect to p', m'_i, m'_o. But then $H = (T, F' + st)$ satisfies the requirements with respect to p, m_i, m_o. ∎

Remark   The proof of the theorem gives rise to a (strongly) polynomial time algorithm to find the desired digraph H provided that the following oracles are available for p. For any pair of nodes x, y and vector $m : V \rightarrow Z_+$, minimize $m(X) - p(X)$ over the sets X (A) containing both x and y, (B) neither x nor y.

Corollary 3.6   *In Theorem 3.5 H may be chosen loopless if and only if (3.6) holds and*

$$m(v) := m_i(v) + m_o(v) \leq \gamma \text{ for every } v \in T. \tag{3.7}$$

Proof   If there is a loopless H satisfying (3.5), then every edge entering v leaves $T - v$ and hence $m_i(v) \leq m_o(T - v)$, that is, $m_i(v) + m_o(v) \leq m_o(T - v) + m_o(v) = \gamma$ and the necessity of (3.7) follows.

To see the sufficiency, let us start with a solution $H = (T, F)$ provided by Theorem 3.5 and assume that H has a minimum number of loops. If this minimum is zero, we are done. Suppose that at a node v there is a loop e in H. If there is an edge $f = xy$ of H with end-nodes different from v then we can replace e and f by xv and vy. The revised digraph clearly satisfies (3.4) and (3.5) and has one less loop than H, a contradiction. Therefore v is one of two end-nodes of each edge of H. But then v violates (3.7). ∎

Theorem 3.7   *Let p be the same as in Theorem 3.5. There exists a digraph $H = (T, F)$ satisfying (3.4) so that H has at most γ edges if and only if*

$$\sum p(X_i) \leq \gamma \tag{3.7a}$$

and

$$\sum p(T - X_i) \leq \gamma \tag{3.7b}$$

holds for every sub-partition $\{X_1, \ldots, X_t\}$ of T.

Proof   The necessity of (3.7) is straightforward. We prove the sufficiency with the following idea. Determine first two functions m_i and m_o satisfying (3.6) and apply then Theorem 3.5. To this end let m_i and m_o be integer-valued function on T satisfying (3.6) (but not necessarily $m_o(T) = m_i(T)$) and assume that m_i and m_o are minimal with respect to this (3.6a) and (3.6b), respectively. (That is, (3.6a), say, is destroyed if we reduce m_i by one on any element v where $m_i(v) > 0$.)

Claim $m_i(T) \leq \gamma$ and $m_o(T) \leq \gamma$.

Proof   By symmetry we may assume that $m_i(T) \geq m_o(T)$. Increase m_o so that $m_o(T) = m_i(T)$ (this way we may loose the minimality of m_o but it does not matter). Since m_i is minimal, every element $v \in T$ for which $m_i(v) > 0$ belongs to an in-tight set. Let $\mathcal{F} := \{X_1, \ldots, X_t\}$ be a family of in-tight sets so that each v with positive $m_i(v)$ belongs to a member of $\mathcal{F}$ and $|\mathcal{F}|$ is minimum. There are no two crossing members of $\mathcal{F}$ since, by Lemma 4, their union is in-tight, contradicting the minimality of $\mathcal{F}$. If $\mathcal{F}$ have two members X, Y for which $T = X \cup Y$, then by Lemma 2, $m(X \cap Y) = 0$. Applying (3.7b) to $\{T - X, T - Y\}$ we have $m_i(T) = m_i(X) + m_i(Y) = p(X) + p(Y) \geq \gamma$. Finally, if $\mathcal{F}$ consists of disjoint subsets, then by (3.7a) we get $m(T) = \sum_j m_i(X_j) = \sum_j p(X_j) \geq \gamma$. Now the theorem directly follows from Theorem 3.5. ∎

Theorem 3.7 implies the following generalization of Theorem 3.4. Let $D = (V, E)$ be a directed graph and T a subset of nodes. We say that D is k-**edge-connected in** T if $\lambda(u, v) \geq k$ for every pair of nodes $u, v \in T$.

Theorem 3.8 *Given a digraph D and a subset T of nodes, it is possible to make D k-edge-connected in T by adding at most γ new edges connecting elements of T if and only if*

$$\sum_i (k - \varrho(X_i)) \leq \gamma \text{ and } \sum_i (k - \delta(X_i)) \leq \gamma \qquad (3.8)$$

holds for every family $\mathcal{F} = \{X_1, \dots, X_t\}$ of subsets V for which $\emptyset \subset X_i \cap T \subset T$ and $\mathcal{F}|T$ is a sub-partition of T.

Proof For every subset X of T define $p(X) := \max((k - \varrho(X \cup Z))^+ : Z \subseteq V - T)$. This p satisfies the hypothesis of Theorem 3.5 and (3.7) transforms to (3.8) and hence Theorem 3.7 implies Theorem 3.8. ∎

More can be said if D is di-Eulerian outside T, that is, $\varrho(v) = \delta(v)$ for every $v \in V - T$.

Corollary 3.9 *Suppose that D is di-Eulerian outside T. It is possible to make D k-edge-connected in T by adding at most γ new edges connecting elements of T if and only if (3.8) holds for every sub-partition $\mathcal{F} = \{X_1, \dots, X_t\}$ of subsets V for which $\emptyset \subset X_i \cap T \subset T$.*

Proof If the conditions of Theorem 3.8 are satisfied, we are done. Suppose indirectly that there is a family $\mathcal{F} = \{X_1, \dots, X_t\}$ for which $\mathcal{F}|T$ is a sub-partition and $\mathcal{F}$ violates (3.8). We may assume that $\sum |X_i|$ is minimum. Since a sub-partition of V satisfies (3.8), there are two members X, Y of $\mathcal{F}$ whose intersection is non-empty. By the hypothesis every node in $X \cap Y$ is di-Eulerian, therefore $\varrho(X) + \varrho(Y) \geq \varrho(X - Y) + \varrho(Y - X)$. Replacing X and Y by $X - Y$ and $Y - X$ we obtain a family $\mathcal{F}'$ which also violates (3.8), contradicting the minimal choice of $\mathcal{F}$. ∎

Since the condition in Corollary 3.9 is necessary even if new edges are allowed to have end-nodes outside T, it also follows that the minimum number of new edges whose addition makes a digraph k-edge-connected in T does not depend on whether we may only add edges with end-nodes in T or arbitrary new edges are allowed, provided that D is di-Eulerian outside T.

The following generalization of Corollary 3.9 is due to [1]. Let $D = (V, E)$ be a digraph and let $T(D) := \{v \in V : \varrho_D(v) \neq \delta_D(v)\}$ be the set of non-di-Eulerian nodes. Let k be a positive integer and $r(x, y)$, $(x, y \in V)$ a non-negative integer-valued demand function satisfying

$$r(x, y) = r(y, x) \leq k \text{ for every } x, y \in V \text{ and} \qquad (3.9a)$$

$$r(x, y) = k \text{ for every } x, y \in T(D). \qquad (3.9b)$$

Let $R(\emptyset) = R(V) = 0$ and for $X \subseteq V$ let $R(X) := \max(r(x, y) : X \text{ separates } x$ and $y)$. Let us define $q_i(X) := R(X) - \varrho_D(X)$, $q_o(X) := R(X) - \delta_D(X)$.

Theorem 3.10 (Bang-Jensen, Frank and Jackson[1]) *Given a digraph* $D = (V, E)$, *positive integers* k, γ, *and a demand function* $r(x, y)$ *satisfying (3.9)*, D *can be extended to* D^+ *by adding* γ *new directed edges so that*

$$\lambda(x, y; D^+) \geq r(x, y) \text{ for every } x, y \in V \qquad (3.10)$$

if and only if both

$$\sum_j q_i(X_j) \leq \gamma \qquad (3.11a)$$

and

$$\sum_j q_o(X_j) \leq \gamma \qquad (3.11b)$$

hold for every sub-partition $\{X_1, \ldots, X_t\}$ *of* V.

Corollary 3.11 *Given an Eulerian digraph* $D = (V, E)$, *and a symmetric demand function* $r(x, y)$, D *can be extended to an Eulerian digraph* D^+ *by adding* γ *new edges so that (3.10) holds if and only if (3.11) is satisfied.*

Our next problem is to find a k-edge-connected augmentation of minimum cardinality if upper and lower bounds are imposed both on the in-degrees and on the out-degrees of the digraph of newly added edges. Let $f_i \leq g_i$ and $f_o \leq g_o$ be four non-negative integer-valued functions on V (infinite values are allowed for g_i and g_o). The following two results appeared in [18, 19].

Theorem 3.12 *Given a directed graph* $D = (V, E)$ *and a positive integer* k, D *can be made* k-edge-connected by adding a set F of precisely γ new edges so that both

$$f_i(v) \leq \varrho_F(v) \leq g_i(v) \qquad (3.12a)$$

and

$$f_o(v) \leq \delta_F(v) \leq g_o(v) \qquad (3.12b)$$

hold for every node v *of* D *if and only if both*

$$k - \varrho(X) \leq g_i(X) \qquad (3.13a)$$

and

$$k - \delta(X) \leq g_o(X) \tag{3.13b}$$

hold for every subset $\emptyset \subset X \subset V$ *and both*

$$\sum_j (k - \varrho(X_j) : j = 1, \ldots, t) + f_i(X_0) \leq \gamma \tag{3.14a}$$

and

$$\sum_j (k - \delta(X_j) : j = 1, \ldots, t) + f_o(X_0) \leq \gamma \tag{3.14b}$$

hold for every partition $\{X_0, X_1, X_2, \ldots, X_t\}$ *of* V *where only* X_0 *may be empty.*

One may be interested in degree-constrained augmentations when there is no requirement on the number of new edges.

Theorem 3.13 *Given a directed graph* $D = (V, E)$ *and a positive integer* k, D *can be made* k-edge-connected by adding a set F of new edges satisfying (3.12) if and only if (3.13) holds and and

$$\sum_j (k - \varrho(X_j) : j = 1, \ldots, t) + f_i(X_0) \leq \alpha \tag{3.15a}$$

and

$$\sum_j (k - \delta(X_j) : j = 1, \ldots, t) + f_o(X_0) \leq \alpha \tag{3.15b}$$

hold for every partition $\{X_0, X_1, X_2, \ldots, X_t\}$ *of* V *where only* X_0 *may be empty and* $\alpha := \min(g_o(V), g_i(V))$.

We close this section by another generalization of Theorem 3.8. Let $D = (V, A)$ be a digraph with two specified non-empty subsets S, T of nodes (which may or may not be disjoint). We say that D is k-**edge-connected from** S **to** T if there are k edge-disjoint paths from every node of S to every node of T. (When $S = T$ we are back at k-edge-connectivity.) We say that a family of subsets of nodes is (S, T)-**independent** if it contains at most one $t\bar{s}$-set for every pair $s \in S, t \in T$.

Theorem 3.14 *A digraph* $D = (V, E)$ *can be made* k-edge-connected from S to T by adding at most γ new edges with tails in S and heads in T if and only if

$$\sum_j (k - \varrho(X_j)) \leq \gamma$$

holds for every choice of (S, T)-*independent family of subsets* $X_j \subseteq V$ *where* $T \cap X_j \neq \emptyset, S - X_j \neq \emptyset$ *for each* X_j.

In all other theorems in this section (except the Lucchesi-Younger theorem) subpartitions played the main role in the characterization in question. In Theorem 3.14 the characterization is more complicated. In fact, its proof goes along a line completely different from the approach applied for proving the previous theorems. The theorem is a consequence of a general result of [20] on crossing bi-supermodular functions, which among others, gives rise to a solution to the node-connectivity augmentation problem of directed graphs. This is the topic of the next section.

4 Node-Connectivity Augmentation of Digraphs

Given a directed graph $D = (V, E)$, how many new edges have to be added to D to make it k-node-connected, in short, k-connected. Recall that a digraph is called k-connected if it remains strongly connected after deleting at most $k - 1$ nodes. That is, k-connectivity is defined only if $k \leq n - 1$. If $k = n - 1$, then in a k-connected digraph xy is an edge for every ordered pair $\{x, y\}$ of nodes. This case is uninteresting so we will assume that $k \leq n - 2$. Also, when $k = 1$, edge-connectivity and node-connectivity coincide (:strong connectivity) so we will assume that $k \geq 2$.

In Section 2 we indicated that a related augmentation problem, when the goal is to reach k-connectivity from a specified node, could be solved [23], including the minimum cost version. The general minimum cost k-connectivity augmentation problem is NP-complete so we concentrate only on the minimum cardinality case. Masuzawa, Hagihara and Tokura (1987) solved it when the starting digraph is an arborescence (a directed tree so that every node is reachable from a source-node). Their result easily extends to branchings:

Theorem 4.1 (Masuzawa et al. 1987) *The minimum number of edges whose addition makes a branching* $D = (V, E)$ *k-connected is* $(\sum (k - \delta(v))^+ : v \in V)$*, that is, the sum of out-deficiencies of the nodes.*

For more general digraphs stronger lower bounds are required. One natural idea is to mimic the approach applied successfully in Theorem 3.4. For a subset X of nodes (with $|V - X| \geq k + 1$) let $I(X)$ (respectively, $O(X)$) denote the set of nodes in $V - X$ from which there is an edge to X (into which there is an edge from X). In a k-connected digraph the cardinality of $I(X)$ and $O(X)$ must be at least k. Therefore, if the digraph is not k-connected, we may call the quantity $Q_I(X) := (k - |I(X)|)^+$ the **in-deficiency** and $Q_O(X) := (k - |O(X)|)^+$ the **out-deficiency** of set X. Clearly, if there is a family of disjoint sets (each having cardinality at most $|V| - 1 - k$)), then the sum of the in-deficiencies and the sum of out-deficiencies are both lower bounds for the necessary number of new edges. Theorem 3.4 asserted that the maximum of the analogous lower bounds in the edge-connectivity augmentation provides the correct minimum for the number of new edges. Unfortunately, this is not the case for node-connectivity even if the starting digraph is $k - 1$ connected. An example in [36] shows that the minimum of the required new edges may be $k - 1$ larger than the maximum sum of out- or in-deficiencies of a sub-partition. (On the other hand, in a recent paper[21] we can derive from the general min-max theorem below that this gap actually can never get bigger than $k - 1$.)

Theorem 2.1 however suggests that, instead of single sets, it might be helpful to consider pairs of disjoint sets. Let us call an ordered pair (A, B) of non-empty disjoint subsets of nodes **one-way** if there is no edge in D from A to B. The deficiency $p_{def}(A, B)$ of a one-way pair is defined by $(k - (|V - (A \cup B)|))^+$. Clearly in a k-connected augmentation of D at least that many edges from A to B must be added to D. Finally, call two pairs $(A_i, B_i)(i = 1, 2)$ **independent** if at least one of $A_1 \cap A_2$ and $B_1 \cap B_2$ is empty.

Theorem 4.2 (Frank and Jordán [20]) *A digraph $D = (V, E)$ can be made k node-connected by adding at most γ new edges if and only if*

$$\sum (p_{def}(X, Y) : (X, Y) \in \mathcal{F}) \le \gamma \qquad (4.1)$$

holds for every family $\mathcal{F}$ of pairwise independent one-way pairs.

Since this is a characterization for a general starting digraph, one may expect that Theorem 4.1, where the starting digraph is a branching, can be derived from it. So far we were not able to do that. The following conjecture, if true, is a generalization of Theorem 4.2.

Conjecture *If D is a simple acyclic digraph, then the minimum number of new edges whose addition makes D k-connected is equal to the maximum of the sum of out-deficiencies and the sum of in-deficiencies of nodes.*

Note that M. Bussieck [3] pointed out that the the corresponding statement for edge-connectivity easily follows from Theorem 3.4.

Actually, Theorem 4.2 is a special case of a more general result. Let V be a ground-set and S, T two (not-necessarily disjoint) subsets of V. Let A denote the set of all directed edges st for which $s \in S, t \in T$.

Let $\mathcal{A}$ denote the set of all ordered pairs (X, Y) with $X \subseteq S, Y \subseteq T$. We call X and Y the **tail** and the **head** of the pair, respectively. A directed edge xy **covers** a pair $(X, Y) \in \mathcal{A}$ if $x \in X, y \in Y$. We say that a sub-family $\mathcal{F}$ of $\mathcal{A}$ is **independent** if every edge of A covers at most one member of $\mathcal{F}$. This is equivalent to requiring that there are no two members $(X_i, Y_i)(i = 1, 2)$ of $\mathcal{F}$ for which $X_1 \cap X_2 \ne \emptyset$ and $Y_1 \cap Y_2 \ne \emptyset$.

Let p be a non-negative, integer-valued function defined on $\mathcal{A}$ for which $p(X, \emptyset) = p(\emptyset, Y) = 0$. We say that p is **crossing bi-supermodular** if

$$p(X, Y) + p(X', Y') \le p(X \cap X', Y \cup Y') + p(X \cup X', Y \cap Y') \qquad (4.2)$$

holds whenever $p(X, Y), p(X', Y') > 0, X \cap X', Y \cap Y' \ne \emptyset$.

For a non-negative function x defined on A, define $\delta_x(A, B) := \sum (x(s, t) : s \in S, t \in T)$. We say that x **covers** p or that x is a **covering** of p if $\delta_x \ge p$. The main result in [20, 21] is:

Theorem 4.3 *For an integer-valued crossing bi-supermodular function p the following min-max equality holds. $\tau_p := \min(z(A) : z$ an integer-valued covering of $p) = \nu_p := \max(p(\mathcal{F}): \mathcal{F} \subseteq \mathcal{A}, \mathcal{F}$ independent).*

Theorem 4.2 as well as Theorems 3.4 and 3.7 are special cases of this result and Theorem 3.5 can also easily be derived from it. (In [20, 21] a difficult min-max theorem on intervals of E. Győri was also shown to be a consequence but this has nothing to do with connectivity augmentation). Having so many corollaries, it is indeed surprising that the proof of Theorem 4.3 is short and is rather standard (demonstrating nicely that finding the right notions and formulation of results might subsume sophisticated proofs.) This proof is, however, not constructive! Though the theorem may be used to develop a polynomial time algorithm for finding a minimum k-connected augmentation of a digraph, the algorithm is based on the ellipsoid method. Designing a combinatorial algorithm for this task is one of the most challenging algorithmical problems of the area. We do not know such an algorithm even if D is $(k - 1)$-connected, that is, the goal is to increase the connectivity of D only by 1 (For $k = 1$, Eswaran and Tarjan have such an algorithm. In a recent paper we developed an algorithm for $k = 2$ [21]).

In the same paper we were able to show that in Theorem 4.2, if D is $(k - 1)$-connected, then the optimal digraph of new edges may be chosen to consist of disjoint directed paths and circuits. This implies

Theorem 4.4 *Let D be a k-connected digraph for which the in-degree and out-degree of every node is k. Then it is possible to add disjoint directed circuits to D so that the resulting digraph D^+ is $(k + 1)$-connected.*

Note that it is not always possible to increase the connectivity of a digraph by adding a directed Hamiltonian circuit. (Take a digraph arising from $K_{3,3}$ by replacing each edge by two oppositely directed edges.)

5   Edge-Connectivity Augmentation with Undirected Edges

The purpose of the present chapter is to review results concerning edge-connectivity augmentation when only undirected edges are allowed to be added. Typically the starting graph $G = (V, E)$ is also undirected but many results extend to mixed starting graphs as well.

Let $r(x, y)$ be a symmetric, non-negative, integer-valued function defined on the pairs of nodes. Add a minimum number of new edges to G so that $\lambda(x, y; G^+) \geq r(x, y)$ holds for every pair of nodes x, y of the augmented graph G^+.

The first results concerned the special case when the starting graph has no edges. For the fractional version of this case R.E. Gomory and T.C. Hu [28] provided an elegant solution and proved that the optimal (fractional) augmentation can be realized by half-integers. W. Chou and H. Frank [5] solved the integer-valued version. In another paper Frank and Chou[5] solved the restricted problem when no parallel edges are allowed to be used. J. Edmonds [6] proved that if there exists a simple graph with a specified degree sequence and each degree is at least k, then there is a k-edge-connected simple graph with the given degree sequence.

The first papers concerning general starting graphs appeared in 1976. K. Eswaran and R. E. Tarjan and J. Plesnik solved the 2-edge-connectivity augmentation problem.

Eswaran and Tarjan also provided a linear time algorithm while Plesnik's paper is the first where the idea of splitting off technique appears. In a tiny note at the end of his paper, Plesnik remarks that the 2-edge-connectivity augmentation theorem also follows from a (then recent) theorem of Lovász on splitting off edges. It turned out that this approach has far reaching consequences. The general k-edge-connectivity augmentation problem was solved by T. Watanabe and A. Nakamura[54]. In their solution there is no restriction on the number of copies a new edge may be added. It is an important open problem to find algorithms that does not add parallel edges. Very recently this task was solved for the special cases $k \leq 5$ by Taoka, Takafuji and Watanabe[51].

The fundamental min-max theorem of Watanabe and Nakamura is as follows.

Theorem 5.1 *The minimum number of edges whose addition makes an undirected graph $G = (V, E)$ k-edge-connected ($k \geq 2$) is equal to*

$$\max\lceil(\sum_i(k - d(X_i)))/2\rceil \qquad (5.1)$$

where the maximum is taken over all sub-partitions $\{X_1, \ldots, X_t\}$ of V.

The proof of Watanabe and Nakamura is based on the recognition that the augmentation problems for different k's are strongly related. They prove various exciting structural properties of edge-connectivity augmentations. Below we cite two of them. These are not only the basis of their augmentation algorithm but serve as a framework for subsequent improved algorithms as well [47, 27, 2].

Let us first study how an optimal sub-partition for (5.1) may be found. It is an easy observation that for any fixed integer l the relation "$\lambda(x, y) \geq l$" on the node-set of a graph $G = (V, E)$ is an equivalence relationship. An equivalence class may be called an **edge-connectivity component** (in short, **ec-component**) or an l-**component**. (That is, an l-component is a maximal subset of nodes for which $\lambda(x, y) \geq l$.) From the definition it is straightforward that the family $\mathcal{F}_{ec}$ of all ec-components is a laminar family (and hence it has at most $2n$ members). For $l = 0$ the node-set V is an ec-component and for $l = |E| + 1$ each single node is a ec-component.

Call a subset $X \subseteq V$ **extreme** if $d(X') > d(X)$ holds for every proper, non-empty subset X' of X. If we choose an optimal sub-partition in (5.1) so that the union of its members has minimum cardinality, then the sub-partition consists of extreme sets. In other words it suffices to restrict (5.1) on sub-partitions consisting of extreme sets. The following lemma is basic to explore the structure of extreme sets.

Lemma 5.2 *Each extreme set is an edge-connectivity component.*

Proof Let C be an extreme set and let $l := \min(\lambda(x, y) : x, y \in C)$. Since V and each singleton is an ec-component, we may assume that $1 < |C| < |V|$. We have to show that $\lambda(x, y) < l$ for any pair $x \in C, y \in V - C$. This is clearly the case if $d(C) < l$ so suppose that $d(C) \geq l$. By Menger's theorem there is a set M for which $M \cap C$ and $C - M$ are non-empty and $d(M) = l$. By taking the complement if necessary, we may assume that $x \in M$. If $y \notin M$, then we have $d(C) + d(M) \geq d(C \cap M) + d(C \cup M) > d(C) + d(C \cup M)$ and therefore $\lambda(x, y) \leq d(C \cup M) < d(M) = l$. If $y \in M$, then we have $d(C) + d(M) \geq d(C - M) + d(M - C) > d(C) + d(M - C)$ and therefore $\lambda(x, y) \leq d(M - C) < d(M) = l$. ∎

By this lemma the family $\mathcal{F}^*$ of extreme sets is a sub-family of $\mathcal{F}_{ec}$ and therefore it is a laminar family. Since the ec-components of G can be computed with the help of a Gomory-Hu tree, $\mathcal{F}^*$ can also be computed. The nice thing is that the family of extreme sets includes all information which is required to determine the optimum in (5.1) for **any** k.

This may be done recursively. For each extreme set X let us define the **recursive k-deficiency** $R_k(X)$ as follows. For singletons let $R_k(v) := (k - d(v))^+$. If for all maximal extreme subsets X' of an extreme set X (which form, incidentally, a subpartition of X) $R_k(X')$ has already been determined, then define $R_k(X) := \max((k - d(X))^+, \sum(R_k(X') : X'$ is a maximal extreme subset of $X))$. Parallel to this we may store a sub-partition $\mathcal{R}_k(X)$ of X. It consists of the single set X if the maximum in the definition of $R_k(X)$ is attained on the first term. If the maximum is attained on the second term, then let $\mathcal{R}_k(X) := \cup(\mathcal{R}_k(X') : X'$ is a maximal extreme subset of $X)$. From the definition, it is clear that, among the sub-partitions $\mathcal{R}$ of V consisting of extreme sets, $\mathcal{R}_k(V)$ maximizes the sum $\sum(k - d(X) : X \in \mathcal{R})$ and therefore $\mathcal{R}_k$ is an optimal solution to (5.1).

Not only the best sub-partitions for different values of k may be encoded into a single laminar family (of the extreme sets) but Watanabe and Nakamura proved that the optimal edge-connectivity augmentations for increasing k may be chosen as a sequence of ever increasing supergraphs:

Theorem 5.3 *Suppose that the edge-connectivity of the starting graph G is l. There is a sequence $G_l := G, G_{l+1}, G_{l+2}, \ldots$ of graphs so that for each $i \geq l$, G_{i+1} is a supergraph of G_i and G_i is an i-edge-connected augmentation of G using a minimum number of new edges.*

Watanabe and Nakamura described how to compute this sequence in polynomial time. Gusfield, Naor and Martel [47] and Gabow [27] found improvements for the complexity. One apparent disadvantage of this approach is that the resulting algorithm is not strongly polynomial if the target edge-connectivity k is very big. This is clearly so since the approach uses one-by-one augmentations. A. Benczúr [2] however devised a clever grouping technique to make the algorithm of Watanabe and Nakamura strongly polynomial.

The first strongly polynomial algorithm [18, 19] for the k-edge-connectivity augmentation problem followed a different line. One of its basic ideas, the use of the splitting off technique, was suggested by Plesnik [48] when $k = 2$ and by Cai and Sun [4] for arbitrary $k \geq 2$. Using splitting off is equivalent to using degree-prescribed augmentation problems.

Theorem 5.4 *Let $G = (V, E)$ be an undirected graph and m a modular non-negative integer-valued function on V. G can be made k-edge-connected $(k \geq 2)$ by adding a* *set F of new edges so that*

$$d_F(v) = m(v) \qquad (5.2)$$

holds for every node v if and only if $m(V)$ is even and

$$d(X) + m(X) \geq k \qquad (5.3)$$

holds for every non-empty proper subset X of V.

This theorem is an immediate consequence of Lovász's theorem on splitting off:

Theorem 5.5 *(Lovász [25, 42]) Suppose that in a graph $G' = (V + s, E')$ $d'(s) > 0$ is even and*

$$\lambda(x, y) \geq k \qquad (5.4)$$

holds for every pair of nodes $x, y \in V$. Then the edges incident to s can be paired into $d'(s)/2$ pairs so that after splitting off these pairs the resulting digraph on node-set V is k-edge-connected.

To derive Theorem 5.4, add a new node s to D and $m(v)$ parallel edges between v and s for every $v \in V$. Then (5.3) is equivalent to (5.4) and the edge-set F arising in Theorem 5.5 from the splitting off operations satisfies the requirements in Theorem 5.4.

An equivalent form of Theorem 5.1 of Watanabe and Nakamura is:

Theorem 5.1' *An undirected graph $G = (V, E)$ can be made k edge-connected by adding at most γ new edges if and only if*

$$\sum_i (k - d(X_i)) \leq 2\gamma \qquad (5.5)$$

holds for every sub-partition $\{X_1, \ldots, X_t\}$ of V.

Proof   Let m be an integer-valued function for which (5.3) holds and m is minimal with respect to this property. Call a set X tight if it satisfies (5.3) with equality. We claim that if X and Y are intersecting tight sets, then both $X - Y$ and $Y - X$ are tight and $m(X \cap Y) = 0$. Indeed, $k - m(X) + k - m(Y) = d(X) + d(Y) \geq d(X - Y) + d(Y - X) \geq k - m(X - Y) + k - m(Y - X) = k - m(X) + k - m(Y) + 2m(X \cap Y)$, from which the claim follows. ∎

By the minimality of m there is a family $\mathcal{F}$ of tight sets so that each v for which $m(v)$ is positive belongs to a member of $\mathcal{F}$. Choose $\mathcal{F}$ so that $\sum(|X| : X \in \mathcal{F})$ is minimum. This choice and the claim shows that $\mathcal{F}$ is laminar. Therefore the maximal members of $\mathcal{F}$ form a sub-partition $\{X_1, \ldots, X_t\}$ of V covering all elements v with $m(v) > 0$.

It follows from (5.5) that $m(V) = \sum_i m(X_i) = \sum_i (k - d(X_i)) \leq 2\gamma$. By increasing m if necessary, we may assume that $m(V) = 2\gamma$. Finally, by applying Theorem 5.4 to this m, Theorem 5.1' follows. ∎

Let us turn to the general augmentation problem when we are given an arbitrary starting graph $G = (V, E)$, an arbitrary (symmetric, non-negative, integer-valued) demand function $r(x, y)$ and the goal is to determine the minimum number of new edges to be added to G so as to obtain a graph G^+ for which

$$\lambda(x, y; G^+) \geq r(x, y) \tag{5.6}$$

holds for every pair of nodes x, y in the augmented graph G^+. We may call an augmentation satisfying (5.6) **feasible**. If F is the set of new edges in a feasible augmentation, a vector in Z^V defined by $(d_F(v) : v \in V)$ is called an **augmentation vector**.

Recall that the corresponding augmentation problem for directed graphs is NP-complete already for very special demand functions (e.g., if $r(x, y) = 1$ when both x and y belong to a subset T of nodes and 0 otherwise.) In this light it is especially surprising that the undirected augmentation problem is tractable for any starting graph and for any demand function [18, 19].

In order to formulate the general augmentation result, let us define a set-function R on the subsets of V so that $R(\emptyset) := R(V) := 0$ and

$$R(X) := \max(r(x, y) : x \in X, y \in V - X). \tag{5.7}$$

By Menger's theorem $\lambda(x, y) \geq r(x, y)$ holds in a graph for every pair of nodes x, y if and only if $d(X) \geq R(X)$ for every subset $X \subseteq V$. That is, $R(X)$ serves as a lower bound for the number of edges in a cut $[X, V - X]$ and $q(X) := (R(X) - d(X))^+$ may be considered as the deficiency of X. Now

$$\max(\lceil (\sum_i q(X_i))/2 \rceil : \{X_1, X_2, \dots, X_t\} \text{ a sub-partition of } V) \tag{5.8}$$

is a lower bound for the minimum number $\gamma(G, r)$ of new edges. Theorem 5.1 asserts that this lower bound is achievable if $r \equiv k (\geq 2)$.

On the way to generalize Theorem 5.1 we have to prepare, however, to overcome a little anomaly indicated already by the fact that Theorem 5.1 is not true for $k = 1$ (take a starting graph on four nodes with no edges) while the augmentation problem when $k = 1$ is trivial. This distinction must be handled in the general case, as well. To this end let $C (\neq V)$ be the node-set of a component of G and call C a **marginal component** (with respect to the demand function r) if $q(C) \leq 1$ and $q(X) = 0$ for every proper subset of C.

The solution in [18] to find a minimal feasible augmentation of G consists of two parts. In the first part the marginal components are eliminated while the second one consists of proving (algorithmically) that $\gamma(G, r)$ is equal to the maximum in (5.8) when there are no marginal components. (This is, by the way, the case if G is connected).

Let C be a marginal component, $G_1 := G - C$ and let r_1 denote the demand function restricted on the node set of G_1. It is proved in [18] that $\gamma(G, r) = \gamma(G_1, r_1) + q(C)$. It is also shown how an optimal feasible augmentation of G_1 can be extended to an optimal feasible augmentation of G by adding $q(C)$ (which is 0 or 1) edge. This way we can eliminate the marginal components one by one.

Theorem 5.6 *If G has no marginal components, there is a feasible augmentation of G using at most γ new edges if and only if*

$$\sum_i q(X_i) \le 2\gamma \qquad (5.9)$$

holds for every sub-partition $\{X_1, X_2, \ldots, X_t\}$ of V. Or, equivalently, the minimum number of new edges $\gamma(G, r) = \max(\lceil (\sum_i q(X_i))/2 \rceil \; : \; \{X_1, X_2, \ldots, X_t\}$ a sub-partition of V).

Corollary 5.6' *Let $G = (V, E)$ be an undirected graph, $r(u, v)$ an integer-valued demand-function such that G has no marginal components, and g an integer-valued capacity function on E. There is an optimal solution to the undirected max-flow augmentation problem which is half integral. Furthermore, an optimal integer-valued solution is either optimal among the real-valued augmentations or its total increment is one half bigger than that of a (real-valued) optimal solution.*

The key to the proof of Theorem 5.6 is the following deep splitting off theorem of W. Mader.

Theorem 5.7 (Mader [44]) *Let $G' = (V + s, E')$ be a (connected) undirected graph in which $0 < d_G(s) \ne 3$ and there is no cut-edge incident with s. Then there exists a pair of edges $e = su, f = st$ so that*

$$\lambda(x, y; G) = \lambda(x, y; G^{ef})$$

holds for every $x, y \in V$.

(For a relatively simple proof, using submodularity, see [18, 19]) In Section 3 it was pointed out that Mader's directed splitting off theorem is equivalent to Theorem 3.2 on the existence of a k-edge-connectivity augmentation of a digraph satisfying pre-scriptions of the in-degree and out-degree. Analogously, Theorem 5.7 is equivalent to:

Theorem 5.7' *Let $m : V \to Z_+$ be an integer-valued function so that $m(V)$ is even and $m(C) \ge 2$ for each component C of $G = (V, E)$. There is a set F of new edges so that $G^+ = (V, E + F)$ is a feasible augmentation of G and $d_F(v) = m(v)$ for every node v (that is, m is an augmentation vector) if and only if*

$$m(X) \ge R(X) - d_G(X) \qquad (5.10)$$

for every $X \subseteq V$.

The material of the closing part of this section is taken from a recent work of [1]. They proved an extension of Mader's theorem when the graph is a mixed one but all edges incident to s are undirected. This was used to derive a generalization of Theorem 5.6.

Let $N = (V, E + A)$ be a mixed graph composed from an undirected graph $G = (V, E)$ and a directed graph $D = (V, A)$ in which $T(D) := \{v \in V : \varrho_D(v) \neq \delta_D(v)\}$ is the set of non-di-Eulerian nodes. Let $k \geq 2$ be an integer and $r(x, y)$ $(x, y \in V)$ a non-negative, integer-valued demand function satisfying

$$r(x, y) = r(y, x) \leq k \text{ for every } x, y \in V \text{ and} \tag{5.11a}$$

$$r(x, y) \equiv k \text{ for every } x, y \in T(D). \tag{5.11b}$$

Let $R(\emptyset) = R(V) = 0$ and for $X \subseteq V$ let $R(X) := \max(r(x, y) : X$ separates x and $y)$. We say that a component C of N is **marginal** (with respect to r) if $r(u, v) \leq \lambda(u, v; N)$ for every $u, v \in C$ and $r(u, v) \leq \lambda(u, v; N) + 1$ for every u, v separated by C. Let $\beta_N(X) := \min(\varrho_D(X) + d_G(X), \delta_D(X) + d_G(X))$.

Theorem 5.8 (Bang-Jensen, Frank and Jackson [1]) *Given a mixed graph N, integers $k \geq 2$, $\gamma \geq 0$, and a demand function $r(x, y)$ satisfying (5.11) so that there is no marginal components, N can be extended to a mixed graph N^+ by adding γ new undirected edges so that*

$$\lambda(x, y; N^+) \geq r(x, y) \text{ for every } x, y \in V \tag{5.12}$$

if and only if

$$\sum (R(X_i) - \beta_N(X_i)) \leq 2\gamma \tag{5.13}$$

holds for every sub-partition $\{X_1, \ldots, X_t\}$ of V.

Before Theorem 5.6 we indicated how to eliminate marginal components when the starting graph is undirected. A similar reduction works for mixed undirected graphs as well. When N is an undirected graph, Theorem 5.8 specializes to Theorem 5.6. When $r \equiv k$ for an integer $k \geq 2$ Theorem 5.8 specializes to:

Corollary 5.9 *Let $N = (V, A \cup E)$ be a mixed graph and $k \geq 2, \gamma \geq 1$ integers. N can be made k-edge connected by adding γ new undirected edges if and only if*

$$\sum (k - \beta_N(X_i)) \leq 2\gamma$$

holds for every sub-partition $\{X_1, .., X_t\}$ of V.

This corollary is not true for $k = 1$. (Let N be a digraph with 4 nodes and 3 edges so that the heads of the edges are distinct but their tails coincide) However, the following can be proved.

Theorem 5.10 *A mixed graph N with connected underlying graph can be made strongly connected by adding γ new undirected edges if and only if $(*)$ for any family $\mathcal{F}$ of $\gamma + 1$ disjoint subsets of nodes contains (not necessarily distinct) members X, Y for which $\varrho_N(Y) > 0$ and $\delta_N(Y) > 0$.*

(A mixed graph is **strongly connected** if every node is reachable from every other node along a path not using oppositely oriented edges.)

6 Node-Connectivity Augmentation of Undirected Graphs

In this section we want to make an undirected graph $G = (V, E)$ k-connected by adding a minimum number of new edges. In previous sections we pointed out that the k-edge-connectivity augmentation problem is tractable for both directed and undirected graphs. It turned out that undirected edge-connectivity augmentation behaves better than the directed one in the sense that even the general-demand augmentation is solvable in the undirected case. Section 4 described a nice min-max theorem for the directed k-connectivity augmentation. From these data one hopes that the undirected k-connectivity augmentation problem also has a solution. Unfortunately, at present, it is not known if the problem is **NP**-complete or is perhaps in co-**NP**∩**NP** or even in **P**.

For general k, F. Harary [31] found the solution when the starting graph has n nodes but no edges. Wang and Kleitman [53] determined a necessary and sufficient condition for the existence of a k-connected graph with specified degree sequence.

For general starting graphs, solutions are known only for small k. When $k = 1$ the problem is obvious. Plesnik [48] and Eswaran and Tarjan [11] proved a min-max formula for $k = 2$ and the latter paper described a linear-time algorithm, as well, to construct the optimal augmentation. For a subset X of nodes let $N(X)$ denote the set of nodes in $V - X$ which have a neighbor in X. In a k-connected graph $|N(X)| \geq k$ whenever $|X| \leq |V| - k - 1$. Therefore, in a k connected augmentation of G at least $q_\kappa(X) := (k - N(X))^+$ new edges must be in the cut $[X, V - X]$. Hence

$$\max(\lceil (\sum_i q_\kappa(X_i))/2 \rceil \; : \{X_i\} \text{ a sub-partition of } V) \tag{6.1}$$

is a lower bound for the minimum number of new edges. Theorem 5.1 asserted that in the edge-connectivity augmentation an analogous lower bound is achievable, except the (otherwise trivial) case of $k = 1$. The difficulty of the node-connectivity augmentation arises from the fact that this kind of trouble may occur for any big k. For example, let G be a star, that is, G is a simple graph on n nodes in which every edge is incident to a node s. The bound of (6.1) when $k = 2$ is $\lceil (n-1)/2 \rceil$. But to make G 2-connected one needs $n - 2$ edges. From this example we may extract another lower bound for $k = 2$. Let $c(X)$ denote the number of components of the graph arising from G by deleting the node-set X. Then at least $c(v) - 1$ new edges have to be added to G to make it 2-connected,

Theorem 6.1 (Eswaran and Tarjan [11], Plesnik [48]) *An undirected graph G can be made 2-connected by adding at most γ new edges if and only if*

$$\sum_i (2 - |N(X_i)|) \leq 2\gamma \text{ for every sub-partition } \{X_i\} \text{ of } V \tag{6.2}$$

where $(|X_i| \leq |V| - 3)$ *and*

$$c(v) \leq \gamma + 1 \tag{6.3}$$

for every node $v \in V$.

For the 3-connectivity augmentation problem an analogous theorem holds:

Theorem 6.2 (Watanabe and Nakamura [54, 55]) *An undirected graph G can be made 3-connected by adding at most γ new edges if and only if*

$$\sum_i (3 - |N(X_i)|) \leq 2\gamma \text{ for every sub-partition } \{X_i\} \text{ of } V \qquad (6.2')$$

where $(|X_i| \leq |V| - 4)$ and

$$c(X) \leq \gamma + 1 \qquad (6.3')$$

for every 2-element subset X of V.

Watanabe and Nakamura also developed a polynomial-time algorithm to find the optimal augmentation. A linear time algorithm (and another proof) was given by Hsu and Ramachandran [33]. Jordán [37] solved the degree-constrained 3-connectivity augmentation problem when the starting graph is 2-connected. In particular, he proved that G can be made optimally 3-connected by adding disjoint paths.

More generally, if the starting graph G is $k-1$ connected, then both $\max\lceil(\sum_i q_\kappa(X_i))/2\rceil$ and $\max(c(X)-1)$ are lower bounds for the number of new edges where $\{X_i\}$ is a sub-partition of V consisting of subsets of at most $|V| - k - 1$ elements and X is a cut-set of precisely $k - 1$ elements. Theorems 6.1 and 6.2 are equivalent to saying that the maximum of these two lower bounds are achievable when $k = 2, 3$. But not so if $k = 4$ as is shown by $K_{3,3}$. Here the optimum is 4 while the first maximum is 3 and the second is 2. For this case (that is, to find a minimal 4-connected augmentation of a 3-connected graph) a polynomial algorithm was developed by T. Hsu [3]. He also showed that the maximum of the two lower bounds is achievable except for some well-described small graphs. A related nice result of Jordán [37] asserts that a 3-regular 3-connected graph on at least 8 nodes can be made 4-connected by adding a perfect matching.

For higher k the complete bipartite graph $K_{k-1,k-1}$ shows that the gap between the maximum and the minimum may be as big as $k - 3$. T. Jordán [35] proved that this is the largest possible gap for any $(k - 1)$-connected starting graph. Relying on this theorem, he developed a polynomial time algorithm to make an arbitrary $(k-1)$-connected graph k-connected and the solution provided by the algorithm uses at most $(k - 3)^+$ more edges than the optimum. To compute this optimum exactly seems to be out of reach at the time of writing this paper.

There is a fractional version of the connectivity augmentation problem. We say that a weighting on the edge-set of a graph G is **connected** if the total weight of every cut is at least 1. We say that a weighting is k-**connected,** if leaving out any set of at most $k - 1$ nodes leaves a graph with connected weighting. Clearly if all weights are 1, then this definition gives back the original definition of k-connectivity.

The fractional node-connectivity augmentation problem is the following. Let $G = (V, E)$ be a simple undirected graph. Define $w_G(x, y) = 1$ if xy is an edge of G and $w_G(x, y) = 0$ otherwise. The goal is to determine a non-negative weighting w on the edge set of the complete graph on V so that $w + w_G$ is k-connected and $1w$ is minimum.

By an elementary construction this problem can be reduced to the directed node-connectivity augmentation problem. As a consequence of Theorem 4.2 one may derive that the optimal fractional augmentation may be chosen half-integral. For example, if the starting graph is $k_{k-1,k-1}$, then an optimum fractional augmentation is 1/2 on the edges of a circuit in both parts, that is, its total value is $k - 1$. Compare this with the integer optimum which is $2(k - 2)$.

7 Polyhedral and Algorithmic Aspects

In Sections 3 and 5 we saw that the cardinality version of the k-edge-connectivity augmentation problem is tractable for both directed and undirected graphs, while the minimum cost version is **NP**-complete. But there is a restricted class of cost functions, the node-induced costs, when even the minimum cost augmentation problem is solvable. The idea is based on the fact that the degree sequences belonging to possible graphs of new edges in feasible augmentations of the starting graph span a g-polymatroid. G-polymatroids generalize polymatroids which are generalization of matroid polyhedra. The matroid greedy algorithm may be extended to g-polymatroids. This feature makes it possible to solve minimum node-cost augmentation problems.

Another important property of g-polymatroids is that their intersection with a box is again a g-polymatroid. This is why the degree-constrained augmentation problems can be handled (see, for example Theorem 3.12).

Here we illustrate some details of this polymatroidal approach concerning the undirected case. The basic framework is similar for directed augmentation. Details may be found in [18, 19].

Let V be a finite ground-set and $b : 2^V \to Z \cup \{\infty\}$ an integer-valued set-function which is zero on the empty set. We call b **fully (intersecting) submodular** if

$$b(X) + b(Y) \geq b(X \cap Y) + b(X \cup Y) \tag{7.1}$$

holds for every (intersecting) $X, Y \subseteq V$. A set function p is called **supermodular** if $-p$ is submodular. We say that p is **skew supermodular** if for every $X, Y \subseteq V$ at least one of the following inequalities holds:

$$p(X) + p(Y) \leq p(X \cap Y) + p(X \cup Y),$$

$$p(X) + p(Y) \leq p(X - Y) + p(Y - X).$$

Note that intersecting supermodular functions are skew supermodular.

We say that a pair (p, b) of set-functions is a **strong pair** if p (resp. b) is fully supermodular (submodular) and they are **compliant,** that is,

$$b(X) - p(Y) \geq b(X - Y) - p(Y - X) \tag{7.2}$$

holds for every $X, Y \subseteq S$. If p and b are intersecting super- and submodular functions and (7.2) holds for intersecting X, Y, then (p, b) is called a **weak pair.**

Given a strong pair (p, b), the polyhedron

$$Q(p, b) := \{x \in R^S : p(A) \le x(A) \le b(A) \text{ for every } A \subseteq S\}.$$

is called a **generalized polymatroid** (in short, **g-polymatroid.**) For technical reasons the empty set is also considered as a g-polymatroid. Properties of g-polymatroids were extensively studied in [22]. It can be proved that $Q := Q(p, b)$ is non-empty and that Q uniquely determines p and q, that is, different strong pairs define different g-polymatroids. On the other-hand (p', b') is a g-polymatroid for any weak pair. Let f and g be two integer-valued functions on V with $f \le g$ and let $B := \{x : f \le x \le g\}$ a box.

Theorem 7.1 *For a weak pair (p', b') the intersection $M := Q(p', b') \cap B$ is an integral g-polymatroid. M is non-empty if and only if*

$$g(Z_0) + \sum_{i \ge 1} b'(Z_i) \ge p'(\cup_{i \ge 0} Z_i) \text{ and}$$

$$f(Z_0) + \sum_{i \ge 1} p'(Z_i) \le b'(\cup_{i \ge 0} Z_i)$$

holds for every sub-partition $\{Z_i\}$ of V where only part Z_0 may be empty.

This theorem is in the background of connectivity augmentation results (such as Theorem 3.12) when upper and lower bounds are imposed on the degrees of the augmented graph.

Contra-polymatroids form a special class of g-polymatroids. Given a fully super-modular, monotone increasing function p,

$$C(p) := \{x \in R^S : x \ge 0, x(A) \ge p(A) \text{ for every } A \subseteq S\}$$

is called a **contra-polymatroid.** Such a p is uniquely determined by the polyhedron but weaker functions may also define contra-polymatroids.

Theorem 7.2 *Let p^* be a skew supermodular function. Then $C(p^*)$ is a contra-polymatroid whose unique (monotone, fully supermodular) defining function p is given by $p(X) := \max(\sum p^*(X_i))$ where the maximum is taken over all sub-partitions $\{X_i\}$ of X.*

This theorem (proved in [18, 19]) made it possible for minimum node-cost and degree-constrained augmentation problems to become tractable. The link between augmentations and contra-polymatroids is revealed by the observation that the function $q(X) := R(X) - d_G(X)$ is skew supermodular. (For the notation, see Theorem 5.6.)

Recall the notion of an augmentation vector. Combining Theorem 7.12 and Theorem 5.6 together we obtain:

Corollary 7.3 *For a connected graph G, an integer vector z is an augmentation vector if and only if $z(V)$ is even and $z \in C(q)$.*

Suppose we are given a non-negative cost-function c on V. For each possible new edge xy define a cost by $c(x) + c(y)$. We call such a function a node-induced cost function. With a slight modification of the greedy algorithm one can find an integer-valued element z of $C(q)$ with $z(V)$ even which minimizes cz. Therefore, for node-induced cost-functions, the minimum cost edge-connectivity augmentation problem can be solved in polynomial time.

References

[1] J. Bang-Jensen, A. Frank, and B. Jackson, "Preserving and increasing local edge-connectivity in mixed graphs," *SIAM J. Discrete Mathematics*, to appear.

[2] A. Benczúr, "Augmenting undirected connectivity in RNC and in randomized $O(n^3)$ time," Proceedings of STOC, 1994.

[3] M. Bussieck, "Augmenting capacities of networks," preprint (TU Braunschweig, Ab-teilung für Math. Opt, Germany).

[4] G. R. Cai and Y. G. Sun, "The minimum augmentation of any graph to k-edge-connected graph," *Networks* **19**(1989) 151-172.

[5] W. Chou and H. Frank, "Survivable communication networks and the terminal capacity matrix," *IEEE Trans. Circuit Theory*, **CT-17** (1970), 192-197.

[6] J. Edmonds, "Existence of k-edge-connected ordinary graphs with prescribed degrees," *J. Research of the National Bureau of Standards*, **68B**, 73-74.

[7] J. Edmonds, "Edge-disjoint branchings," in: *Combinatorial Algorithms*, B. Rustin (editor), Acad. Press, New York, 1973, 91-96.

[8] J. Edmonds, "Submodular functions, matroids, and a certain polyhedra", *Combinatorial Structures and their applications* R. Guy, H. Hanani, N. Sauer, and J. Schonheim (editors), Gordon and Breach, New York, 1970, 69-87.

[9] J. Edmonds, "Matroid intersection," *Annals of Discrete Math*, **4** (1979), 39-49.

[10] J. Edmonds and R. Giles, "A min-max relation for submodular functions on graphs," *Annals of Discrete Mathematics*, **1** (1977), 185-204.

[11] K. P. Eswaran and R.E. Tarjan, "Augmentation problems," *SIAM J. Computing*, **5** (1976), 653-665.

[12] L. R. Ford and D. R. Fulkerson, *Flows in Networks*, Princeton Univ. Press, Princeton NJ, 1962.

[13] A. Frank, "Kernel systems of directed graphs," *Acta Scientiarum Mathematicarum (Szeged)*, **41** (1979), 63-76.

[14] A. Frank, "How to make a digraph strongly connected," *Combinatorica* **1** (1981), 145-153.

[15] A. Frank, "A weighted matroid intersection algorithm," *J. Algorithms* **2** (1981), 328-336.

[16] A. Frank, "Submodular flows," *Combinatorial Optimization*, W. Pulleyblank (editor), Academic Press, 1984, 147-165.

[17] A. Frank, "On connectivity properties of Eulerian digraphs," *Annals of Discrete Math*, **41** (1989), 179-194.

[18] A. Frank, "Augmenting graphs to meet edge-connectivity requirements," *SIAM J. on Discrete Mathematics*, **5** (1992), 22-53.

[19] A. Frank, "On a theorem of Mader," Proceedings of the Graph Theory Conference held in Denmark, 1990, *Annals of Discrete Mathematics* **101** (1992), 49-57.

[20] A. Frank and T. Jordán, "Minimal edge-coverings of pairs of sets," *J. Combinatorial Theory, Ser. B*, submitted.

[21] A. Frank and T. Jordán, "Increasing connectivity of digraphs," in preparation.

[22] A. Frank and É. Tardos, "Generalized polymatroids and submodular flows," *Mathematical Programming, Ser. B*, **42** (1988), 489-563.

[23] A. Frank and É. Tardos, "An application of submodular flows," *Linear Algebra and its Applications*, **114/115** (1989), 329-348.

[24] H. Frank and W. Chou, "Connectivity considerations in the design of survivablle networks," *IEEE Trans. Circuit Theory*, **CT-17** (1970), 486-490.

[25] D. R. Fulkerson, "Packing rooted directed cuts in a weighted directed graph," *Math. Programming*, **6** (1974), 1-13.

[26] D. R. Fulkerson and L. S. Shapley, "Minimal k-arc-connceted graphs," *Networks*, **1** (1971), 91-98.

[27] H. N. Gabow, "A matroid approach to finding edge-connectivity and packing arborescences," *Proc. 23rd Annual ACM Symp. on Theory of Comp.*, 1991, 112-122.

[28] R. E.Gomory and T. C. Hu, "Multi-terminal network flows," *SIAM J. Appl. Math.*, **9** (1961), 551-570.

[29] D. Gusfield, "Optimal mixed graph augmentation," *SIAM J. Computing*, **16** (1987), 599-612.

[30] M. Grötschel, C.L. Monma, and M. Stoer, "Design of survivable networks," *Handbook of Operations Research and Management Science*, (Volume on Networks), to appear.

[31] F. Harary, "The maximum connectivity of a graph," *Proc. of 1962 Nat. Acad. of Sci.* **48**, 1962.

[32] T. S. Hsu, "On four-connecting a triconnected graph," *Proc. 33'rd Annual IEEE Symp. on FOCS*, 1992.

[33] T. S. Hsu and V. Ramachandran, "A linear time algorithmn for triconnectivity augmentation," *Proc. 32nd Annual IEEE Symp. on Foundations of Comp.Sci.*, 548-559.

[34] B. Jackson, "Some remarks on arc-connectivity, vertex splitting, and orientation in digraphs," *J. Graph Theory*, **12** (1988), 429-436.

[35] T. Jordán, "On the optimal vertex-connectivity augmentation," *J. Combinatorial Theory, Ser. B*, to appear.

[36] T. Jordán, "Increasing the vertex connectivity in directed graphs," *Proceedings of the first ESA Conference*, Springer Lecture Notes on Computer Sciences **726** (1993), 236-247.

[37] T. Jordán, "Triconnectivity augmentation with degree constraints," in preparation (also in Connectivity in graphs, Master Thesis, 1991, Eötvös University, Budapest (in Hungarian)).

[38] T. Jordán, "Connectivity augmentation of regular graphs," in preparation.

[39] L. Lovász, "Conference on Graph Theory," lecture, Prague, 1974.

[40] L. Lovász, "On two min-max theorems in graph theory," *J. Combinatorial Theory, Ser. B*, **21** (1976), 26-30.

[41] L. Lovász, "On some connectivity properties of Eulerian graphs," *Acta Mat. Akad. Sci. Hungaricae.*, **28** (1976), 129-138.

[42] L. Lovász, *Combinatorial Problems and Exercises*, North-Holland, 1979.

[43] C. L. Lucchesi and D.H. Younger, "A minimax relation for directed graphs," *J. London Math. Soc.*, **17** (2) 1978, 369-374.

[44] W. Mader, "A reduction method for edge-connectivity in graphs," *Ann. Discrete Math.*, 3 (1978), 145-164.

[45] W. Mader, "Konstruktion aller n-fach kantenzusammenhängenden Digraphen," *Europ. J. Combinatorics*, **3** (1982), 63-67.

[46] W. Mader, "Konstruktion aller n-fach kantenzusammenh angenden Digraphen," *Europ. J. Combinatorics*, **3** (1982), 63-67.

[47] D. Naor, D. Gusfield, and Ch. Martel, "A fast algorithm for optimally increasing the edge-connectivity," *31st Annual Symposium on Foundations of Computer Science* 1990, 698-707.

[48] J. Plesnik, "Minimum block containing a given graph," *Archiv der Mathematik*, **XXVII** (1976) Fasc.6, 668-672.

[49] A. Schrijver, "Min-max relations for directed graphs," *Annals of Discrete Mathematics*, **16** (1982) 261-280.

[50] Y. Shiloach, "Edge-disjoint branchings in directed multigraphs," *Inform. Proc. Letters*, **8** (1979), 24-27.

[51] S. Taoka, D. Takafuji, and T. Watanabe, "Simplicity preserving augmentation of the edge-connectivity of a graph," The Institute of Electronics, Information and Communication Engineers, Technical Report of IEICE, COMP93-73 (1994-01) (Department of Circuits and Systems, Faculty of Engineering, Hiroshima University).

[52] E. Tardos, "A strongly polynomial minimum cost circulation algorithm," *Combinatorica*, **5** (1985), 247-255.

[53] D. L. Wang and D. J. Kleitman, "On the existence of *n*-connected graphs with prescribed degrees *(n ≥ 2),"* *Networks,* **3** (1973), 225-239.

[54] T. Watanabe and A. Nakamura, "Edge-connectivity augmentation problems," *Computer and System Siences,* **35** (1987), 96-144.

[55] T. Watanabe and A. Nakamura, "3-connectivity augmentation problems," *Proceedings IEEE International Symposium on Circuits and Systems,* 1988, 1847-1850.

[56] T. Watanabe and A. Nakamura, "A minimum 3-connectivity augmentation of a graph," *J. Computer and System Sciences,* **46** (1993), 91-128.

[57] D. P. Williamson, "On the design of approximation algorithms for a class of graph problems," Ph. D. Thesis, Massachusetts Institute of Technology.

Tabu Search: Improved Solution Alternatives

Fred Glover
US West Chair in Systems Science
Graduate School of Business
University of Colorado
Boulder, Colorado

1 Background

Tabu Search (TS) is a *metaheuristic* that guides a local heuristic search procedure to explore the solution space beyond local optimality. Widespread successes in practical applications of optimization have spurred a rapid growth of tabu search in the past few years. New "records" have been set by TS, and by hybrids of TS with other heuristic and algorithmic procedures, in finding better solutions to problems in scheduling, sequencing, resource allocation, investment planning, telecommunications and many other areas. Some of the diversity of tabu search applications is shown in Table 1. (See also the survey of Glover and Laguna (1993), and the volume edited by Glover, Laguna, Taillard and de Werra (1993).)

Tabu search is based on the premise that problem solving, in order to qualify as intelligent, must incorporate *adaptive memory* and *responsive exploration*. The use of adaptive memory contrasts with "memoryless" designs, such as those inspired by metaphors of physics and biology, and with "rigid memory" designs, such as those exemplified by branch and bound and its AI-related cousins. The emphasis on responsive exploration (and hence purpose) in tabu search, whether in a deterministic or probabilistic implementation, derives from the supposition that a bad strategic choice can yield more information than a good random choice. (In a system that uses memory, a bad choice based on strategy can provide useful clues about how the strategy may profitably be changed. Even in a space with significant randomness which fortunately is not pervasive enough to extinguish all remnants of order in most real world problems a purposeful design can be more adept at uncovering the imprint of structure, and thereby at affording a chance to exploit the conditions where randomness is not all-encompassing.)

These basic elements of tabu search have several important features, summarized in Table 2.

Tabu search is concerned with finding new and more effective ways of taking advantage of the concepts embodied in Table 2, and with identifying associated principles that can expand the foundations of intelligent search. As this occurs, new strategic mixes of the basic ideas emerge, leading to improved solutions and better practical implementations. This makes TS a fertile area for research and empirical study.

The remainder of this paper is divided into two main parts. Section 2 and its subsections are devoted to presenting the main concepts and strategies of tabu search, and to showing how they interrelate. Section 3 focuses on specific aspects of implementation, with illustrations of useful ways to organize memory processes to enhance the efficiency of the search. Finally, implications for future developments are discussed in the concluding section.

2   Tabu Search Foundations

The basis for tabu search may be described as follows. Given a function $f(x)$ to be optimized over a set X, TS begins in the same way as ordinary local search, proceeding iteratively from one point (solution) to another until a chosen termination criterion is satisfied. Each $x \in X$ has an associated *neighborhood* $N(x) \subset X$, and each solution $x^* \in N(x)$ is reached from x by an operation called a *move*.

TS goes beyond local search by employing a strategy of modifying $N(x)$ as the search progresses, effectively replacing it by another neighborhood $N^*(x)$. As our previous discussion intimates, a key aspect of tabu search is the use of special memory structures which serve to determine $N^*(x)$, and hence to organize the way in which the space is explored.

The solutions admitted to $N^*(x)$ by these memory structures are determined in several ways. One of these, which gives tabu search its name, identifies solutions encountered over a specified horizon (and implicitly, additional related solutions), and forbids them to belong to $N^*(x)$ by classifying them *tabu*. (The tabu terminology is intended to convey a type of restraint that embodies a "cultural" connotation; i.e., one that is subject to the influence of history and context, and capable of being surmounted under appropriate conditions.)

The process by which solutions acquire a tabu status has several facets, designed to promote a judiciously aggressive examination of new points. A useful way of viewing and implementing this process is to conceive of replacing original evaluations of solutions by tabu *evaluations*, which introduce penalties to significantly discourage the choice of tabu solutions (i.e., those preferably to be excluded from $N^*(x)$, according to their dependence on the elements that compose tabu status). In addition, tabu evaluations also periodically include inducements to encourage the choice of other types of solutions, as a result of aspiration levels and longer term influences. The following subsections describe how tabu search takes advantage of memory (and hence learning processes) to carry out these functions.

Explicit and Attributive Memory: The memory used in TS is both explicit and attributive. Explicit memory records complete solutions, typically consisting of elite

Table 1: Illustrative Tabu Search Applications

Scheduling

Flow-Time Cell Manufacturing

Heterogeneous Processor Scheduling

Workforce Planning

Classroom Scheduling

Machine Scheduling

Flow Shop Scheduling

Job Shop Scheduling

Sequencing and Batching

Design

Computer-Aided Design

Fault Tolerant Networks

Transport Network Design

Architectural Space Planning

Diagram Coherency

Fixed Charge Network Design

Irregular Cutting Problems

Lay-Out Planning

Location and Allocation

Multicommodity ocation/Allocation

Quadratic Assignment

Quadratic Semi-Assignment

Multilevel Generalized assignment

Telecommunications

Call Routing

Bandwidth Packing

Hub Facility Location

Path Assignment

Network Design for Services

Customer Discount Planning

Failure Immune Architecture

Synchronous Optical Networks

Production, Inventory and Investment

Flexible Manufacturing

Just-in-Time Production

Capacitated MRP

Part Selection

Multi-item Inventory Planning

Volume Discount Acquisition

Fixed Mix Investment

Routing

Vehicle Routing

Capacitated Routing

Time Window Routing

Multi-Mode Routing

Mixed Fleet Routing

Traveling Salesman

Traveling Purchaser

Table 1 Continued

Logic and Artificial Intelligence

    Maximum Satisfiability

    Probabilistic Logic

    Clustering

    Pattern Recognition/Classification

    Data Integrity

    Neural Network Training

    Neural Network Design

Technology

    Seismic Inversion

    Electrical Power Distribution

    Engineering Structural Design

    Minimum Volume Ellipsoids

    Space Station Construction

    Circuit Cell Placement

    Off-Shore Oil Exploration

Graph Optimization

    Graph Partitioning

    Graph Coloring

    Clique Partitioning

    Maximum Clique Problems

    Maximum Planner Graphs

    P-Median Problems

General Combinational Optimization

    Zero-One Programming

    Fixed Charge Optimization

    Nonconvex Nonlinear Programming

    I-or-None Networks

    Bilevel Programming

    General Mixed Integer Optimization

Table 2: Principal Tabu Search Features

Adaptive Memory

    Selectivity (including strategic forgetting)

    Abstraction and decomposition (through explicit and attributive memory)

    Timing:

        recency of events

        frequency of events

        differentiation between short term and long term

    Quality and impact:

        relative attractiveness of alternative choices

        magnitude of changes in structure or constraining relationships

    Context:

        regional interdependence

        structural interdependence

        sequential interdependence

Responsive Exploration

    Strategically imposed restraints and inducements
    tabu conditions and *aspiration levels*)

    Concentrated focus on good regions and good solution features
    (*intensification processes*)

    Characterizing and exploring promising new regions
    (*diversification processes*)

    Non-montonic search patterns
    (*strategic oscillation*)

    Integrating and extending solutions (*path relinking*)

solutions visited during the search (or highly attractive but unexplored neighbors of such solutions). These special solutions are introduced at strategic intervals to enlarge $N^*(x)$, and thereby provide useful options not in $N(x)$.

TS memory is also designed to exert a more subtle effect on the search through the use of attributive memory, which records information about solution attributes that change in moving from one solution to another. For example, in a graph or network setting, attributes can consist of nodes or arcs that are added, dropped or repositioned by the moves executed. In more abstract problem formulations, attributes may correspond to values of variables or functions. Sometimes attributes are also strategically combined to create other attributes, as by hashing procedures or by AI related chunking or "vocabulary building" methods (Hansen and Jaumard (1990), Woodruff and Zemel (1992), Battiti and Tecchioli (1992a), Woodruff (1993), Glover and Laguna (1993)).

2.1 Short Term Memory and its Accompaniments

An important distinction in TS arises by differentiating between short term memory and longer term memory. Each type of memory is accompanied by its own special strategies. The most commonly used short term memory keeps track of solution attributes that have changed during the recent past, and is called *recency-based* memory. To exploit this memory, selected attributes that occur in solutions recently visited are designated *tabu-active*, and solutions that contain tabu-active elements, or particular combinations of these attributes, are those that become tabu. This prevents certain solutions from the recent past from belonging to $N^*(x)$ and hence from being revisited. Other solutions that share such tabu-active attributes are also similarly prevented from being revisited. The use of tabu evaluations, with large penalties assigned to appropriate sets of tabu-active attributes, can allow tabu status to vary by degrees.

Managing Recency-Based Memory: The process is managed by creating one or several tabu lists, which record the tabu-active attributes and implicitly or explicitly identify their current status. The duration that an attribute remains tabu-active (measured in numbers of iterations) is called its *tabu tenure*. Tabu tenure can vary for different types or combinations of attributes, and can also vary over different intervals of time or stages of search. This varying tenure makes it possible to create different kinds of trade offs between short term and longer term strategies. It also provides a dynamic and robust form of search. (See, e.g., Taillard (1991), Dell'Amico and Trubian (1993), Glover and Laguna (1993).) Illustrations of recency-based memory structures and associated rules for implementing them are given in Section 3.1.

Aspirations Levels: An important element of flexibility in tabu search is introduced by means of aspiration criteria. The tabu status of a solution (or a move) can be overruled if certain conditions are met, expressed in the form of aspiration levels. In effect, these aspiration levels provide thresholds of attractiveness that govern whether the solutions may be considered admissible in spite of being classified tabu. Clearly a solution better than any previously seen deserves to be considered admissible. Similar criteria of solution quality provide aspiration criteria over subsets of solutions that belong to common regions or that share specified features (such as a particular functional value or level of infeasibility). Additional examples of aspiration criteria are provided later.

Candidate List Strategies: The aggressive aspect of TS is reinforced by seeking the best available move that can be determined with an appropriate amount of effort. It should be kept in mind that the meaning of best is not limited to the objective function evaluation. (As already noted, tabu evaluations are affected by penalties and inducements determined by the search history.) For situations where $N^*(x)$ is large or its elements are expensive to evaluate, candidate list strategies are used to restrict the number of solutions examined on a given iteration.

Because of the importance TS attaches to selecting elements judiciously, efficient rules for generating and evaluating good candidates are critical to the search process. Even where candidate list strategies are not used explicitly, memory structures to give efficient updates of move evaluations from one iteration to another, and to reduce the effort of finding best or near best moves, are often integral to TS implementations. Intelligent updating can appreciably reduce solution times, and the inclusion of explicit candidate list strategies, for problems that are large, can significantly magnify the resulting benefits. Useful kinds of candidate list strategies are indicated in Section 3.2.

The operation of these short term elements is illustrated in Figure 1.

The representation of penalties in Figure 1 either as "large" or "very small" expresses a thresholding effect: either the tabu status yields a greatly deteriorated evaluation or else it chiefly serves to break ties among solutions with highest evaluations. Such an effect of course can be modulated to shift evaluations across levels other than these extremes. If all moves currently available lead to solutions that are tabu (with evaluations that normally would exclude them from being selected), the penalties result in choosing a "least tabu" solution.

It may be noted that the sequence of the Tabu Test and the Aspiration Test in Figure 1 can be interchanged (that is, by employing the tabu test only if the aspiration threshold is not satisfied). Also, the tabu evaluation can be modified by creating inducements based on the aspiration level, just as it is modified by creating penalties based on tabu status. In this sense, aspiration conditions and tabu conditions can be conceived roughly as "mirror images" of each other.

The TS variant called *probabilistic tabu search* follows a corresponding design, with a short term component that can be represented by the same diagram. The approach additionally keeps track of tabu evaluations generated during the process that results in selecting a move. Based on this record, the move is chosen probabilistically from the pool of those evaluated (or from a subset of the best members of this pool), weighting the moves so that those with higher evaluations are especially favored. Fuller discussions of probabilistic tabu search are found in Glover (1989, 1993), Soriano and Gendreau (1993) and Crainic et al. (1993).

Tabu status is often permitted to serve as an all-or-none threshold, without explicit reference to penalties and inducements, by directly excluding tabu options from

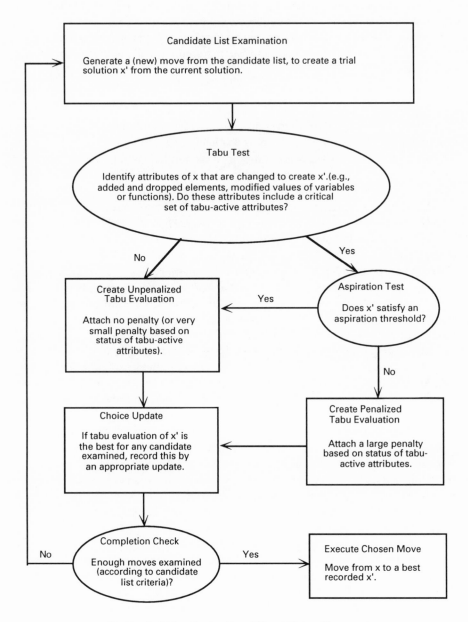

Figure 1: Tabu Evaluation (Short Term Memory)

being selected (subject to the outcome of aspiration tests). Whether or not modified evaluations are explicitly used, the selected move may not be the one with the best objective function value, and consequently the solution with the best objective function value encountered throughout the search history is recorded separately. (A common form of aspiration criterion dictates that such a solution will be chosen as the next one to visit.)

2.2 Longer Term Memory

In some applications, the short term TS memory components are sufficient to produce very high quality solutions. However, in general, TS becomes significantly stronger by including longer term memory and its associated strategies.

Special types of *frequency-based* memory are fundamental to longer term considerations. These operate by introducing penalties and inducements determined by the relative span of time that attributes have belonged to solutions visited by the search, allowing for regional differentiation. *Transition frequencies* keep track of how often attributes change, while *residence frequencies* keep track of relative durations that attributes occur in solutions generated. (These memories are also sometimes accompanied by extended forms of recency-based memory).

Perhaps surprisingly, the use of longer term memory does not require long solution runs before its benefits become visible. Often its improvements begin to be manifest in a relatively modest length of time, and can allow solution efforts to be terminated somewhat earlier than otherwise possible, due to finding very high quality solutions within an economical time span. The fastest methods for job shop and flow shop scheduling problems, for example, are based on including longer term TS memory. (This includes explicit memory of a type subsequently described.) On the other hand, it is also true that the chance of finding still better solutions as time grows–in the case where an optimal solution is not already found–is enhanced by using longer term TS memory in addition to short term memory. Section 3.1 describes forms of frequency-based memory that provide a basis for useful longer term strategies.

Intensification and Diversification: Two highly important longer term components of tabu search are *intensification strategies* and *diversification strategies*. Intensification strategies are based on modifying choice rules to encourage move combinations and solution features historically found good. They may also initiate a return to attractive regions to search them more thoroughly. A simple instance of this second type of intensification strategy is shown in Figure 2.

The strategy for selecting elite solutions is italicized in Figure 2 due to its importance. Two variants have proved quite successful. One, due to Voss (1993), introduces a diversification measure to assure the solutions recorded differ from each other by a desired degree, and then erases all short term memory before resuming from the best of the recorded solutions. The other variant, due to Nowicki and Smutniki (1993), keeps a bounded length sequential list that adds a new solution at the end only if it is better than any previously seen. The current last member of the list is always the one chosen (and removed) as a basis for resuming search. However, TS short term memory that accompanied this solution also is saved, and the first move also forbids

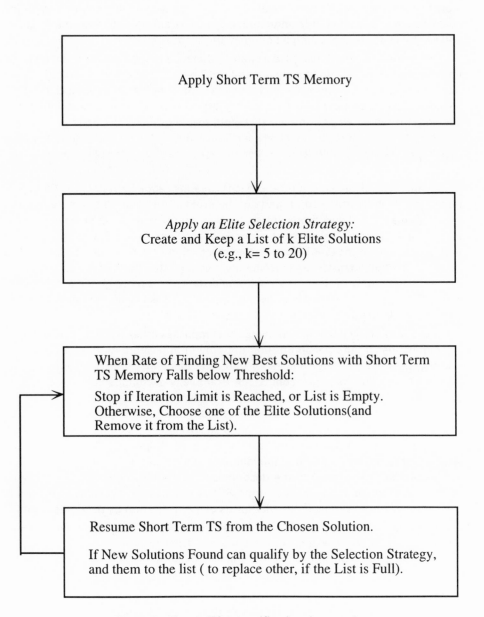

Figure 2: Simple TS Intensification Approach

the move previously taken from this solution, so that a new solution path will be launched. (Sometimes this approach of recovering selected elite solutions is called "backtracking." However, it is unrelated to the backtracking of tree search methods and simply consists of creating a bounded priority queue.)

This second variant is related to a strategy that resumes the search from unvisited neighbors of solutions previously generated (Glover, 1990). Such a strategy keeps track of the quality of these neighbors to select an elite set, and restricts attention to specific types of solutions, such as neighbors of local optima or neighbors of solutions visited on steps immediately before reaching such local optima. This type of "unvisited neighbor" strategy has been little examined. It is noteworthy, however, that the two variants previously indicated have provided solutions of remarkably high quality.

Another type of intensification approach is *intensification by decomposition*, where restrictions may be imposed on parts of the problem or solution structure in order to generate a form of decomposition that allows a more concentrated focus on other parts of the structure. A classical example is provided by the traveling salesman problem, where edges that belong to the intersection of elite tours may be "locked into" the solution, in order to focus on manipulating other parts of the tour. The use of intersections may be seen as an extreme instance of a more general strategy for exploiting frequency information, by a process that seeks to identify and constrain the values of *strongly determined* and *consistent variables*.

Intensification by decomposition also encompasses other types of strategic considerations, basing the decomposition not only on indicators of strength and consistency, but also on opportunities for particular elements to interact productively. Within the traveling salesman context, for example, a decomposition may be based on identifying subchains of an elite tour, where two or more subchains may be assigned to a common set if they contain nodes that are "strongly attracted" to be linked with nodes of other subchains in the set. An edge disjoint collection of subchains can be treated by an intensification process that operates in parallel on each set, subject to the restriction that the identity of the endpoints of the subchains will not be altered. As a result of the decomposition, the best new sets of subchains can be reassembled to create a new tour. Such a process can be applied to multiple alternative decompositions in broader forms of intensification by decomposition.

Diversification Strategies: TS diversification strategies, as their name suggests, are designed to drive the search into new regions. Often they are based on modifying choice rules to bring attributes into the solution that are infrequently used. Alternatively, they may introduce such attributes by partially or fully re-starting the solution process.

The same types of memories previously described are useful as a foundation for such procedures, although these memories are maintained over different (generally larger) subsets of solutions than those maintained by intensification strategies. A simple diversification approach that keeps a frequency-based memory over all solutions previously generated, and that has proved very successful for machine scheduling problems, is shown in Figure 3.

Significant improvements over the application of short term TS memory have been achieved by the procedure of Figure 3 (see Laguna and Glover (1993)). The TS local optima reached by this approach, and used as a basis for launching a sequence of diversifying steps, naturally may differ from true local optima since tabu search choice rules may exclude some improving moves. The success of this approach suggests the merit of incorporating a TS variant that always continues to a true local optimum once an improving move becomes an acceptable choice based on an aspiration criterion that is activated only after executing an improving move. In this approach, as long as additional improving moves exist, the aspiration criterion allows one of them to be selected, by a tabu evaluation rule that penalizes choices based on their tabu status (restricting attention to the improving set). Once a true local optimum is reached, the special aspiration criterion is discontinued until a new improving move is selected by standard TS rules. This approach embodies an instance of *aspiration by search direction*, and can be usefully refined by taking *spheres of influence* into account (Glover and Laguna (1993)).

The precise manner in which frequency-based memories are used to implement strategies of intensification and diversification (apart form defining these memories over different subsets) provides a fertile area for investigation. Two different general patterns for exploiting these memories are illustrated in Figure 4.

A variety of additional alternatives can be inferred from natural variations in these patterns. Diversification strategies that create partial or full restarts are important for problems and neighborhood structures where a solution trajectory can become isolated from worthwhile new alternatives unless a radical change is introduced. Special forms of diversification in these cases have been developed by Hertz and de Werra (1991), Gendreau, Hertz and Laporte (1991), Soriano and Gendreau (1993), Porto and Ribeiro (1993), and Hubscher and Glover (1993).

Diversification strategies can also utilize a long term form of recency-based memory, which results by increasing the tabu tenure of solution attributes. A simple version of this approach that has produced good results (Kelly et al. (1992)) is shown in Figure 5.

The determination of effective ways to balance the concerns of intensification and diversification represents a promising research area. These concerns also lie at the heart of effective parallel processing implementations. The goal from the TS perspective is to design patterns of communication and information sharing across subsets of processors in order to achieve the best tradeoffs between intensification and diversification functions. General analyses and studies of parallel processing with tabu search are given in Taillard (1991, 1993), Battiti and Tecchioli (1992b), Chakrapani and Skorin-Kapov (1991, 1993), Crainic, Toulouse and Gendreau (1993a, 1993b), and Voss (1994).

2.3 Strategic Oscillation

Strategic oscillation is closely linked to the origins of tabu search, and provides a means to achieve an effective interplay between intensification and diversification

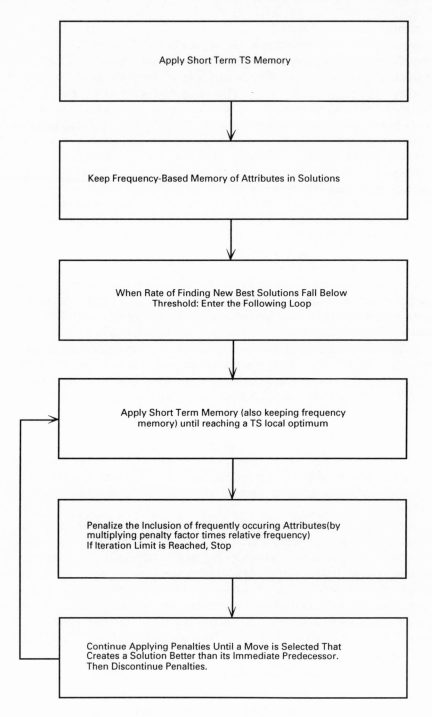

Figure 3: Simple TS Diversification Approach

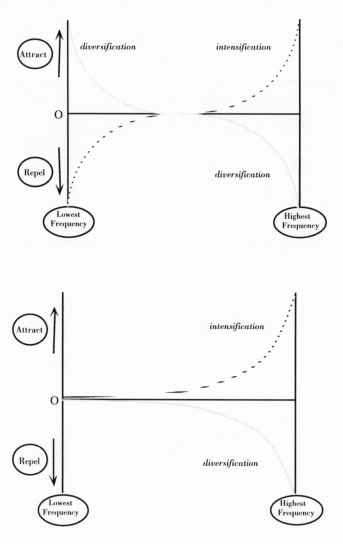

Figure 4: Intensification and Diversification: Degree of Attraction or Repulsion Based on Frequency

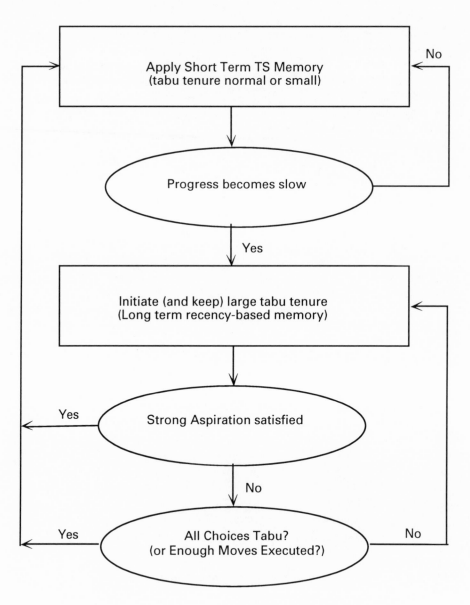

Figure 5: Diversification Using Long Term Recency-Based Memory

over the intermediate to long term. The approach operates by orienting moves in relation to a *critical level*, as identified by a stage of construction or a chosen interval of values for a functional.

Such a critical level often represents a point where the method would normally stop. Instead of stopping when this level is reached, however, the rules for selecting moves are modified, to permit the region defined by the critical level to be crossed. The approach then proceeds for a specified depth beyond the critical level, and turns around. The critical level again is approached and crossed, this time from the opposite direction, and the method proceeds to a new turning point.

The process of repeatedly approaching and crossing the critical level from different directions creates an oscillatory behavior, which gives the method its name. Control over this behavior is established by generating modified evaluations and rules of movement, depending on the region navigated and the direction of search. The possibility of retracing a prior trajectory is avoided by standard tabu search mechanisms.

A simple example of this approach occurs for the multidimensional knapsack problem, where values of zero-one variables are changed from 0 to 1 until reaching the boundary of feasibility. The method then continues into the infeasible region using the same type of changes, but with a modified evaluator. After a selected number of steps, the direction is reversed by choosing moves that change variables from 1 to 0. Evaluation criteria to drive toward improvement vary according to whether the movement occurs inside or outside the feasible region (and whether it is directed toward or away from the boundary), accompanied by associated restrictions on admissible changes to values of variables. An implementation of such an approach by Freville and Plateau (1986, 1992) has generated particularly high quality solutions for multidimensional knapsack problems.

A somewhat different type of application occurs for graph theory problems where the critical level represents a desired form of graph structure, capable of being generated by progressive additions (or insertions) of basic elements such as nodes, edges, or subgraphs. One type of strategic oscillation approach for this problem results by a constructive process of introducing elements until the critical level is reached, and then introducing further elements to cross the boundary defined by the critical level. The current solution may change its structure once this boundary is crossed (as where a forest becomes transformed into a graph that contains loops), and hence a different neighborhood may be required, yielding modified rules for selecting moves. The rules again change in order to proceed in the opposite direction, removing elements until again recovering the structure that defines the critical level. Such rule changes are typical features of strategic oscillation, and provide an enhanced heuristic vitality. The application of different rules may be accompanied by crossing a boundary to different depths on different sides. An option is to approach and retreat from the boundary while remaining on a single side, without crossing (i.e., electing a crossing of "zero depth").

Both of these examples constitute a constructive/destructive type of strategic oscillation, where constructive steps "add" elements (or set variables to 1) and destructive steps "drop" elements (or set variables to 0). In these approaches it is frequently important to spend additional search time in regions close to the critical level, and especially to spend time at the critical level itself. This may be done by inducing a sequence of tight oscillations about the critical level, as a prelude to each larger oscillation that proceeds to a greater depth. Alternately, if greater effort is permitted for evaluating and executing each move, the method may use "exchange moves" (broadly interpreted) to stay at the critical level for longer periods. A simple option, for example, is to use such exchange moves to proceed to a local optimum each time the critical level is reached. A strategy of similarly applying exchanges at additional levels is suggested by a *proximate optimality principle*, which states roughly that good constructions at one level are likely to be close to good constructions at another. A simple version of a constructive/destructive form of strategic oscillation is illustrated in Figure 6.

As observed in the table accompanying Figure 6, the oscillation can also operate by increasing and decreasing bounds for a function $g(x)$. Such an approach has been the basis for a number of effective applications, where $g(x)$ has represented such items as workforce assignments, objective function values, and feasibility/infeasibility levels, to guide the search to probe at various depths with the associated regions. (See, for example, Freville and Plateau (1986), Gendreau, Hertz and Laporte (1993), Kelly, Golden and Assad (1993), Osman (1993), Osman and Christofides (1993), Rochat and Semet (1993), and Voss (1993).)

When the levels refer to degrees of feasibility and infeasibility, $g(x)$ is a vector-valued function associated with a set of problem constraints (which may summarized, for example, by $g(x) \leq b$). In this case, controlling the search by bounding $g(x)$ can be viewed as manipulating a parameterization of the selected constraint set. A preferred alternative is often to make $g(x)$ a Lagrangean or surrogate constraint penalty function, avoiding vector-valued functions and allowing tradeoffs between degrees of violation of different component constraints.

2.4 Path Relinking

A useful integration of intensification and diversification strategies occurs in the approach called *path relinking* (Glover (1989, 1993)). This approach generates new solutions by exploring trajectories that "connect" elite solutions by starting from one of these solutions, called an *initiating solution*, and generating a path in neighborhood space that leads toward the other solutions, called *guiding solutions*. This is accomplished by selecting moves that introduce attributes contained in the guiding solutions.

The approach may be viewed as an extreme (highly focused) instance of a strategy that seeks to incorporate attributes of high quality solutions, by creating inducements to favor these attributes in the moves selected. However, instead of using an inducement that merely encourages the inclusion of such attributes, the path relinking approach subordinates all other considerations to the goal of choosing moves that introduce the attributes of the guiding solutions, in order to create a "good attribute

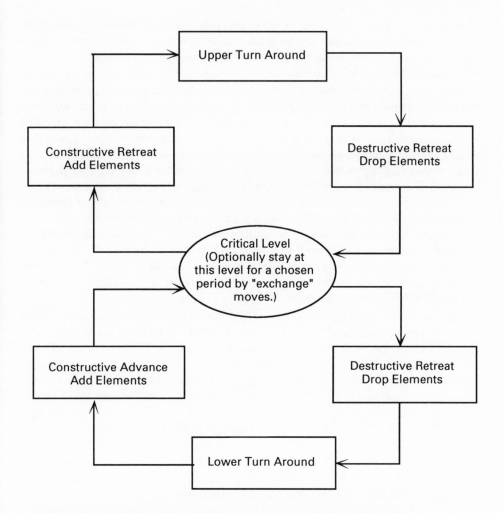

Move types	Example Alternatives	
Constructive	Set $x_j = 1$	increase bounds (L,U) for g(x)*
Destructive	Set $x_j = 0$	decrease bounds (L,U) for g(x)
Exchange	Set $x_p = 1$, $x_q = 0$	maintain bounds for g(x)

* g(x) can be scalar or vector valued function of solution x

Figure 6: Strategic Oscillation Using Constructive/Destructive Moves

composition" in the current solution. The composition at each step is determined by choosing the best move, using customary choice criteria, from the restricted set of moves that incorporate a maximum number (or a maximum *weighted value*) of the attributes of the guiding solutions.

Specifically, upon identifying a collection of one or more elite solutions to guide the path of a given solution, the attributes of these guiding solutions are assigned preemptive weights as inducements to be selected. Larger weights are assigned to attributes that occur in greater numbers of the guiding solutions, allowing bias to give increased emphasis to solutions with higher quality or with special features (e.g., complementing those of the solution that initiated the new trajectory). More generally, it is not necessary for an attribute to occur in a guiding solution in order to have a favored status. In some settings attributes can share degrees of similarity, and in this case it can be useful to view a solution vector as providing "votes" to favor or discourage particular attributes (Glover (1991)). Only the strongest forms of aspiration criteria are allowed to overcome this type of choice rule, so that the path does not deviate unless a neighboring solution is better than any of the guiding solutions (and the initiating solution), whereupon the preemptive guidance rule is resumed after the deviation runs its course.

In a given collection of elite solutions, the role of initiating solution and guiding solutions can be alternated. That is, a set of current solutions may be generated simultaneously, extending different paths, and allowing an initiating solution to be replaced (as a guiding solution for others) whenever its associated current solution satisfies a sufficiently strong aspiration criterion. Because their roles are interchangeable, the initiating and guiding solutions are collectively called *reference solutions*.

An idealized form of such a process is shown in Figure 7. The chosen collection of reference solutions consists of the three members, A, B and C. Paths are generated by allowing each to serve as initiating solution, and by allowing either one or both of the other two solutions to operate as guiding solutions. Intermediate solutions encountered along the paths are not shown. The representation of the paths as straight lines of course is oversimplified, since choosing among available moves in a current neighborhood will generally produce a considerably more complex trajectory.

As Figure 7 indicates, at least one path continuation is allowed beyond each initiating/guiding solution. Such a continuation can be accomplished by penalizing the inclusion of attributes dropped during a trajectory, including attributes of guiding solutions that may be compelled to be dropped in order to continue the path. (An initiating solution may also be repelled from the guiding solutions by penalizing the inclusion of their attributes from the outset.) Probabilistic TS variants operate in the path relinking setting, as it does in others, by translating evaluations for deterministic rules into probabilities of selection, strongly biased to favor higher evaluations.

Promising regions may be searched more thoroughly in path relinking by modifying the weights attached to attributes of the guiding solutions, and by altering the bias associated with solution quality and selected solution features. Figure 8 depicts the type of variation that can result, where the point X represents an initiating solution

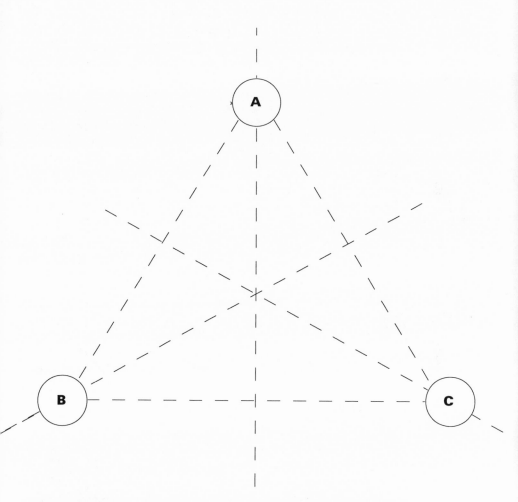

Intensification: Generate paths from similar solutions
Diversification: Generate paths from dissimilar solutions
Aspiration: Explore deviations from paths at attractive neighbors

Figure 7: Path Relinking in Neighborhood Space

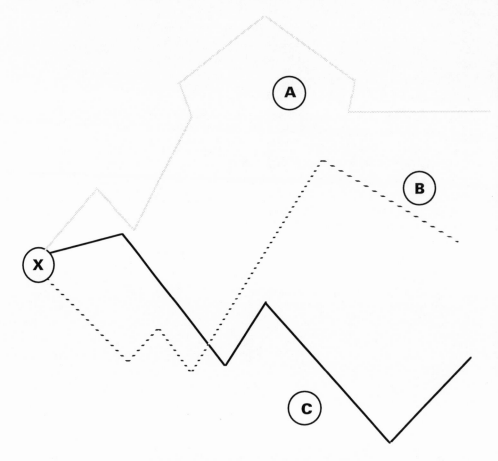

Figure 8: Path Relinking by Attribute Bias

and the points A, B and C represent guiding solutions. Variations of this type within a promising domain are motivated by the proximate optimality principle discussed in connection with strategic oscillation. For appropriate choices of the reference points (and neighborhoods for generating paths from them), this principle suggests that additional elite points are likely to be found in the regions traversed by the paths, upon launching new searches from high quality points on these paths. Evidence that combinatorial solution spaces often have topologies that may be usefully exploited by such an approach is provided by findings of Moscato (1993), Moscato and Tinetti (1994), and Nowicki and Smutnicki (1993, 1994).

3 Illustrative Tabu Search Memory Structures and Strategies

This section focuses on considerations relevant for implementing tabu search, with an emphasis on examples to illustrate main ideas.

3.1 Recency-Based and Frequency-Based Memory Structures

Recency-Based Memory Structures: We begin by indicating some commonly used recency-based memory structures for identifying attributes that are tabu-active, and for determining the tabu status of solutions containing these attributes. Let $\{S = 1, 2, \ldots, s\}$ denote an index set for a collection of solution attributes. For example, the indexes $i \in S$ may correspond to indexes of zero-one variables x_i, or they may be indexes of edges that may be added or deleted from a graph. (More precisely, attributes referenced by S in these two cases consist of the specific values assigned to the variables or the specific add/drop states adopted by the edges.) In general, an index $i \in S$ can summarize more detailed information; e.g., by referring to an ordered pair (j, k) that summarizes a value assignment $x_j = k$. Hence, the index i may be viewed as a notational convenience for representing a pair or a vector, etc. (Consideration can often be limited to move representations in which only a very small number of attributes–or critical attributes–change at a time. E.g., a pivot step, which changes the values of many variables, can be recorded by indicating that just two variables change their states, one entering and one leaving a basis.)

For the present illustration, suppose that each $i \in S$ corresponds to a 0-1 variable x_i. We will not bother to write $(i, 0)$ and $(i, 1)$ to identify the two associated attributes $x_i = 0$ and $x_i = 1$ (since by knowing the current value of x_i we also know its unique alternative value). To record recency-based TS information for each variable, we keep track of iterations by an iteration counter denoted *current-iteration*, which starts at 0 and increases by 1 each time a move is made.

When a move is executed that causes a variable x_i to change its value, we record *tabu-start(i) = current-iteration* immediately after updating the iteration counter. This means that if the move has resulted in $x_i = 1$, then the attribute $x_i = 0$ becomes tabu-active at the iteration *tabu-start(i)*. Further, we stipulate that this attribute will remain tabu-active for a number of iterations equal to *tabu-tenure(i)*, whose value will be determined in a manner soon to be indicated. Thus, in particular, the recency-based tabu criterion says that the previous value of x_i is tabu-active throughout all iterations such that

$$tabu\text{-}start(i) + tabu\text{-}tenure(i) \geq current\text{-}iteration.$$

Once *current-iteration* increases to the point where this inequality no longer holds, x_i will no longer be tabu-active at its previous value and hence will not be discouraged from receiving this value again.

The value *tabu-start(i)* can be set to 0 before initiating the method, as a convention to indicate no prior history exists. Then we automatically avoid assigning a tabu-active status to any variable with *tabu-start(i) = 0* (since the starting value for variable x_i has not yet been changed).

For convenience in the following we will refer to a *variable* x_i as tabu-active with the understanding that the tabu-active condition applies to a specific associated *attribute* the attribute $x_i = k$ where k is the last value previously assigned to x_i. If only one variable changes its value on an iteration, a move may be classified tabu whenever it changes the value of a tabu-active variable. However, if two variables change their values, as where one is set to 0 and the other to 1, then there are several choices. For example, the move can be designated tabu if:

(a) both variables are tabu-active
(b) either variable is tabu-active
(c) the variable that changes from 0 to 1 is tabu-active

The possibility that the tabu status should depend on a particular change of values, as in (c), can also be reflected by giving *tabu-tenure(i)* a different value according to the value assigned to x_i.

The choice of a preferred value for *tabu-tenure(i)* is customarily based on empirical test, starting by considering a common value for all attributes (or for all attributes in a specific class). Experience shows that options can then be quickly narrowed to a range where every value in the range gives good results, particularly if the value is treated as the center of a small interval in which *tabu-tenure(i)* is varied, either systematically or randomly. For example, the approximate outlines of such a range can be quickly inferred by investigating values that are multiples of 5 or 7.

Tabu tenure values for given classes of problems typically can be expressed as a simple function of the total number of attributes (such as fraction or a multiple of the square root of s). For increased refinement, such values then can be differentiated according to types of attributes as for example, according to assignments $x_i = 0$ or 1, and according to specific types of variables. These types of refinements can be made adaptively within the solution process itself, by monitoring the consequences of chosen alternatives. For example, Laguna et al. (1992) monitor the quality of moves associated with particular attribute changes, and vary the tabu tenure of the attributes as they participate in moves of greater or lesser attractiveness. In another type of approach, Kelly et al. (1992) keep track of patterns of objective function value changes, and modify the tabu status of moves when the pattern suggests the possibility of cycling. Battiti and Tecchioli (1992a) provide an effective method that uses a hashing function as a cycling indicator, and directly modifies an overall tabu tenure value as the search process continues, until this value is just large enough to eliminate traces of cycling. (This type of approach can be extended by taking advantage of the chunking ideas of Woodruff (1993).)

A dynamic strategy with a somewhat different foundation determines tabu status without relying on a tabu tenure at all, but by accounting for logical relationships in the sequence of attribute changes. Appropriate reference to these relationships makes it possible to determine in advance if a particular current change can produce cycling, and thus to generate tabu restrictions that are both necessary and sufficient to keep from returning to previous solutions (Glover (1990)). A small tabu tenure introduces extra vigor into the search, since the avoidance of cycling is not the only goal of recency-based memory. (In addition, a "bounded memory span" reduces over-head and provides increased flexibility, as where it may sometimes be preferable to revisit solutions previously encountered.) This means of exploiting logical inter-dependencies also provides information that is useful for diversification strategies. Innovative implementations have been developed by Dammeyer and Voss (1991) and Voss (1992, 1993).

While interesting opportunities exist for applying advanced forms of recency-based memory in tabu search, it is to be noted that simpler forms often work quite well. This motivates the use of straightforward types of memory as a basis for devel-oping initial TS implementations. Experience with such implementations can then suggest the basis for productive elaborations. This feature of tabu search, which makes it possible to introduce refinements by natural stages, is particularly useful for progressing to designs that incorporate longer term memory.

Frequency-Based Memory Structures: Again we consider the setting of a zero-one optimization problem, and make reference to an attribute set $S = 1, \ldots, s$ that consists of indexes of 0-1 variables x_i. The form of transition memory to record the number of times x_i changes its value consists simply of keeping a counter for x_i that is incremented at each move where such a change occurs. Since x_i is a zero-one variable, such a memory also discloses the number of times x_i changes to and from each of its possible assigned values. (In more complex situations, a matrix memory can be used to determine numbers of transitions involving assignments such as $x_j = k$.) However, in using this memory, penalties and inducements are based on *relative* numbers (rather than absolute numbers) of transitions, hence requiring that recorded transition values are divided by the total number of iterations (or the total number of transitions).

Residence memory requires only slightly more effort to maintain than transition memory, by taking advantage of the recency-based memory stored in *tabu-start(i)*. The following approach can be used to track the number of solutions in which $x_i = 1$, thereby allowing the number of solutions in which $x_i = 0$ to be inferred from this. Start with *residence(i) = 0* for all i. Then, whenever x_i changes from 1 to 0, after updating *current-iteration* but before updating *tabu-start(i)*, set

$$residence(i) = residence(i) + current\text{-}iteration - tabu\text{-}start(i).$$

Then, during iterations when $x_i = 0$, *residence(i)* correctly stores the number of earlier solutions in which $x_i = 1$. During iterations when $x_i = 1$, the "true" value of *residence(i)* is the right hand side of the preceding assignment, but the update only has to be made at the indicated points when x_i changes from 1 to 0.

As with transition memory, residence memory should be translated into a relative measure (dividing by the total number of iterations, hence solutions generated), as a basis for creating penalties and inducements. The preferred magnitude of penalties and inducements, when not preemptive, is established by empirical test. (See, for example, Laguna and Glover (1993) and Gendreau, Soriano and Salvail (1993).)

3.2 Considerations for Candidate List Strategies

Both solution speed and quality can be significantly influenced by the use of appropriate candidate list strategies. Perhaps surprisingly, the importance of such approaches is often overlooked, though they are fundamental to the TS emphasis on making judicious choices. We give examples of a few candidate list strategies that are particularly useful, and that give a basis for understanding the relevant concerns.

The first of these approaches, called the *Aspiration Plus* strategy, establishes on aspiration threshold for the quality of move to be selected, based on the history of the search pattern. The Aspiration Plus strategy then operates by examining moves until finding one that satisfies this threshold. At this point, an additional number of moves are examined, equal to a selected value *Plus* (e.g., for *Plus* in the interval from 20 to 100), and the best move overall is selected. To assure that neither too few nor too many moves are examined in total, this rule is qualified to require that at least *Min* moves and at most *Max* moves are examined, for chosen values of *Min* and *Max*. (When the upper limit of *Max* moves is reached, before satisfying other conditions, the approach simply selects the best of the moves seen. The values of *Min* and *Max* can be modified as a function of the number of moves required to meet the threshold.)

The aspiration threshold for this approach can be determined in several ways. For example, during a sequence of improving moves, the aspiration may specify that the next move chosen should likewise be improving, at a level based on other recent moves and the current objective value. During a nonimproving sequence the aspiration will typically be lower, but rise toward the improving level as the sequence lengthens. The quality of currently examined moves can shift the threshold, as by encountering moves that significantly surpass or that uniformly fall below the threshold. As an elementary option, the threshold can simply be a function of the quality of the initial *Min* moves examined on the current iteration.

This Aspiration Plus strategy includes many other strategies as special cases. For example, a "first improving" strategy results by setting *Plus = 0* and directing the aspiration threshold to accept moves that qualify as improving, while ignoring the values of *Min* and *Max*. A slightly more advanced strategy can determine *Min* and *Max* to assure some specified additional number of moves will be examined after first satisfying an improving threshold. In general, in applying the Aspiration Plus strategy, it is important to assure that new moves are examined on each iteration that are not among those just reviewed (as by starting where the previous examination left off).

A second useful candidate list approach is the Elite Candidate List strategy. This approach first builds a Master List by examining all (or a relatively large number of) moves, selecting the k best moves encountered, where k is a parameter of the process. (e.g., for $k = 15$ to 50). Then at each subsequent iteration, the current best move from the Master List is chosen to be executed, continuing until such a move falls below a given quality threshold, or until a given number of iterations have elapsed. Then a new Master List is constructed and the process repeats.

The evaluation and precise identity of a given move on the list must be appropriately monitored, since one or both may change as result of executing other moves from the list. Such an Elite Candidate List strategy can be advantageously extended by a variant of the Aspiration Plus strategy, allowing some additional number of moves outside the Master List to be examined at each iteration (where those of sufficiently high quality may replace elements of the Master List).

Another useful candidate list approach is the Successive Filter strategy. Moves can often be broken into component operations, and the set of moves examined can be reduced by restricting consideration to those that yield high quality outcomes for each operation separately. For example, the choice of an exchange move that includes an "add component" and a "drop component" may restrict attention only to exchanges created from a relatively small subset of "best add" and "best drop" components. In such applications, the evaluations of the separate components often will give only approximate information about their combined evaluation. Nevertheless, if this information is good enough to insure a significant number of the best complete moves will result by combining these *apparently best* components, then the approach can be quite effective. Improved information may be obtained by sequential evaluations, as where the evaluation of one component is conditional upon the prior (restricted) choices of another.

Finally we note that a Bounded Change candidate list strategy can be worth considering, provided an improved solution can be found by restricting the domain of choices so that no solution component changes by more than a limited degree on any step. A bound on this degree, expressed by a distance metric appropriate to the context, is selected large enough to encompass possibilities considered strategically relevant. (The metric may allow large changes along one dimension, but limit the changes along another so that choices can be reduced and evaluated more quickly.) Such an approach offers particular benefits as part of an intensification strategy based on decomposition, as discussed in Section 2, where the decomposition itself suggests the limits for bounding the changes considered.

In constructing candidate lists such as the foregoing, the concept of *move influence* is important to longer term considerations. Thus, for example, evaluation criteria should be periodically modified (especially where no improving moves exist) to encourage moves that create significant structural changes. A limit is required on the number of influential moves allowed in a given interval, and more particularly on their cumulative interacting effects, since moves of high influence can also be mutually incompatible as a foundation for generating solutions of the best quality.

An expanded discussion of these considerations, and of additional candidate list procedures is given in Glover, Taillard and de Werra (1993) and Glover (1993).

4 Conclusion

The practical successes of tabu search have promoted useful research into ways to exploit its underlying ideas more fully. At the same time, many facets of these ideas remain to be explored. The issues of identifying best combinations of short and long term memory and best balances of intensification and diversification strategies still contain many unexamined corners, and some of them undoubtedly harbor important discoveries for developing more powerful solution methods in the future.

There are evident contrasts between TS perspectives and the views currently favored by the artificial intelligence and neural network communities, particularly concerning the role of memory in search. However, there are also useful complementarities among these views, which raise the possibility of creating systems that integrate their fundamental concerns. Advances are already underway in this realm, with the creation of *tabu training and learning* models (de Werra and Hertz (1989), Beyer and Orgier (1991), Battiti and Tecchioli (1993), Gee and Prager (1994)), *tabu machines* (Chakrapani and Skorin-Kapov (1993), Nemati and Sun (1994)) and *tabu design* procedures (Kelly and Gordon (1994)). The outcomes from this work have shown promising consequences for supplementing customary connectionist models and paradigms as by yielding levels of performance notably superior to that of models based on *Boltzmann machines*, and by yielding processes for modifying network linkages that give more reliable mappings of inputs to outputs.

Recent years have undeniably witnessed significant gains in solving difficult optimation problems, but it must also be acknowledged that a great deal remains to be learned. Research in these areas is full of uncharted and inviting landscapes.

References

[1] R. Battiti and G. Tecchiolli, "The reactive tabu search," IRST Technical Report 9303-13, to appear in *ORSA Journal on Computing*.

[2] R. Battiti and G. Tecchiolli, "Parallel biased search for combinatorial optimization: Genetic algorithms and tabu," *Microprocessors and Microsystems*, **16** (1992), 351-367.

[3] R. Battiti and G. Tecchioli, "Training neural nets with the reactive tabu search," Technical Report UTM 421, Univ. of Trento, Italy, November 1992.

[4] D. Beyer and R. Ogier, "Tabu learning: A neural network search method for solving nonconvex optimization problems," *Proceedings of the International Joint Conference on Neural Networks*, IEEE and INNS, Singapore, 1991.

[5] J. Chakrapani and J. Skorin-Kapov, "Massively parallel tabu search for the quadratic assignment problem," Working Paper Harriman School for Management and Policy, State University of New York at Stony Brook, 1991.

[6] J. Chakrapani and J. Skorin-Kapov, "Connection machine implementation of a tabu search algorithm for the traveling salesman problem," *Journal of Computing and Information Technology* CIT(1) 1993, 29-36.

[7] T. G. Crainic, M. Gendreau, P. Soriano, and M. Toulouse, "A tabu search procedure for multicommodity location/allocation with balancing requirements," *Annals of Operations Research*, **41**(1-4) 1993, 359-383.

[8] T. G. Crainic, M. Toulouse, and M. Gendreau, *A Study of Synchronous Parallelization Strategies for Tabu Search*, Publication 934, Centre de recherche sur les transports, Universite de Montreal, 1993.

[9] T. G. Crainic, M. Toulouse, and M. Gendreau, *Appraisal of Asynchronous Parallelization Approaches for Tabu Search Algorithms*, Publication 935, Centre de recherche sur les transports, Universite de Montreal, 1993.

[10] F. Dammeyer and S. Voss, "Dynamic tabu list management using the reverse elimination method," Working Paper TH Darmstadt, to appear.

[11] R. L. Daniels and J. B. Mazzola, "A tabu search heuristic for the flexible-resource flow shop scheduling problem," *Annals of Operations Research*, **41** (1993), 207-230.

[12] M. Dell'Amico and M. Trubian, "Applying tabu search to the job-shop scheduling problem," *Annals of Operations Research*, **41** (1993), 231-252.

[13] A. Freville and G. Plateau, "Heuristics and reduction methods for multiple constraint 0-1 linear programming problems," *European Journal of Operational Research*, **24** (1986), 206-215.

[14] A. Freville and G. Plateau, *Methodes Heuristiques Performantes Pour les Problemes in Variables 0-1 a Plussieurs Constraintes en inegalite*, Publication ANO-91, Universite de Sciences et Techniques de Lille, 1992.

[15] A. H. Gee and R.W. Prager, "Polyhedral combinatorics and neural networks," *Neural Computation*, **6** (1994), 161-180.

[16] M. A. Gendreau, Hertz, and G. Laporte, "A tabu search heuristic for vehicle routing," CRT-777, Centre de Recherche sur les transports, Universite de Montreal, to appear in *Management Science*.

[17] M. Gendreau, P. Soriano, and L. Salvail, "Solving the maximum clique problem using a tabu search approach," *Annals of Operations Research*, **41** (1993), 385-404.

[18] F. Glover, "Heuristics for integer programming using surrogate constraints," *Decision Sciences*, **8**(1) January 1977, 156-166.

[19] F. Glover, "Tabu search: Part I," *ORSA Journal on Computing*, **1**(3) 1989, 190-206.

[20] F. Glover, "Tabu search: Part II," *ORSA Journal on Computing*, **2** (1990), 4-32.

[21] F. Glover, "Tabu search for nonlinear and parametric optimization (with links to genetic algorithms)," to appear in *Discrete Applied Mathematics*.

[22] F. Glover, "Tabu thresholding: Improved search by nonmonotonic trajectories," to appear in *ORSA Journal on Computing*.

[23] F. Glover and M. Laguna, "Tabu search," *Modern Heuristic Techniques for Combinatorial Problems*, C. Reeves (editor), Blackwell Scientific Publishing, 1993, 70-141.

[24] F. Glover, M. Laguna, E. Taillard, and D. de Werra (editors), "Tabu Search," special issue of the *Annals of Operations Research*, **41** (1993), J. C. Baltzer.

[25] F. Glover, E. Taillard and D. de Werra, "A user's Guide to tabu search," *Annals of Operations Research*, **41** (1993), 3-28.

[26] P. Hansen, "The steepest ascent, mildest descent heuristic for combinatorial programming," presented at the Congress on Numerical Methods in Combinatorial Optimization, Capri, Italy, 1986.

[27] P. Hansen and B. Jaumard, "Algorithms for the maximum satisfiability problem," *Computing*, **44** (1990), 279-303.

[28] P. Hansen, B. Jaumard, and Da Silva, "Average linkage divisive hierarchical clustering," to appear in *Journal of Classification*.

[29] A. Hertz and D. de Werra, "The tabu search metaheuristic: how we used it," *Annals of Mathematics and Artificial Intelligence*, **1** (1991), 111-121.

[30] R. Hubscher and F. Glover, "Applying tabu search with influential diversification to multiprocessor scheduling," Graduate School of Business, University of Colorado, to appear in *Computers and Operations Research*.

[31] J. P. Kelly and K. Gordon, "Predicting the rescheduling of world debt: A neural network-based approach that introduces new construction and evaluation techniques," Working Paper, College of Business and Administration, University of Colorado, Boulder, CO 80309, 1994.

[32] J. P. Kelly, B. L. Golden, and A. A. Assad, "Large-scale controlled rounding using tabu search with strategic oscillation," *Annals of Operations Research*, **41** (1993), 69-84.

[33] J. P. Kelly, M. Laguna, and F. Glover, "A study of diversification strategies for the quadratic assignment Problem," to appear in *Computers and Operations Research*.

[34] M. Laguna and F. Glover, "Integrating target analysis and tabu search for improved scheduling systems," *Expert Systems with Applications*, 6 (1993), 287-297.

[35] M. Laguna, J. P. Kelly, J. L. Gonzalez-Velarde, and F. Glover, "Tabu search for the multilevel generalized assignment problem," Graduate School of Business and Administration, University of Colorado at Boulder, to appear in *European Journal of Operations Research*.

[36] P. Moscato, "An introduction to population approaches for optimization and hierarchical objective functions: A discussion on the role of tabu search," *Annals of Operations Research*, 41 (1993), 85-122.

[37] P. Moscato and F. Tinetti, "Blending heuristics with a population-based approach: A memetic algorithm for the traveling salesman problem," to appear in *Discrete Applied Mathematics*.

[38] H. Nemati and M. Sun, "A tabu machine for connectionist methods," Joint National ORSA/TIMS Meeting, Boston, MA, 1994.

[39] E. Nowicki and C. Smutnick, "A fast taboo search algorithm for the job shop problem," Report 8/93, Institute of Engineering Cybernetics, Technical University of Wroclaw, 1993.

[40] E. Nowicki and C. Smutnicki, "A fast tabu search algorithm for the flow shop problem," Institute of Engineering Cybernetics, Technical University of Wroclaw, 1994.

[41] I. H. Osman, "Metastrategy simulated annealing and tabu search algorithms for the vehicle routing problem," *Annals of Operations Research*, 41 (1993), 421-451.

[42] I. H. Osman and N. Christofides, "Capacitated clustering problems by hybrid simulated annealing and tabu search," Report No. UKC/IMS/OR93/5. Institute of Mathematics and Statistics, University of Kent, Canterbury, UK, 1993. Forthcoming in: *International Transactions in Operational Research*, 1994.

[43] S. C. Porto and C. Ribeiro, "A tabu search approach to task scheduling on heterogeneous processors under precedence constraints," Monographia em Ciêcia da Computaiço, No. 03/93, Pontificia Universidade Catalica do Rio de Janeiro, 1993.

[44] C. R. Reeves, "Diversification in genetic algorthms: Some connection with tabu search," Conventry University, U.K., 1993.

[45] V. Rochat and A. Semet, "Tabu search approach for delivering pet food and flour in Switzerland," ORWP92/9, 1992.

[46] P. Soriano and M. Gendreau, "Diversification strategies in tabu search algorithms for the maximum clique problem," Publication #940, Centre de Recherche sur les Transports, Universite de Montreal, 1993.

[47] E. Taillard, "Parallel tabu search technique for the job shop scheduling problem," Research Report ORWP 91/10, Departement de Mathematiques, Ecole Polytechnique Federale de Lausanne, 1991.

[48] E. Taillard, "Parallel iterative search methods for vehicle routing problems," *Networks*, 23 (1993), 661-673.

[49] V. Verdejo, R. M. Cunquero, and P. Sarli, "An application of the tabu thresholding technique: Minimization of the number of arc crossings in an acyclic digraph," Departamento de Estadistica e Investigacion Operative, Universidad de Valencia, Spain, 1993.

[50] S. Voss, "Tabu search: Applications and prospects," Technical report, Technische Hochshule Darmstadt, 1992.

[51] S. Voss, "Solving quadratic assignment problems using the reverse elimination method," Technische Hochschule Darmstadt, Germany, 1993.

[52] S. Voss, "Concepts for parallel tabu search," Technische Hochschule Dormstadt, Germany, 1994.

[53] D. de Werra and A. Hertz, "Tabu search techniques: A tutorial and an application to neural networks," *OR Spectrum*, 11 (1989), 131-141.

[54] D. L. Woodruff, "Tabu search and chunking," working paper, University of California, Davis, 1993.

[55] D. L. Woodruff and E. Zemel, "Hashing vectors for tabu search," *Annals of Operations Research*, 41 (1993), 123-138.

Interior Point Path Following Algorithms

Clovis C. Gonzaga

Department of Systems Engineering
and Computer Science
COPPE-Federal University of Rio de Janeiro
Rio de Janeiro, Brazil

1 Introduction

In the last few years the research on interior point methods for linear programming has been dominated by the study of primal-dual algorithms. These algorithms start from the non linear system given by optimality conditions and solve the system by clever applications of Newton's method. These optimality conditions are naturally extended to linear complementarity problems, and it has been found that most algorithms designed for linear programming preserve its properties when applied to certain LCP formulations.

It seems that the most general LCP formulation that preserves both the properties of algorithms for LP and the simplicity of the convergence proofs is the monotone horizontal LCP. For this reason most of this talk will describe methods for this problem.

The literature on interior point methods is huge. Over a thousand references are available, covering a wide range of problems. Our choice for this talk was the following: we stress the basic mechanisms of algorithms for the simplest case, in which an initial interior feasible solution is available. We develop in some detail the algorithms that follow the central path and give some hints on their convergence properties. This development is described in this abstract. In the last part of the talk we comment on how these techniques are extended to the case in which no initial feasible point is available, and make comments on several extensions.

Barrier and Potential Functions

Algorithms may be either based on the use of auxiliary functions, or based on the direct application of Newton's method to the optimality conditions. In many cases there is a close relationship between these two ways of working, as we comment below. Most algorithms for LCP's can be stated in either of these "languages", with similar results. Although it may be said that methods based on auxiliary functions are more general, the formulations based on the direct application of Newton's method are very appealing for their simplicity and will be the choice in this talk.

References
The results described in this abstract are due to too many authors to be cited in an abstract. Our choice is not to include any references at all. This development and an extensive discussion of the literature will be in a book that we are presently writing.

2 Path Following Algorithms

In this text we use the following conventions: Given a vector x, d, the corresponding upper case symbol denotes as usual the diagonal matrix X, D defined by the vector. The symbol e will represent the vector of all ones, with dimension given by the context. We denote component-wise operations on vectors by the usual notations for real numbers. Thus, given two vectors u, v of the same dimension, uv, u/v, etc. will denote the vectors with components $u_i v_i$, u_i/v_i, etc.

The monotone horizontal linear complementarity problem is the following: Find $(x, s) \in \mathbb{R}^{2n}$ such that

$$(P) \qquad \begin{aligned} xs &= 0 \\ Qx + Rs &= b \\ x, s &\geq 0 \end{aligned}$$

where $b \in \mathbb{R}^n$, and $Q, R \in \mathbb{R}^{n \times n}$ are such that for any $u, v \in \mathbb{R}^n$,

$$\text{if } Qu + Rv = 0 \text{ then } u^T v \geq 0.$$

The feasible set for (P) and the set of interior solutions are respectively

$$\begin{aligned} F &= \{(x, s) \in \mathbb{R}^{2n} \mid Qx + Rs = b \,, \, x, s \geq 0\}, \\ F^0 &= \{(x, s) \in F \mid x > 0, s > 0\}. \end{aligned}$$

The set of optimal solutions and the set of strictly complementary optimal solutions are respectively

$$\begin{aligned} \mathcal{F} &= \{(x, s) \in F \mid xs = 0\}, \\ \mathcal{F}^0 &= \{(x, s) \in \mathcal{F} \mid x + s > 0\}. \end{aligned}$$

Example: The Quadratic Programming Problem
This format is quite general. For instance, the convex quadratic programming problem is [*]

$$\begin{aligned} \text{minimize} \quad & c^T x + \frac{1}{2} x^T H x \\ \text{subject to} \quad & Ax = b \\ & x \geq 0, \end{aligned}$$

where $c \in \mathbb{R}^n$, $b \in \mathbb{R}^m$, $A \in \mathbb{R}^{m \times n}$, and $H \in \mathbb{R}^{n \times n}$ is a positive semi-definite matrix.

---

[*]The notation in the example is local, and shall not be used in the remainder of the abstract.

The necessary and sufficient optimality conditions for this problem are

$$
\begin{aligned}
x^T s &= 0 \\
-Hx + A^T w + s &= c \\
Ax &= b \\
x, s &\geq 0
\end{aligned}
$$

Let B be a matrix whose rows span the null space of A. Multiplying the second equation by B, one obtains the equivalent relation $-BHx + Bs = Bc$, so that (x, w, s) satisfies the first-order optimality system if and only if

$$
x^T s = 0
$$
$$
\begin{bmatrix} A & 0 \\ -BH & B \end{bmatrix} \begin{bmatrix} x \\ s \end{bmatrix} = \begin{bmatrix} b \\ Bc \end{bmatrix}
$$
$$
x, s \geq 0
$$

Now let $u, v \in \mathbb{R}^n$ be such that $Au = 0$ and $-Hu + A^T w + v = 0$. Multiplying this equation by u^T, we obtain $u^T v = u^T H u \geq 0$, and conclude that the optimality conditions constitute a monotone linear complementarity problem. This is also trivially true for the linear programming problem, where $H = 0$.

The Optimal Face

The set of optimal solutions $\mathcal{F}$ is a face of the polyhedron F. This means that there are sets of indices $N, B \subset \{1, \ldots, n\}$ such that

$$
\mathcal{F} = \{(x, s) \in F \mid x_N = 0, \ s_B = 0\}.
$$

The complementarity condition $xs = 0$ implies that $N \cup B = \{1, \ldots, n\}$. For simplicity, we assume that there exists $(x, s) \in \mathcal{F}$ such that $x + s > 0$, and this implies that $N \cap B = \emptyset$. Such a solution is called *strictly complementary*: its existence is guaranteed in the case of linear programming by a theorem due to Goldman and Tucker, and it is a hypothesis in the general case. This hypothesis is not needed for proving the complexity results below, but it is essential for establishing superlinear convergence results.

With the hypothesis, $\{N, B\}$ is a partition of the set of indices, called *optimal partition*, and any pair (x, s) in the relative interior of the optimal face satisfies

$$
\begin{bmatrix} x_B \\ s_N \end{bmatrix} > 0 \ , \quad \begin{bmatrix} x_N \\ s_B \end{bmatrix} = 0.
$$

These vectors are called respectively vectors of *large* and *small* variables.

A particular optimal solution has much interest: the central optimum (x^*, s^*), the analytic center of the optimal face. It is given by

$$
(x^*, s^*) = \underset{(x,s) \in \mathcal{F}^0}{argmax} \sum_{i \in B} \log x_i + \sum_{i \in N} \log s_i.
$$

The Central Path

The difficulty in solving (P) is due to the combinatorial problem of determining the optimal partition. A much easier problem is solving the following perturbed system, where $\mu > 0$ and e is the vector of ones:

$$(P_\mu) \qquad \begin{aligned} xs &= \mu e \\ Qx + Rs &= b \\ x, s &> 0 \end{aligned}$$

This system has a unique solution $(x(\mu), s(\mu))$ for any $\mu > 0$, and defines a smooth curve $\mu > 0 \mapsto (x(\mu), s(\mu))$ called *central path*. The central path runs through the set of interior points, keeping a comfortable distance from the non-optimal faces of F, and ends at the central optimum (x^*, s^*).

Adding the equations $x_i(\mu)s_i(\mu) = \mu$, we see that $x(\mu)^T s(\mu) = n\mu$. Taking $x^T s$ as an objective to minimize (it equals the duality gap in linear or quadratic programming), it is enough to find (approximately) points on the central path with $\mu \to 0$. This is what path following algorithms do, by following a homotopy approach:

Algorithm 2.1 *Data:* $\epsilon > 0$, $\mu^0 > 0$.

$\quad k := 0$

$\quad$ repeat

$\qquad$ Find *approximately* $x(\mu^k)$, $s(\mu^k)$.

$\qquad$ Choose $\mu^{k+1} < \mu^k$.

$\qquad k := k + 1$.

$\quad$ until $\mu^k < \epsilon$

The Proximity Measure

The word 'approximately' in the algorithm model above will be our concern now. Given $\mu > 0$, we want to find (x, s) such that $xs/\mu = e$. The deviation of this condition for a given feasible pair (x, s) is given by the proximity measure

$$\delta(x, s, \mu) = \left\| \frac{xs}{\mu} - e \right\|.$$

If $\delta(x, s, \mu) < 0.5$, then it related to the distance from (x, s) to the central point for μ in a certain norm. By adding components as we did above, we conclude that

$$x^T s \le (n + 0.5\sqrt{n})\mu,$$

and again the measure of complementarity $x^T s$ is well related to μ for such *nearly central points*.

In a moment we shall describe the Newton iteration for solving (P_μ). We can advance now what seems to be the most important result in interior point methods:

Theorem 2.2 _Given_ $(x, s) \in F$ _and_ $\mu > 0$ _such that_ $\delta(x, s, \mu) = \delta \leq 0.7$, _let_ (x^+, s^+) _be the result of a Newton step for_ (P_μ) _from_ (x, s). _Then_

$$\delta(x^+, s^+, \mu) \leq \delta^2.$$

This means that Newton's method is very efficient for solving P_μ in the sense that it reduces the proximity measure quadratically in a large region. The region in which efficient centering steps can be used is a neighborhood of the central path, defined as follows.

Let $\alpha \in (0, 1)$ be given (usually taken as $\alpha = 0.5$). Our neighborhood of the central path will be defined as

$$\mathcal{N}_\alpha = \bigcup_{\mu \in (0, \infty)} \{(x, s) \mid \delta(x, s, \mu) \leq \alpha\}.$$

Path following algorithms will generate sequences in $\mathcal{N}_\alpha$.

The Newton Step

Let $(x, s) \in F^0$ and $\mu > 0$ be given. Following the homotopy scheme above, our goal is approaching the central point $(x(\gamma\mu), s(\gamma\mu))$ for some $\gamma \in [0, 1]$. Ideally, we want to find

$$x^+ = x + u \quad , \quad s^+ = s + v$$

such that $(x^+, s^+) \in F$ and $x^+ s^+ = \gamma\mu e$. The Newton step solves this approximately by linearizing P_μ. The linearization is straightforward, and leads to the following system, which has a unique solution:

$$(P_N) \qquad \begin{aligned} xv + su &= -xs + \gamma\mu e \\ Qu + Rv &= 0 \end{aligned}$$

Before we describe algorithms based on the Newton step, let us examine how it depends on the choice of γ. There are two obvious choices, $\gamma = 0$ and $\gamma = 1$, which we describe now.

(i) $\gamma = 0$: the affine-scaling step (u^a, v^a).

Choosing $\gamma = 0$, we try to solve (P) directly. This is the boldest possible choice of γ, and the resulting pair

$$(x^a, s^a) = (x, s) + (u^a, v^a)$$

is in general infeasible. When this direction is used, it is always necessary to do a line search to avoid leaving the proximity of the central path.

(ii) $\gamma = 1$: the centering step (u^c, v^c).

This choice is the most conservative one. Its importance stems from the centering theorem 2.2. It gives the fastest possible approach to the central path. If (x, s) is nearly centered, then the resulting pair

$$(x^c, s^c) = (x, s) + (u^c, v^c)$$

will be much nearer the central path.

Observing now that the right hand side of the first equation in (P_N) can be written as

$$-xs + \gamma\mu e = \gamma(-xs + \mu e) + (1-\gamma)(-xs),$$

it follows by superposition that for any $\gamma \in [0,1]$,

$$(u, v) = \gamma(u^c, v^c) + (1-\gamma)(u^a, v^a),$$

and also

$$(x^+, s^+) = \gamma(x^c, s^c) + (1-\gamma)(x^a, s^a).$$

The Algorithms

Now, we describe the general scheme of path following algorithms by refining the homotopy model above. Each iteration computes a Newton step and possibly uses a line search along it. This is still a model, with an open choice of parameters, which will be specified below.

Algorithm 2.3 *Data:* $\epsilon > 0$, $(x^0, s^0) \in \mathcal{N}_\alpha$, $\mu^0 > 0$ *such that* $\delta(x^0, s^0, \mu^0) \leq \alpha$.

$k := 0$

repeat

    Choose $\gamma \in [0,1]$.

    Solve (P_N) from $(x, s) := (x^k, s^k)$.

    Choose $\lambda \in (0,1]$ and set $(x^{k+1}, s^{k+1}) := (x^k, s^k) + \lambda(u, v)$.

    Choose $\mu^{k+1} \in (0, \mu^k)$

    $k := k + 1$.

until $\mu^k < \epsilon$

The choice of γ, λ and μ^{k+1} in each iteration must be such that for all k, $\delta(x^k, s^k, \mu^k) \leq \alpha$.

Now, we describe three interesting ways of choosing these parameters.

(i) The Short Step Algorithm

At each iteration, choose

$$\gamma = 1 - \frac{0.2}{\sqrt{n}}, \quad \lambda = 1, \quad \mu^{k+1} = \gamma\mu^k.$$

In the beginning of the iteration k, assume that $\delta(x^k, s^k, \mu^k) \leq 0.5$. We now show that $\delta(x^k, s^k, \mu^{k+1}) \leq 0.7$.

We must prove that

$$\|x^k s^k - (1 - \frac{0.2}{\sqrt{n}})\mu e\| \leq (1 - \frac{0.2}{\sqrt{n}})\mu.$$

We have:

$$\|x^k s^k - (1 - \frac{0.2}{\sqrt{n}})\mu e\| \leq \|x^k s^k - \mu e\| + \|\frac{0.2}{\sqrt{n}}\mu e\| \leq 0.5\mu + 0.2\mu,$$

and the proof is complete.

Now, directly from the centering theorem,

$$\delta(x^{k+1}, s^{k+1}, \mu^{k+1}) \leq 0.7^2 < 0.5.$$

We conclude that since $\delta(x^0, s^0, \mu^0) \leq 0.5$, all points satisfy this same inequality, and the algorithm is well defined.

The short step algorithm is obviously not practical, but it gives an immediate proof of polynomiality: we have

$$\mu^k \leq (1 - \frac{0.2}{\sqrt{n}})^k \mu^0,$$

and hence

$$log_2(\frac{\mu^k}{\mu^0}) \leq k \, log_2(1 - \frac{0.2}{\sqrt{n}}) \leq -0.2\frac{k}{\sqrt{n}}.$$

Given an integer $L > 0$, we conclude that after no more than $5\sqrt{n}L$ iterations, $\mu^k \leq 2^{-L}\mu^0$. This essentially means that (we are omitting some simple details) $x^T s$ can be reduced to the value 2^{-L} in $O(\sqrt{n}L)$ iterations. For linear programming, if L is the bit length of the input data and $\mu^0 = O(2^L)$, then an exact optimal solution can be obtained by purification (rounding) after $O(\sqrt{n}L)$ iterations.

(ii) The Largest Step Algorithm

At each iteration, choose $\lambda = 1$ (no line searches) and $\mu^{k+1} = \gamma\mu^k$. We want to determine the minimum possible γ such that the result of the pure Newton step is in $\mathcal{N}_\alpha$. This is done by the following computation:

Solve (P_N) for $\gamma = 0$ and $\gamma = 1$, obtaining respectively the pairs (x^a, s^a) and (x^c, s^c).

Compute γ such that $\delta(\gamma x^c + (1 - \gamma)x^a, \gamma s^c + (1 - \gamma)s^a, \gamma\mu^k) = 0.5$.

Set $(x^{k+1}, s^{k+1}) := \gamma(x^c, s^c) + (1 - \gamma)(x^a, s^a)$.

The computation of (x^a, s^a) and (x^c, s^c) needs one factorization and two back substitutions. The computation of γ is a problem in one variable, and hence easy. Developing the definition of the proximity δ, it is shown that γ is a root of a quadratic equation; if there is more than one positive root, one should choose the smallest one.

This provides the largest possible step using perfect Newton iterates that stay in the neighborhood $\mathcal{N}_\alpha$. The polynomial bound of $O(\sqrt{n}L)$ iterations is obviously kept, since at all iterations γ is smaller than the value used in the short step method.

(iii) The Predictor-Corrector Algorithm

This algorithm is similar to the one above, but it uses the affine-scaling direction and the centering step in alternate iterations. These are the parameter choices:

Odd iterations (corrector step): given (x^k, s^k) and μ^k such that $\delta(x^k, s^k, \mu^k) \leq 0.5$, take a centering step ($\gamma = 1$, $\lambda = 1$, $\mu^{k+1} = \mu^k$), to obtain by the centering theorem

$\delta(x^{k+1}, s^{k+1}, \mu^{k+1}) \leq 0.25$.

Even iterations (predictor step): given (x^k, s^k) and μ^k such that $\delta(x^k, s^k, \mu^k) \leq 0.25$, compute the affine-scaling step ($\gamma = 0$) and compute $\lambda \in (0, 1)$ such that

$$\delta(x^k + \lambda u^a, s^k + \lambda v^a, (1 - \lambda)\mu^k) = 0.5.$$

Set $(x^{k+1}, s^{k+1}, \mu^{k+1}) := (x^k + \lambda u^a, s^k + \lambda v^a, (1 - \lambda)\mu^k)$.

Each predictor-corrector cycle needs two complete solutions of (P_N), and the polynomial bound of $O(\sqrt{n}L)$ iterations is preserved.

Iterate Convergence

The sequences generated by the methods described above always converge to an optimal solution. The short step and the predictor-corrector algorithms generate sequences that converge to the central optimum (x^*, s^*), while the largest step algorithm finds in general a point in the proximity of the central optimum. Whenever the optimal face is not a vertex, the limit point in this method is likely to differ from (x^*, s^*).

Asymptotic Rates of Convergence

This study depends on the separate behaviour of small and large variables along the sequences generated by the algorithms. The short steps method obviously leads to linear convergence of the objective $x^T s$. The predictor-corrector algorithm has quadratic convergence when we consider each pair of steps as one iteration. The largest step algorithm is superlinearly convergent in objective values, but it achieves quadratic convergence with the addition of a computationally trivial safeguard.

3 Concluding Remarks

To conclude this abstract, we make some remarks on infeasible interior point algorithms.

In general, no initial feasible point is available. The algorithms should be able to either find an optimal solution or to certify that no optimal solution exists and explain why (it should detect infeasibility and unboundedness in the case of linear or quadratic programming).

Given an initial arbitrary positive pair (x^0, s^0), let

$$r = Qx^0 + Rs^0 - b.$$

Here r measures the infeasibility at the initial point. Path following algorithms are constructed by perturbing the problem (P) in the following way:

$$(P_{\mu,\theta}) \qquad \begin{aligned} xs &= \mu e \\ Qx + Rs - b &= \theta r \\ x, s &> 0 \end{aligned}$$

Note that if (x^0, s^0) is feasible, this reduces to (P_μ). Otherwise the solutions of this system form a two-dimensional surface of analytic centers. Trajectories in this surface are obtained by relating the parameters μ and θ, for instance by setting $\mu = \theta$ (the most popular choice).

The relationship between the parameters is still not well understood, but much research on this topic is in progress now.

A different way of approaching infeasible point methods for linear programming starts from the construction of a homogeneous self-dual problem with a bounded optimal set, whose optimal solutions lead either to optimal solutions to the original problem or to the detection of infeasibility. The solution of the homogeneous problem is done by the feasible point methods discussed above, and low polynomial complexity is achieved.

In the talk, we summarize these results, and comment on several extensions and variants.

Finally, I regret that it is infeasible to include in this talk any comments on the generalizations made by Nesterov and Nemirovsky, using their concept of self-concordant barriers, since this would take the whole lecture (maybe to the advantage of the public).

Computational Differentiation and Optimization

Andreas Griewank

Institute of Scientific Computing
Technical University Dresden
and
Mathematics and Computer Science Division
Argonne National Laboratory
Argonne, Illinois

1 Introduction

Optimization should be a prime application area for automatic differentiation. Unfortunately, this chain-rule-based technique has not even approached its full potential. Yet, its basic variants — the forward and the reverse mode — have been known in some form for more than twenty and thirty years, respectively. For a historical survey of automatic differentiation and related areas one may consult the article by Masao Iri in the Proceedings of the Breckenridge conference in 1991 [22]. The same volume also contains several applications studies and a survey by David Juedes of the-then existing software packages. M.C Bartholomew et. al. have recently published a careful numerical comparison of dense and sparse versions of the forward and reverse mode implemented in ADA and applied to optimization test problems [2].

The efficient use of function and derivative information has long been a chief concern of algorithm design in nonlinear optimization. Since the gradients of objective and constraint functions enter directly into the Kuhn-Tucker-Karush conditions, their accurate evaluation is virtually necessary for achieving good solution accuracy. It is also well understood that some of the curvature information represented by the Hessian of the Lagrangian needs to be evaluated or approximated if one wishes to achieve superlinear convergence of any kind.

Except for rather limited efforts to design derivative-free optimization methods in the unconstrained case (see, e.g., Nelder-Mead, Powell, Brent), it has usually been understood that gradient values are indispensable for efficient optimization calculations but that Hessians may be approximated unless they can be formed and factored quite cheaply. For unstructured problems of moderate size, this assessment can be confirmed in the light of recent results from automatic, or computational, differentiation. For large and structured problems, however, secant approximations to Hessians and Jacobians have proven, in general, quite unsatisfactory. The reason is that low-rank corrections require many iterations to make up for high-dimensional derivative discrepancies and that (sparse) high-rank updates destroy the lower linear algebra cost per step, which classical secant methods enjoy in comparison with Newton's method.

On discretizations of operator equations and other regularly sparse problems, Jacobians and Hessians can often be evaluated by hand-coded derivative routines, which significantly reduce the run time compared with the divided difference alternative. To facilitate the hand coding of derivatives with reasonable human effort and reliability, computational scientists sometimes eschew more sophisticated models and discretizations in favor of algebraically simpler ones. With the assistance of suitable automatic differentiation software, this self-imposed impediment to accurate computer modeling can be avoided, and reliable derivative codes can be generated with little human effort.

Apart from being useful as a tool for optimization software, automatic differentiation poses a number of challenging discrete optimization problems. One problem is the efficient calculation of Jacobians by a process of vertex elimination in a directed graph, which is closely related to the problem of sparse Gaussian elimination that has received considerable attention in the literature. Adapting the original proof of Rose and Tarjan [35], Herley [26] has recently shown that the minimum fill-in problem for Jacobian accumulation is also NP-complete. While promising results have been obtained using the Markowitz criterion [24], there are important classes of problems where the resulting elimination order is much less efficient than the obvious ones that correspond to the classical forward and reverse mode. Hence there is considerable scope for the development of practical heuristics and suitable implementations in the future. A related task is that of computing Newton steps or their approximations directly, that is, without forming the Jacobian at all [18]. Finally, if one wishes to reduce the potentially large storage requirement of the reverse mode by recalculating certain intermediate results from suitable check points, minimizing the number of operations becomes also a difficult discrete optimization task. This scheduling problem can be solved optimally [20] if the evaluation consists of a sequence of computational steps that are homogeneous with regards to their temporal and spatial complexity, as can be expected for time-dependent problems. In general, this problem is also very difficult, especially since a practical preprocessor implementation would have to be based on the partial information about the computational graph that is discernible at compile time.

In this survey article we cover both aspects of automatic differentiation: its utility for providing derivatives in nonlinear programming and its role as a source of combinatorial optimization tasks. In Section 2 we discuss the basic functionality of automatic differentiation in comparison to other techniques for evaluating or estimating derivatives. In Section 3 we survey the basic forward and reverse modes as well as their complexities. The key conclusion is that gradients have essentially the same operations count and memory requirements as the underlying scalar function.

In Section 4 we discuss the ramifications for optimization with regards to gradient methods, truncated Newton schemes, and their preconditioning. In Section 5 we consider the evaluation of Jacobians and one-sided projection of the Lagrangian Hessian, which turn out to be no more expensive than the two-sided projection. In Section 6 we briefly examine the differentiation of evaluation procedures that contain iterative processes and other nonsmooth or potentially nonsmooth aspects. Section 7 describes two combinatorial optimization problems associated with automatic differentiation in terms of the computational graph.

2 Functions and Arguments IN ⇒ Derivative Values OUT

A major stumbling block for the application of automatic differentiation techniques and tools for the benefit of optimization and other numerical purposes has been a serious confusion in terminology. Depending on their educational background and professional interests, people have very different notions of what the terms "function", "derivative", "explicit formula", "expression tree", "analytical differentiation", "real variable" etc., mean. The meanings of such terms in a certain context usually become established through common usage, but so far there simply has not been enough discussion on the topic of constructive derivative calculations. On the other hand, the common interpretation of the term "numerical differentiation" is rather unfortunate, since it tends to be understood as the natural alternative to "symbolic differentiation", a dichotomy that apparently leaves no room for anything else. Depending on one's preferences, one may view automatic differentiation as numerical because it "only" yields real values of derivatives, or one can view it as symbolic because there is no truncation error, or both. The term "automatic" differentiation itself is at best nondescriptive and at worst misleading, since it suggests that users can simply plug in some CD software into their application without learning anything about the underlying principles. As one might suspect, that is not quite the case, especially when codes involve iterative procedures or more general branching. The here preferred term "computational" differentiation is no more descriptive, but probably less misleading.

2.1 Vector Functions as Evaluation Procedures

Since computational differentiation is based on variations of the chain rule, it cannot be applied to functions as abstract mathematical objects but only to their realization in form of an evaluation procedure, usually of a computer program. The same mathematical mapping may have several realizations with widely varying properties regarding run-time efficiency and numerical stability. In contrast to symbolic manipulators, automatic differentiation tools generally do not try to simplify or otherwise improve the user supplied evaluation code but merely enhance it with additional instructions for computing derivatives.

We strongly believe that the issue of code quality is largely independent of the differentiation task. In other words, the derivative-enhanced procedure will be fast and accurate exactly if the original evaluation procedure is. Our main justification for this claim is that the code extensions needed for automatic differentiation are local, in that each statement of the original code spawns a small number of derivative assignments or adjoint statements. The latter extensions occur in the so-called adjoint or reverse mode of automatic differentiation, which also involve some additional save and restore operations.

From now on we will assume that the vector function consisting of all objectives and constraints is user supplied in the form of a procedural evaluation program written in Fortran(90), C(++), ADA or some other programming language, including in principle even assembly language. We believe that this assumption is valid in the overwhelming majority of optimization applications, even though speakers at mathematical programming meetings occasionally allude to the possibility that the function evaluation process involves physical experiments, demoscopic surveys, or actual pseudo-random number generation. In the latter cases it seems rather unlikely that a calculus-based optimization scheme could be of much use, so differentiation would not be an issue anyway. Note that we do not require the evaluation procedure to be a straight-line program. Subroutine calls and branching are quite acceptable, even though the latter may destroy differentiability and even continuity in the strict mathematical sense.

2.2 Common Intermediates, Graphs, and Locations

Sequential function evaluation procedures have at least two key ingredients that are of no concern from the strictly mathematical point of view: the occurrence of common intermediates (sometimes subexpressions) and the allocation of storage space for scalar and vector quantities.

Common intermediates are of central importance for the relative efficiency of various differentiation modes and imply, in particular, that for an evaluation procedure defining a vector function

$$F : I\!R^n \to I\!R^m,$$

the ratio of operation counts

$$\gamma\{F\} \equiv \left(\sum_{i=1}^{m} \text{OPS}\{F_i\} \right) / \text{OPS}\{F\} \tag{1}$$

may be much greater than one. In that sense a vector function is not just the concatenation of its components. Interpreting all intermediate quantities generated by an evaluation program as vertices connected by edges representing data dependencies, one obtains an acyclic graph as displayed in Figure 1 for an example with two independents, two dependents, and one common intermediate.

This *computational graph* can reduce to a collection of trees only if there are no common intermediates. Yet, this exceptional case of an expression forest seems to be what many people have in mind when they talk about functions being given "explicitly" or "analytically". They then are likely to ask for similarly "symbolic" formulas for the derivatives, that is, character strings representing the trees in in-fix notation. Computer algebra people tend to advertise this nonprocedural way of specifying functional relations as natural to the mathematician and engineer. Of course, it is always possible to convert a directed acyclic graph into a forest by replicating intermediate vertices as often as they occur on a distinct path to a dependent vertex, or leaf. This corresponds to the symbolic substitution of common intermediates and tends to expand the number of vertices exponentially.

We believe that the popular notion of explicit or analytic functions and their derivatives is quite inappropriate for problems of any significant size. What we look for instead are procedures for evaluating derivatives at arbitrary arguments that derived more or less automatically from procedures for evaluating the functions itself. For small problems a display of the acyclic graph as shown in Figure 1 may provide analytical insight into the nature of the function. The box associated with each intermediate quantity contains its value v an adjoint value g and its one or two elementary partial derivatives with respect to its one or two predecessor nodes, respectively. These *elementary partials* were multiplied by the reverse mode (to be discussed in Section 3.4) to yield the adjoints g in the independent boxes, which represent the gradient of the weighted average $0.2 * y_1 + 0.3 * y_2 + 0.5 * y_3$ with respect to x_1, x_2, x_3, and x_4. The first common intermediate $v = x_2 * x_3$ would occur six times in the corresponding forest of three expression trees.

For any problem function whose evaluation takes a nonnegligible fraction of a second on a modern workstation, the computational graph will be so large that it can be hardly kept in core, let alone displayed on the screen. Yet the function evaluation procedure itself may require only very little memory space, since long sequences of intermediate quantities can replace each other in the same memory location. The pattern of allocating mathematical quantities into storage places is determined largely by the programmers variable declarations and to a lesser extent the compiler's handling of temporaries. The usage of partitioned work arrays, equivalences and other space-saving tricks makes it hard or even impossible for the compiler to perform a meaningful dataflow analysis for the purposes of code optimization on a target computer architecture. Very similar effects must be considered with regard to differentiation, which requires a tight identification of all intermediate quantities that are *active*, that is, may lie on a dependency path between an independent and a dependent variable. All other variables can remain *passive* in that no derivative data and calculations need to be included for them in the enhanced code. For successful analysis of compile-time activity, it is highly desirable that the user does not assign both active and passive quantities to the same memory locations or mix these two kinds of variable in the same array.

2.3   Points, Directions, and Weights

By its numerical nature, automatic differentiation is not restricted to the evaluation of Cartesian derivatives, that is, first or higher derivatives of certain dependent variables y_i with respect to certain independent variables x_j. Instead of computing in this way full gradients, Jacobians, Hessians, or other derivative arrays, one can immediately restrict the dimensionality of domain and range by differentiating weighted linear combinations $u^T y$ of the dependents with respect to directions v of the independents. In particular, one can directly compute for a vector function F the Jacobian vector

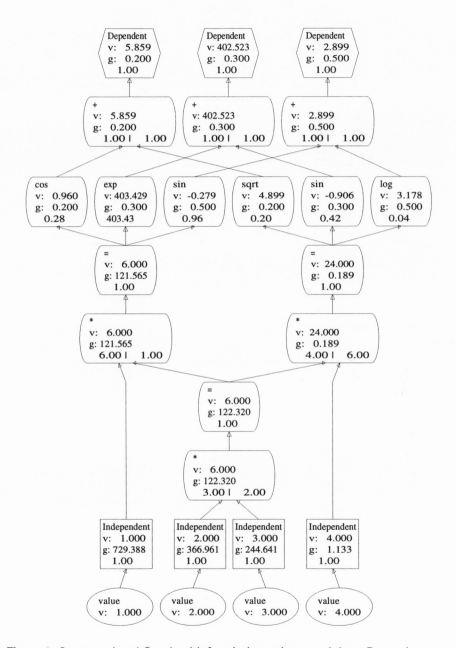

Figure 1: Computational Graph with four Independents and three Dependents

products
$$F'(x)\,w \quad \text{and} \quad u^T F'(x)$$

in the forward and reverse mode, respectively. Both products are obtained at a complexity comparable to that of evaluating F by itself, which may be orders of magnitude cheaper than the cost of computing the full Jacobian.

In general, automatic differentiation tools avoid first evaluating the full derivative tensor and subsequently multiplying it from the left and right with row and column vectors. Instead, they build this contraction implicitly into the differentiation process. A typical application is the evaluation of Lagrangian gradients and Hessians, where the dependents are weighted by Lagrange multipliers and the independent directions can often be restricted to the tangent space of the feasible set. If full Cartesian derivative matrices and tensors are actually needed, they can be obtained by letting u and w range over the m and n Cartesian basis vectors of the range and domain, respectively. This can be done either sequentially over several evaluation sweeps or simultaneously with u or w replaced by the identity matrices of order m and n, respectively. When the Jacobian $F'(x)$ has a known sparsity pattern, these identity matrices can often be compressed to rectangular *seed matrices* with a much smaller number of rows or columns, respectively. The prior determination of the sparsity pattern is in principle no problem for CD tools, but it has to our knowledge not yet been implemented.

2.4   How It Can Be Put Together

Ideally one may imagine the following scenario for using automatic differentiation in an optimization context. The user provides source code for the objective function and the constraints in one or more files listed in a descriptor file. Either there or directly in the function source(s) the user nominates the independent and dependent variable scalars and vectors, where the latter may be of variable dimensions. Together with the optimization library comes a makefile that tells the user how to invoke a suitable automatic differentiation package to process the function sources and generate extensions that can be called upon by the optimizer to evaluate weighted directional derivatives at arbitrary points. These extensions can take the form of compilable sources (see e.g. ADIFOR[4] [1]), scratch files that are interpreted by problem independent CD utilities (see e.g. ADOL-C[23] [2]), or combinations thereof (see e.g. GRESS[27] [3] and PADRE2[30] [4]).

After these preparations, the actual optimization can be started. It could eventually be stopped with the help of evaluation error estimates also provided by the CD tool. There are other pieces of information that the CD tool could extract from the function sources and provide to the optimizer for gains in efficiency and reliability. Such information might include sparsity patterns, partial separability, and linearity

---

[1] Contact: Christain Bischof (adifor@mcs.anl.gov)

[2] Contact: Andreas Griewank (adol-c@math.tu-dresden.de)

[3] Contact: Jim Horwedel (jqh%ornlstc.bitnet)

[4] Contact: Koichi Kubota (kubota@ae.keio.ac.jp)

of certain functions or derivatives with respect to certain variables. Especially on adaptive discretizations of operator equations, providing this redundant information in addition to the function sources can be quite laborious and prone to error. The immediate gain of the envisioned integration of differentiation and optimization would be user convenience, but we can also imagine a significant effect on the algorithmic design of optimization methods. Rather than being restricted to calling up an oracle that provides only function, gradient, and possibly Hessian values, the optimization method can now ask for more selected or structural information regarding the problem at hand and can use it accordingly. Naturally, the exchange of information between the two software packages and the user can happen in an interactive fashion, possibly with the help of a graphical interface.

Actually, full integration of differentiation and optimization has already occurred in some modeling packages like GAMS [10]. However, we think it is sometimes important. that problem functions need not be specified in a certain input format, but can be defined by Fortran or C code. These codes may have been generated automatically by an application specific package with more or less "symbolic" inputs. It is hoped that, eventually differentiation facilities with nearly optimal user-convenience and efficiency will be provided by compilers. Now (and for the forseeable future) however, automatic differentiation tools are based on source generation by compiler-like preprocessors or some form of function and operator overloading. The former approach is mostly applied to Fortran 77 and usually achieves better run-time performance through compilable derivative codes. The latter is comparatively easy to implement in ADA, C++ or Fortran 90 and provides more flexibility to programmer and user. The resulting interpretive overhead can be amortized if more complex derivative objects, such as truncated Taylor series or gradients and Hessians in a sparse storage mode, are propagated in an evaluation sweep. Similar considerations apply to communication costs on multiprocessors. Optimization methods should be designed to bundle evaluation requests whenever possible in order to reduce the overhead. Nevertheless, one must be prepared that, especially on small test problems, the run time needed by general purpose CD packages may be longer by factors of ten or more than those of hand coded derivative programs, which quite often come close to the theoretical complexity ratios discussed in the following section.

2.5   If We Don't Get It All Together

Especially in the design of complex systems such as aircraft, it may be impossible to collect the source codes for the various substructures like wing, fuselage, propulsion, and their aerodynamics on one computing platform. Then the whole multi-layered evaluation procedure cannot be processed together by one CD tool. Apart from the fact that the codes may be written in different languages, they may not even run under the same operating system. It then becomes necessary that, together with intermediate quantities, some corresponding sensitivity information is passed across the interfaces between the various disciplines in a suitable format. Similarly, messages between processes on a distributed system need to be enhanced with sensitivity information, which makes dependency and activity analysis quite hard. For the reverse mode discussed below this situation seems to require message logging between global synchronization levels. In general, their has been very little work in this ares of what one might call distributed differentiation.

2.6 Section Summary

Roughly in agreement with C terminology, we consider a 'function' as a program consisting of one or more routines written in a procedural language that may call each other, possibly in a recursive fashion. From among the input parameters of the top-level routine, the user must distinguish a set of independent variables and similarly select a set of dependent variables from the output parameters. The user should strive to organize variables that need to be active and those that may remain passive into separate data structures and should perform the corresponding calculations in separate subroutines whenever possible. Other than that, the criteria for good and efficient programming are the same as for numerical evaluation procedures without derivative calculations.

3 Basic Assumptions and Techniques

We believe that anybody involved in developing numerical methods for nonlinear problems should have an understanding of the simple complexity results reviewed here. These upper and lower bounds on operations counts and memory usage are sharp up to small constants, and some are not restricted to particular differentiation techniques or implementations. These are mathematical statements about functions that are compositions of elementary functions from a certain finite pool of differentiable functions including the basic arithmetic operations. In the context of optimization we are interested mainly in the cost of evaluating gradients, Jacobians, and Hessians, or projections thereof.

3.1 A Simple Example

To illustrate the results we consider the vector function F and the scalar function f defined by

$$f(x) \equiv \frac{1}{2}\|F(x)\|^2 \quad \text{with} \quad F(x) \equiv B^{-1}(Ax) - b \,, \tag{2}$$

where B is an upper bidiagonal matrix of order m and A a sparse $m \times n$ matrix with $k \geq n$ nonvanishing elements, including all entries in the last row. The parentheses around the matrix-vector product Ax indicate that the evaluation of F uses a back substitution process to solve $Bz = Ax$ for an intermediate vector z, which then immediately yields the value of F as $z - b$. The linking through the bidiagonal matrix B has the effect that the ratio defined in (1) satisfies

$$\gamma\{F\} \quad = \quad \frac{m(k + m/2)}{k + m} \quad \approx \quad m$$

since evaluating the first component of F involves computing all others. For this simple linear least squares example, one can make immediately the following observations:

- The number of operations needed for evaluating the gradient

$$\nabla f^T \quad = \quad F(x)^T F'(x) = \left[F(x)^T B^{-1} \right] A$$

  or any other vector Jacobian product $u^T F'(x)$ together with f is no more than twice that needed for evaluating f by itself.

- For generic bidiagonal B the Jacobian

$$F'(x) \quad = \quad B^{-1} A$$

  is dense and its calculation by back substitution requires nm operations, compared with $m + k$ for evaluating F by itself.

- The Hessian

$$\nabla^2 f(x) \quad = \quad A^T [B^{-T}(B^{-1}A)]$$

  is also dense and can be computed according to the given bracketing using $n(2m + k)$ arithmetic operations, compared with $(2m + k)$ for evaluating f by itself.

- When $n = m$ calculating the diagonal of the Jacobian $F'(x)$ using n partial back substitutions costs about half as many operations as evaluating the full Jacobian.

The last point applies similarly to the Hessian $\nabla^2 f(x)$ and shows that evaluating diagonals for preconditioning purposes requires in general within a constant as much work as evaluating the full Jacobian. This is bad, news since evaluating Jacobians or Hessians can be $min(m, n)$ times as expensive as evaluating the underlying functions and gradients, as is exactly the case for the Hessian in this example, provided $m \geq n$. We will see that $min(m, n)$ is in fact an upper bound, which is not really that surprising. On the other hand, many sparse or otherwise structured Jacobians and Hessians that can be calculated by a suitable variation of the chain rule, at a small multiple of the cost of evaluating the underlying functions or gradients. Probably the key result of automatic differentiation is the *Cheap Gradient* theorem, which holds under the following natural assumptions on the functions and their evaluation procedures.

3.2 Elementary Assumptions

Like fully symbolic manipulation, computational differentiation is based on the realistic assumption that the function $y = f(x)$ is evaluated by a sequence of scalar assignments

$$v_i = \phi_i(v_j, v_k) \quad \text{or} \quad v_i = \phi_i(v_j), \tag{3}$$

where $j, k < i$ and $i = 1, \ldots, p + m$. For notational convenience we may number the scalar variables v_i such that the first n represent the independent variables $v_{i-n} \equiv x_i$ for $i = 1, \ldots, n$, and the last m represent the dependent variables $y_i \equiv v_{p+i}$ for $i = 1, \ldots, m$. The binary or unary functions ϕ_i are usually either arithmetic operations or transcendental intrinsics like *exp, log, sin, cos, abs,* and *max.* The key property of these and other possible elementary functions is that, throughout most of their respective domains, the first and higher derivatives can be cheaply and accurately evaluated. In other words, the elementary functions can be differentiated "analytically". In contrast to a fully symbolic treatment, automatic differentiation (see [34] and [28]) is based on evaluating the elementary partials at the current argument so that ultimately only "numerical" but truncation-error-free derivative values are obtained. In infinite precision these values would be exact, and in practice they are usually obtained with working accuracy.

Of course, one cannot expect to obtain useful derivative information from a chain-rule-based technique if one or more of the elementary functions are evaluated at a point where they are not properly differentiable. For example, this caveat applies to the absolute value and the square root functions at the origin. Such singularities need not show up in the function value itself because they may cancel fortuitously with themselves or another nonsmooth function. In one code we found a loop that summed the Euclidean distance between several pair of grid points in Euclidean space. Because of a wraparound effect, the first and last points were always identical, so that a zero distance was added to the sum without resulting in any numerical difficulties. Upon differentiation, however, a division of zero by zero occurred, and NaN's were propagated throughout the derivative calculation. (Here NaN stands for *Not a Number* according to the IEEE standard for computer arithmetic.) This problem can be bypassed by defining the derivative(s) of $\phi(v) = \sqrt{v}$ at $v = 0$ as 0 or some finite number, but in any case the user should probably be warned. Inaccurate or discontinuous derivatives often arise through branching. For a careful discussion of these issues, see the working note [7].

Assuming that all elementary functions ϕ_i are continuously differentiable on some neighborhood of the arguments of interest, one can show quite easily that gradient evaluation by the reverse mode is backward stable in the sense of Wilkinson. In other words, the numerically computed gradients correspond to the exact derivative values of a composite function whose elementary partials have been perturbed at the level of the working accuracy. This observation from [24] is somewhat unsatisfactory because it conceptually requires the perturbation of the 1's that represent the partials of additions and subtractions. Apparently, a complete error analysis of the automatic differentiation process has not yet been published. Nevertheless, it is abundantly clear that computational differentiation is not an ill-conditioned procedure since it is based on a computer program or other symbolic representation of the function in question. The traditional assertion that differentiation is an ill-conditioned process applies only if the function is given by an oracle that produces values with a given accuracy.

3.3 The Forward or Direct Mode

Denoting by v' the derivative with respect to one or more given directions in the domain of x, one obtains directly from the chain rule the recurrence

$$v_i' = \frac{\partial \phi_i}{\partial v_j} v_j' + \frac{\partial \phi_i}{\partial v_k} v_k' \quad \text{or} \quad v_i' = \frac{\partial \phi_i}{\partial v_j} v_j' . \tag{4}$$

If initially $v_{i-n}' = x_i' = w_i$, the resulting $v_{p+i}' = y_i'$ represent the components of the tangent vector $F'(x)w$. Since each elementary partial $\partial \phi_i / \partial v_j$ occurs exactly once, the total complexity is bounded by

$$\text{OPS}\{F, F'w\} \leq q \, \text{OPS}\{F\}. \tag{5}$$

The factor q is a small number that depends on the exact definition of the operations count OPS{ } and assumptions on the computing platform. If only multiplications are counted, one has $q = 3$, but we prefer the slightly more conservative bound $q = 5$ defined in [17], which takes into account memory accesses and other costs. All these estimates apply to serial machines, where we may assume that

$$\text{OPS}\{f\} = \sum_{i=1}^{p} \text{OPS}\{\phi_i\}.$$

Since the direct derivative recurrence can be carried out simultaneously with the evaluation of F, the storage requirement is at most doubled, since a scalar v_i' is associated with each intermediate v_i. Quite often one wishes to propagate several directional derivatives or even full gradients with respect to all independents simultaneously. Having initialized the $v_{i-n}' = x_i'$ accordingly, one may then apply (4) in a vector mode. The number of operations and storage locations grows linearly with the number of differentiation directions, but the general overhead is constant. On the other hand if the derivative objects get too large the resulting page faults of the virtual memory system can slow down the calculation so much that two or more repeated sweeps would be more efficient. In our experience propagating some $\hat{n} \in [5, 20]$ derivatives at a time seems to be a reasonable compromise on modern workstations.

On the test suite from MINPACK-2, the resulting run-times for evaluating Jacobians by columns with or without compression by sparsity based coloring were typically two times faster than divided differences and up to 50% slower than handcoded derivatives [1]. The results for five test problems with $m = n \approx 15,000$, between 7,000 and 20,000 nonzero Jacobian entries, and a compressed Jacobian width $\hat{n}$ ranging from 7 to 19 are listed in Table 1.

Table 1 compares the time for computing the Jacobian matrices by divided differences (FD), using the Fortran preprocessor ADIFOR, and handcoded derivatives (MINPACK). The first and last column on the right represent the times needed to "color" the columns of the Jacobians (DSM) and to convert the compressed Jacobian into a sparse matrix format (FDJS). The uncompressed and compressed sparsity pattern for the Jacobian of the incompressible elastic rod (IER) problem are displayed in Figure 2, where the number of columns (and thus actual differentiation parameter) is reduced from $n = 15003$ to $\hat{n} = 18$.

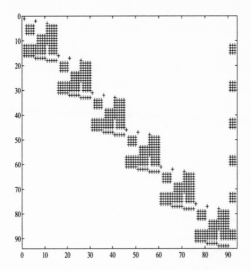 

Figure 2: IER Jacobian Sparsity (left) and IER Compressed Sparsity (right)

The coloring approach of Coleman and Moré [13] is based on the sparsity pattern of the Jacobian F' alone and achieves no savings at all if there is at least one dense row. Nevertheless, most of the gradients $\nabla_x v_i$ may be quite sparse so that significant reductions of the operations count can be achieved by a dynamically sparse implementation (See for example [6] and [2]). Unfortunately, the overhead for maintaining the sparsity information cancels out a significant part of the reduction in floating-point operations. A highly efficient sparse SAXPY option for ADIFOR has recently been completed by P. Khademi.

Problem	n	nnz	$\hat{n}$	Time (seconds)				
				DSM	FD	ADIFOR	MINPACK	FDJS
SFI	14884	73932	7	2.38	1.35	0.93	0.59	.19
FIC	16000	123987	9	2.62	5.23	2.55	1.46	.35
SFD	14000	154981	14	3.79	5.48	2.69	2.08	.43
IER	15003	158000	18	6.14	7.43	2.77	3.16	.48
FDC	14884	191056	19	9.11	11.71	6.53	6.50	.54

Table 1: Solbourne Timings on Minpack2 Test Problems.

The forward mode is easily extended to derivatives of arbitrary order, which have been used extensively for the numerical solution of initial value problems in ordinary differential equations. It is simpler to implement then the reverse mode (to be discussed next), especially for codes with iterations or branches.

3.4 The Reverse or Adjoint Mode

Rather than applying the chain rule to calculate derivatives of intermediates v_i for $i \geq 1$ with respect to independents x_i or linear combinations thereof, the reverse mode propagates for a fixed weight vector u the sensitivities

$$\bar{v}_i \equiv \frac{\partial u^T y}{\partial v_i} \quad \text{for} \quad i = p, p - 1, \ldots, 1 - n$$

backward. To this end, each scalar assignment (3) is transformed to the corresponding adjoint operation(s)

$$\bar{v}_j + = \frac{\partial \phi_i}{\partial v_j} \bar{v}_i \quad \text{and} \quad v_k + = \frac{\partial \phi_i}{\partial v_k} \bar{v}_i, \tag{6}$$

where $a+ = b$ abbreviates $a = a + b$ and the second half applies only for bivariate elementary functions. The adjoint operations are executed in reverse order, which requires that all intermediates v_i are either stored in memory or recalculated during the gradient calculation. Alternatively, one may also save and retrieve the elementary partials $\partial \phi_i / \partial v_j$ instead of the values v_j. In either case one obtains a temporal complexity bound of the form

$$\text{OPS}\{f, \nabla f\} \leq q \, \text{OPS}\{f\}. \tag{7}$$

Here we have scalarized the problem by setting

$$f(x) \equiv u^T F(x)$$

for a fixed weight vector u, which can be set to 1 so that $f = F$ if $m = 1$. With regard to complexity, we consider the premultiplication of F by u as negligible; hence, $\text{OPS}\{f\} \equiv \text{OPS}\{F\}$. Unfortunately, run-time ratios achieved by automatic differentiation packages are often *considerably* larger as a result of various overhead expenses. On the other hand, run-time ratios near $q \simeq 2$ can be achieved if the reverse recurrence (6) is executed by compilable code written by hand or generated by a good precompiler. Currently, we know of two adjoint code generators under development, namely, AMC by Ralph Giering [16] and ODYSSEE by Nicole Rostaing et al. [36]. Both packages accept a subset of Fortran 77 and rely on user-supplied directives to keep the memory requirement within reason. The differentiation tool PCOMP [31] accepts a Fortran-like function specification and generates compilable gradient code for problems of moderate size.

3.5 Checking the Memory

The temporal complexity bound (7) is based on (3) and the assumption that all intermediate values v_i are stored during the forward evaluation and then are retrieved during the reverse sweep (6). Consequently, the memory requirement for ∇f is proportional to the operations count for f, that is,

$$\mathrm{MEM}\{\nabla f\} \sim \mathrm{OPS}\{f\}. \qquad (8)$$

In practice, this relation means that each megaflop during the evaluation of f generates several megabytes of execution trace that need to be stored on disk. Fortunately, since these temporary data are used in exactly the opposite order in which they were generated, the access pattern is strictly sequential. Nevertheless, the proportionality (8) was seen as a severe limitation of the reverse mode in the past. Before showing how the proportionality between the temporal complexity of the function and the spatial complexity can be overcome, let us first stress that this supposed constraint is not active for many practical optimization problems (not to mention the usual suite of toy test problems).

Whenever the aspect ratio

$$h\{F\} \equiv \mathrm{OPS}\{F\}/\mathrm{MEM}\{F\} \qquad (9)$$

is of moderate size, a gradient $\nabla f = u^T F'$ can be calculated by the basic version of the reverse mode with a reasonable growth in memory. This observation applies, for example, to time-independent spatial structures (vehicles, buildings, geological formations, and even molecules) whose long-range atomic interactions are additive. In all these cases the computer model contains data representing local substructures that interact only with a small number of neighbors. Consequently, the number of operations needed to evaluate all constraints and an overall, additive objective is at most a small multiple of the memory locations needed to represent the structure.

In contrast, when a system is time dependent, each state variable is likely to be overwritten many times, and we may view the aspect ratio $h\{F\}$ essentially as the number of time steps. A similar effect occurs when the function evaluation involves an iterative numerical procedure, in which case one may think of $h\{F\}$ as the total number of iterations. The question of how close the iteratively obtained derivatives are to the derivatives of the limiting (implicit) function will be discussed in Section 6. Especially for long-term computer simulations, say of the earth's climate, the ratio $h\{F\}$ can be very large; in such cases, the reverse mode in its basic form is not practical.

3.6 Repeat Performances Save Memory

Fortunately, it is possible to limit the memory growth, $\mathrm{MEM}\{\nabla f\}/\mathrm{MEM}\{f\}$, to a fractional power or even a logarithm of $h\{F\}$, albeit at the cost of a moderate increase in $\mathrm{OPS}\{\nabla f\}$. The key idea is to recalculate some of the v_i repeatedly, starting from carefully selected checkpoints.

The main complication of the reverse mode compared with the forward mode is that the execution of the adjoint operation requires a full reversal of the program execution. The simplest way of doing this is to record a detailed execution trace on the way forward and then to read it on the way back, which corresponds to the basic version of the reverse mode of differentiation discussed earlier. The problem of logical program reversal has been considered in theoretical computer science [3] and is of some interest in the context of debugging tools. The following binomial checkpointing scheme was first published in [20].

Let S be the memory space used by the computer at hand. We may view any program execution as a sequence of state transformations

$$z_{i+1} = T_i(z_i) \quad \text{for} \quad i = 0, 1, \ldots, p-1. \tag{10}$$

Typically, the individual transformations $T_i : S \to S$ are not reversible at all or the recomputation of the previous state z_i from z_{i+1} is numerically poorly conditioned. Yet, the execution of the adjoint recurrence demands that we somehow produce the intermediate states, s_i, in reverse order, starting from $i = p$ down to $i = 0$.

One possibility is to save the initial state z_0 and to restart the forward calculation p times to reach the states $z_p, z_{p-1}, z_{p-2}, \ldots, z_2$, respectively. While this simple scheme minimizes the memory requirement, it essentially squares the operations count, since transition function T_i needs to be evaluated on average $p/2$ times during the repeated forward sweeps. Fortunately, more sensible tradeoffs between spatial and temporal complexity are possible. For example, if p is a perfect square, one may save only every $\sqrt{p}$-th intermediate state and the last $\sqrt{p}$ states during the first forward sweep. Having reversed through the last $\sqrt{p}$ states, one may then recalculate the preceding block of $\sqrt{p}$ states starting from state $p - 2\sqrt{p}$. The maximal memory requirement for this scheme is $2\sqrt{p}$ state vectors, and the operations count has grown by only one extra forward evaluation. Naturally, this idea may be applied recursively, and it was found in [20] that h can be as large as the reciprocal beta function

$$b(t, s) \equiv (t + s)!/(s! \, t!), \tag{11}$$

given a bound t on the number of repetitions for any T_i and a bound s on the number of snapshots in memory at any one time. Using Sterling's formula, one obtains for fixed t and arbitrary $h \leq b(t, s)$

$$s = \mathcal{O}(\sqrt[t]{b}).$$

Because of the remarkable symmetry of b with respect to t and s, one may also fix the spatial complexity factor s and then obtain t proportional to the k-th root of h.

When both t and s are chosen equal, they grow like the logarithm base t of h. This particular choice yields the total cost

$$\text{COST}\{\nabla f\} \equiv \text{OPS}\{\nabla f\} + c \, \text{MEM}\{\nabla f\} = \tilde{O}(\text{COST}\{f\}),$$

where $\tilde{O}(m) = O(m \, P(\log m))$ for some polynomial P, and c is a suitable conversion factor. Hence, gradients can be calculated for functions of virtually any computational complexity.

Provided that one can estimate the aspect ratio h of a given evaluation problem (e.g., by bounding the number of time steps in a dynamical simulation), one may choose various feasible combinations (s, t) of growth factors in spatial and temporal complexity. On a multiprocessor or dedicated machine, one may select the maximal s that can be accommodated in memory (assuming we know MEM$\{F\}$) and then suffer the consequences, that is, the resulting run-time growth factor t. Because of the exponential growth of h in s and t, getting either one of the estimates wrong by a moderate factor will result in a variation of only one or possibly two in the remaining free parameter. Currently, no general-purpose implementation of the checkpointing scheme exists. Such a scheme should, however, be quite easy to implement by hand on explicitly time-dependent problems.

A fully automatic implementation for general problems would have to place checkpoints where only a comparatively small number of variables need to be saved (typically, fairly high up in the calling tree). It must also save regular variables and the program counter so that the calculation can later be restarted with exactly the same results. The problem of performing a reverse sweep using repeated partial forward sweeps to minimize storage has been cast by John Gilbert as a pebble game, which is formulated in Section 7.

4 Ramifications for Optimization

Summing up the observations in the previous two subsections, we conclude that, provided they exist and are smooth, gradients can be evaluated at essentially the same complexity as the underlying function. Thus, one can perform design optimization and parameter estimation by gradient methods on computer models of virtually any complexity.

This conclusion applies not only to the unconstrained case of a single objective function but to the more realistic multicriteria scenario where one has to juggle a handful of $\hat{m}$ objectives and constraints. Since the program reversal effort is independent of the number $\hat{m}$ of dependents, the temporal complexity is then likely to grow by $(\hat{m} + t)$, where t satisfies again $b(t, s) \leq h$ for given h and s. Alternatively, the choice of equal growth factors $s = t$ yields

$$\frac{\text{COST}\{F'\}}{\text{COST}\{F\}} \leq c \left[\hat{m} + ln \left(\frac{\text{OPS}\{F\}}{\text{MEM}\{F\}} \right) \right] , \tag{12}$$

where COST denotes a weighted sum of OPS and MEM that measure temporal and spatial complexity, respectively. Even though this theoretical result is currently not achieved by any practical implementation, it bodes well for the future of optimization on very large problems.

4.1 Affordable Vector-Jacobian Products

If one wishes to solve a nonlinear system $F(x) = 0$ with $m = n$ quite large, the result (12) may still be useful because it allows the calculation of block vector-Jacobian products of the form

$$U^T F(x)' \quad \text{with} \quad U^T \in I\!R^{\hat{m} \times m} .$$

Vector-Jacobian products occur frequently in iterative methods for computing Newton-like steps as approximate solution of a linear system in the Jacobian. They are avoided by so-called transpose-free methods [32] because their calculation was thought to require the prior evaluation of the full Jacobian, which is no longer true if the reverse mode is applied. Bundles of $\hat{m}$ vector-Jacobian products can be used in block iterative methods [11] and reduce the overhead of the reverse mode, especially in a limited storage version with checkpointing. In the symmetric case one can compute

$$\nabla^2 f(x) V \quad \text{with} \quad V \in I\!R^{n \times \hat{n}}$$

at a slightly larger complexity by applying (12) to $F(x) \equiv V^T \nabla f(x)$. This vector function can be evaluated at roughly $2\hat{n}$ times the cost of f in the forward mode of automatic differentiation. The required double sweep [12] for these selected second derivatives has, for example, been implemented in ADOL-C [23] and PADRE2 [30].

Here, "affordable" does not — unfortunately — mean cheap since the ratio between the multiplication counts for $f(x)$ and $\nabla^2 f(x)w$ equals 9 for the simple example

$$f(x) = ((\ldots((x_1 x_2)x_3) \cdots \cdots x_{n-1})x_n) \quad .$$

More specifically each multiplication statement in the original code spawns the following derivative operations

$$c = a\, b; \quad c' = a'\, b + a\, b'; \quad (a, a', b, b') > \text{TAPE}$$

on the way forward and then on the way back

$$(a, a', b, b') < \text{TAPE}; \bar{b}'+ = \bar{c}'a; \bar{a}'+ = \bar{c}'b; \bar{b}+ = \bar{c}'a' + \bar{c}a; \bar{a}+ = \bar{c}'b' + \bar{c}b; \bar{c} = \bar{c}' = 0.$$

Here TAPE is a stack for saving and restoring intermediate data sequentially. If more is known about the code context some of the incremental statements can be replaced by assignments and the resetting of $(\bar{c}, \bar{c}')$ to zero may be avoided. Nevertheless, it is clear that the innocent original multiplication has generated quite a lot of computational work and memory traffic. If only gradients are required all operations involving primed variables can be omitted, which reduces the complexity significantly. Hence we must expect that whenever the evaluation process is dominated by multiplication operations, simultaneously computing $\hat{n}$ Hessian vector products at the same argument requires at least $(1+8\hat{n})$ times as many multiplications as evaluating f by itself. Under the same assumptions a combined function gradient evaluation involves about 3 times as many operations as evaluating the function by itself. Hence shooting $\hat{n}$ divided differences to estimate $\hat{n}$ Hessian vector products costs only $3(1+\hat{n})$ as many multiplications as evaluating f by itself, including the evaluation at the base point. The situation might, however, be more favorable for an analytical approach, when the evaluation process for f involves many linear calculations or transcendental function calls, for which the work ratio is much smaller. It also helps if there are significant data transfers or other *passive* set up costs that arise at each new evaluation but are independent of the number of derivatives being calculated. This question is currently being investigated.

4.2   No Cheap Preconditioning

It is well known that iterate equation solvers are crucially dependent on good scaling, which typically requires preconditioning. Unfortunately, it appears that neither computational differentiation nor any other technique can provide the required scaling information at a cost significantly below that of evaluating the full Jacobian or Hessian. This conclusion follows as a corollary of the cheap gradient result by the following observation of Morgenstern and Kaltofen.

Consider the vector function

$$F(x) = Y^T ABx : I\!R^n \to I\!R^n$$

with Y, A, and B general $n \times n$ matrices. Hence, we have the Jacobian

$$F'(x) = Y^T AB \in I\!R^{n \times n}$$

and the trace

$$Tr(F'(x)) = Tr(Y^T AB),$$

which we may consider as absolutely minimal scaling information about F' and thus F itself. Applying the cheap gradient result to differentiation of the trace with respect to Y, we find that

$$\nabla_Y Tr(Y^T AB) = AB$$

costs no more than q times as much as evaluating $Tr(Y^T AB)$. Disregarding storage costs, we derive that

$$\frac{\text{OPS}\{Tr[F'(x)]\}}{\text{OPS}\{F'(x)\}} \geq \frac{\text{OPS}\{AB\}/q}{\text{OPS}\{Y^T AB\}} \geq \frac{1}{2q}.$$

The last inequality follows from the observation that, whatever scheme for computing dense matrix products is being used, the ternary product $Y^T AB$ can always be obtained at a cost not exceeding that of two binary products. Since the same argument is valid for the gradient $(Y^T AB + B^T A^T Y)x$ of $f(x) = x^T Y ABx$, we can conclude that traces of Jacobians and Hessians are, in general, not much cheaper to evaluate than the full derivative matrices themselves. Clearly, any useful, specific conditioning information is likely to cost even more. It would appear very likely that this "traces are no cheaper than matrices" principle extends to sparse or otherwise structured Jacobians and Hessian, but we are currently not aware of a generalized result in this direction. Nevertheless, we conclude tentatively that preconditioning information can be gained significantly cheaper than the whole Jacobian or Hessian only if structural information apart from the evaluation program is available.

5   Jacobians and Hessians in the Reverse Mode

It is important to realize why the "cheap gradient" result discussed in the preceding section does not mean that first derivatives in general are cheap. By "cheap" we mean here that a certain collection of derivatives can be calculated in the $\tilde{O}$ sense at essentially the same complexity as the underlying scalar or vector function F : $IR^n \to IR^m$. It follows immediately from the application of the cheap gradient result to the component functions F_i of F that

$$\text{OPS}\{F'\} \leq \sum_{i=1}^{m} \text{OPS}\{\nabla F_i\} \leq \sum_{i=1}^{m} \tilde{O}(\text{OPS}\{F_i\})$$

$$\leq \tilde{O}\left(\sum_{i=1}^{m} \text{OPS}\{F_i\}\right) \leq \tilde{O}\left(\gamma\{F\}\text{OPS}\{F\}\right),$$

where $\gamma\{F\}$ is the ratio defined in (1). There are certainly function classes where γ is so small that the "cheap gradient" result does translate into a "cheap Jacobian" result. For example, this is true for the right hand sides of many ordinary differential equations that can be "written down" explicitly and thus without any common intermediates. We believe that the relation $\gamma\{F\} \sim 1$ applies also quite frequently for the constraints in nonlinear programming. For example, if formulated in Cartesian coordinates, the constraints between neighboring components in a multibody system are usually independent of each other, except that some preliminary notation into the lab coordinate frame represents some joint calculation.

5.1 Common Intermediates Hurt

The example of Subsection 3.2 with $n = m$ and $k = \mathcal{O}(n)$ has a dense Jacobian $F'(x)$ and a dense Hessian $\nabla^2 f(x)$, which both cannot be computed by any conceivable method at a small multiple of the $\mathcal{O}(n)$ operations it takes to evaluate the underlying functions F and f. The apparent cause are the vector Ax and other common intermediates, whose presence seems to help much more in evaluating the functions themselves than in calculating their derivative matrices, thus degrading the work ratio $\text{OPS}\{F'\}/\text{OPS}\{F\}$.

It is not hard to see that the number of component functions F_i that depend on any particular intermediate value v_j is bounded by the maximal number of generically nonzero entries in any column of the Jacobian F'. Since this sparsity bound equals the maximal row length of the Jacobian $[F'(x)]^T$, we may denote it by $\rho(F'(x)^T)$. As each intermediate value v_j needs to be (re)evaluated only for those F_i that depend on it, we find that

$$\gamma\{F\} \leq \rho(F'(x)^T).$$

A more detailed analysis of this and related inequalities can be found in [21]. Consequently, the reverse mode yields cheap Jacobians on classes of vector functions whose Jacobians have columns that do not exceed a certain bound on the number of nonzero entries. In other words, all variables are local in that they enter into only a small number of equations. Many discretizations of partial differential equations fall into this category, unless these are variables that enter at each node. Such global parameters may, for instance, represent physical properties or common conditions such as temperature or pressure. They introduce dense columns in the Jacobian, which may drastically slow the reverse mode, as observed in [24].

For general sparsity patterns one may employ the coloring techniques (see, e.g., [14], [13], [38], and [8]) which were developed for approximating derivative matrices by difference quotients. Numerical results using the Fortran preprocessor ADIFOR [4] were already quoted in Section 3.3. In addition to the classical grouping of Jacobian columns, automatic differentiation in the reverse mode also allows the combination of structurally orthogonal rows to one "color". Since the column- or row-wise compressed Jacobians are obtained with working accuracy, the approach of Newsam and Ramsdell [33] looks very promising, despite the less than optimal conditioning of the Vandermonde systems that need to be solved.

5.2 Time Does Tell

Especially when there is explicit time dependence, Jacobians tend to be dense and their evaluation is quite expensive. For example, let us reinterpret (10) as a sequence of smooth transformations T_i on a Euclidean space S with tridiagonal Jacobians T_i'. Then the full Jacobian $\partial z_p/\partial z_0$ is the product of the T_i' and must be structurally dense if $p \gg m \equiv \dim(S)$, as we will assume. Moreover, the gradient of any particular component y_i of $y = z_p$ with respect to all but the last n intermediate states S_i is already dense. Consequently, the first $p - m$ transformation must be executed fully for the evaluation of y_i, and we have

$$\mathsf{OPS}\{y_i = F_i\} \geq \sum_{i=0}^{p-m} \mathsf{OPS}\{T_i\}.$$

Provided that $\mathsf{OPS}\{T_i\}$ is the same for all i, it follows that

$$\gamma = m\left(1 - \frac{m}{p}\right) \approx m.$$

It can be shown that all simple schemes for accumulating $F' = \partial z_p/\partial z_0$ from the T_i require $3pm^2 - O(m^3)$ multiplication. This count applies to multiplying the T_i' with each other in various orders, forward (i.e., for $i = 0, \ldots, p-1$), reverse (i.e., for $i = p-1, \ldots, 0$), or bisection. For $p = 4$, these different orderings correspond to the bracketings

$$T_3'(T_2'(T_1'T_0')) = ((T_3'T_2')T_1')T_0' = (T_3'T_2')(T_1'T_0'). \tag{13}$$

Assuming that $\mathsf{OPS}\{T_i\} \equiv 3m$, one finds that all these methods for computing F' have about m times the complexity of evaluating F. Hence, we may put vector functions obtained by integrating evolution equations over long periods of time in the basket of problems with expensive Jacobians. (Here, long periods mean that signals have enough time to travel throughout the domain S so that F' is dense.)

On discretized evolution equations with local stencils in more than one spatial dimensions the various bracketings in Equation (13) are no longer essentially equivalent, as they may lead to vastly different computational complexities. In general, neither the forward nor the reverse mode is optimal for the computation of Jacobians, and finding the method with minimal space complexity is an NP-complete task [26].

5.3   Lessons for Least Squares

A fairly standard argument in nonlinear least squares is that, since the gradient

$$\nabla \tfrac{1}{2}\|F(x)\|^2 = F(x)^T F'(x)$$

involves the Jacobian $F'(x)$, one might as well utilize this linearization in the algorithm for minimizing the sum of squares $f(x) \equiv \tfrac{1}{2}\|F(x)\|^2$. Here it is tacitly assumed that "being part of the definition" of ∇f means having to be evaluated explicitly. However, we have just argued that $F'(x)$ may be orders of magnitude more expensive than F, whereas the gradient ∇f is not much more costly than f, whose complexity is equivalent to that of F. The answer to this puzzle is, of course, that the reverse mode evaluates ∇f without forming F' explicitly or implicitly. Consequently, Gauss Newton–like methods based on F' may indeed be much more expensive per step than gradient methods relying only on numerical values of ∇f. This observation agrees with the computational practice in four-dimensional data-assimilation, which is used for short- and medium-range weather forecasting. Here, z_i represents the discretized state of the atmosphere at the i-th time step, and one wishes to find the initial state z_0 that minimizes a weighted sum of squared discrepancies between the resulting states z_i and actual meteorological observations. Typically, this calculation is carried out over a window of the past twenty-four hours. Naturally, the Jacobian F' is rather large, and even if one could evaluate it cheaply, its storage and manipulation would be practically impossible. For contrast, the gradient ∇f can be evaluated by hand-coded adjoints with little more than twice the operations count for f itself [37].

5.4   Projections of the Lagrangian Hessian

It is well known that on nonlinearly constrained optimization problems, one-step superlinear convergence can be achieved only if the one-sided projection of the Hessian of the Lagrangian is evaluated or approximated with increasing accuracy. The projection is onto the tangent space of the feasible manifold, which is the orthogonal complement of the active constraint gradients.

Using the double sweep already discussed in Subsection 4.1, one can calculate for $f(x) = u^T F(x)$ with u fixed

$$\nabla^2 f \cdot w = u^T F(x)'' w = \sum_{i=1}^{m} u_i \nabla^2 F_i w$$

at a complexity comparable to that of evaluating $F : I\!R^n \to I\!R^m$. In the optimization context, we think of F_1 as the objective function and of the other components as constraints. With w ranging over all vectors in a basis spanning the orthogonal complement of the gradients, one obtains the one-sided projection of the Lagrangian Hessian for the cost of jointly evaluating the objective function and constraints multiplied by a small constant and the degrees of freedom $n - m + 1$. It should be noted that in computing this $(n - m + 1) \times n$ matrix, the second-derivative tensor F'' is immediately restricted to the null spaces of the Jacobian F' and its transpose as already discussed in Section 2.3 . While this selective derivative calculation is likely to result in savings by orders of magnitude compared with a full evaluation of F'', a second projection yielding the so-called reduced Hessian, a symmetric $(n - m + 1) \times (n - m + 1)$ matrix, does not lower the complexity any further. In that regard analytical differentiation is very different from secant updating, where (probably because of a lack of symmetry) attempts to approximate the one-sided projection rather than the full Hessian or its two-sided projection have not been very successful.

On the other hand, low-rank secant methods can work very well on largely unconstrained problems and even discretizations of mildly nonlinear operator equations. At present, it is not at all clear for what number of variables and constraints a crossover between the efficiency of the two approaches might occur, and naturally many other problem characteristics may impact the relative performance.

6 Pushing the Limits

In many practical applications, objectives and constraints are evaluated by programs that involve discretizations with a variable number of grid points, iterations with a variable number of steps, or some other input-dependent branching. In contrast to those defined by so-called straight-line programs, the input-output relations defined by such adaptive procedures are unlikely to constitute a smooth mathematical mapping. Nevertheless, calculus-based optimization methods are widely employed and may remain the only realistic choice, especially if the number of independent variables is quite large.

In general, one can expect that the input-output relations evaluated by practical codes are at least piecewise smooth and that the jumps in function and derivative values are small enough to yield useful local models up to a realistic resolution. For example, this applies to codes involving the solutions of discretized operator equations that converge in a Sobolev norm including the desired derivatives to the limiting smooth solution function. However, there are important counterexamples [9], where the derivatives of the discretized solutions with respect to problem parameters may grow unbounded rather than converging in the limit. Therefore, only certain discretizations are suitable for use in codes that are to be differentiated for optimization purposes. The difficulty with iterative equation solvers has by now been quite well researched, and we summarize a few basic results from [25].

Let us consider a parameter-dependent system of nonlinear equations

$$G(z, x) = 0 \quad \text{with} \quad G : I\!R^m \times I\!R^n \mapsto I\!R^m$$

where x represents, as before, the vector of independent variables or parameters with respect to what we wish to differentiate. The iterates $z_k \in I\!R^m$ generated by most practical methods for approximating a solution $z_* = z_*(x)$ with $G(z_*, x) = 0$ satisfy a recurrence of the form

$$z_{k+1} = \Phi_k(z_k, x) \equiv z_k - P_k G(z_k, x). \tag{14}$$

Here, the preconditioner P_k is some $m \times m$ matrix that approximates the inverse of the Jacobian G_z, which we assume to be nonsingular and joint with G_x Lipschitz-continuous on some neighborhood of z_*, x.

For the sake of numerical stability we assume furthermore that the P_k satisfy the contractivity condition

$$\|[I - P_k\, G_z(z_k, x)]\| \quad \leq \quad \rho < 1$$

with respect to some induced matrix norm $\| \cdot \|$. Then it follows immediately that when z_k is sufficiently close to z_*, the sequence z_k converges linearly with an R-factor

$$\text{limsup}_k \sqrt[k]{\|z_k - z_*\|} \quad \leq \quad \rho \quad .$$

Differentiating (14) with respect to x, we obtain for the matrices $z_k' \equiv \partial z_k / \partial x$ the recurrence

$$z_{k+1}' \;=\; z_k' - P_k\, \partial G(z_k', z_k, x) - P_k'\, G(z_k, t) \,, \tag{15}$$

where

$$\partial G(z_k', z_k, x) \equiv G_z(z_k, x)z_k' + F_x(z_k, x)$$

and $P_k' \equiv \partial P_k / \partial x$. This last derivative may not even exist, since approximating Jacobians or their inverses are sometimes adjusted discontinuously, and in the formulation (14) they must also absorb the step multipier. Theoretically, the best approach is simply to set P_k' to zero and thus to suppress the dependence of P_k on the variable vector x completely. In practice that may not be so easy, because one has to identify and isolate the preconditioner P_k, which may be a complicated code or may have been written by someone else.

The first term in (15) is the really useful part, as it follows from the implicit function theorem that near z_* for some constant c

$$\|z_k' - z_*'\| \quad = \quad c\left(\|\partial G(z_k', z_k, x) + \|G(z_k, x)\|\right) \,, \tag{16}$$

where z_*' is the unique solution of the linear system $\partial G(z', z_*, x) = 0$. Even though the constant c is usually not known, we have a constructive test to gauge to what extent the tuple (z_k', z_k) has converged to the desired limit (z_*', z_*). Note that $\partial G(z_k', z_k, x)$ can be evaluated for any candidate value z_k' by automatic differentiation without the need to form G_z or G_x explicitly. Since the leading term on the right-hand side of (15) is contractive and G_k converges Q-linearly to zero, one can easily establish R-linear derivative convergence with the same factor, namely,

$$\lim_k \sup \|z_k' - z_*'\|^{\frac{1}{k}} \leq \rho,$$

provided that the P_k' remain at least bounded.

Even though it is not clear whether our assumptions were indeed satisfied, it was found in [5] on a 2D transonic code [15] that the derivatives, which were started only after the iterates had settled down, did converge at roughly the same linear rate. In

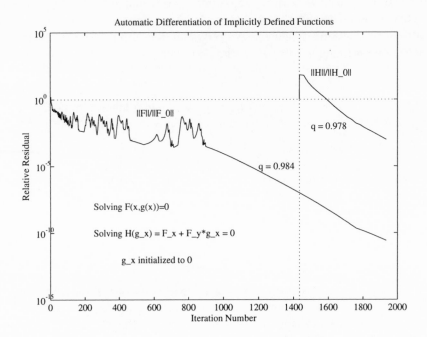

Figure 3: Convergence Behavior of Function and Derivatives in Elbanna and Carlson's Code

Figure 3 F and H represent what we denote here by G and ∂G, respectively. Also $g(x)$ corresponds to our implicit function $z(x)$. The original roughness in the residual stems from the fact that the shock location still moves on the surface of the wing.

For secant updating methods, P'_k may actually grow unbounded, but because of their superlinear convergence rate it can be shown that the perturbation $P'_k G(z_k, x)$ still converges to zero fast enough to ensure derivative convergence with an R-factor not greater than ρ. In such a situation, however, we must expect that the derivative accuracy will significantly lag behind the accuracy achieved for the solution after a certain number of iterations.

7 Combinatorial Games

As we have already noted in Section 5.2 the forward and reverse mode are only particular options in a wide variety of possibilities for applying the chain rule to particular composite function. In terms of a computational graph like the one depicted in Figure 1 applying the chain rule means successively eliminating all intermediate nodes until only the bipartite graph consisting of the independent roots and dependent leaves remains. All arcs have values that represent elementary partials at the beginning and nonvanishing elements of the Jacobian at the end of the elimination process.

The rule for updating arc-values is to compute for each predecessor/successor pair of the intermediate vertex being eliminated the product of the corresponding arc values and to increment the value of the arc connecting them directly by that amount. If no such arc exists, it must be introduced and its value is initialized to the product. One usually wishes to minimize the number of such fill-in arcs and/or the number of multiplications, which is given by the Markowitz degree, i.e. the number of predecessors times the number of successors of the vertex being eliminated. The minimum fill-in problem has been shown to be NP-hard by Herley [26]. It is currently not clear whether the greedy Markowitz criterion really works as well as it seems to in sparse Gaussian elimination.

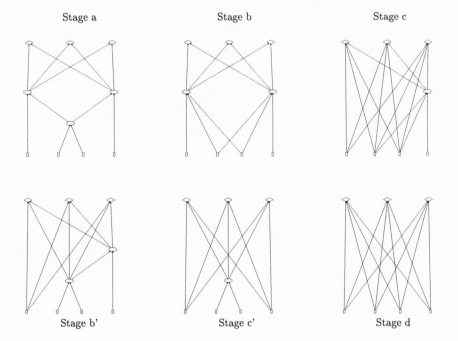

Figure 4: Alternative Elimination Sequences for Jacobian Accumulation on Example from Fig. 1

Some possible elimination sequences for the small problem originally displayed in Figure 1 are displayed in Figures 4. First we note that all vertices in Figure 1 with two incoming and one outgoing arc were introduced somewhat unnecessarily through assignments and can be eliminated at no arithmetic cost since the value of the out going arc is one. Moreover, the elimination of all vertices with Markowitz degree one, i.e. exactly one predecessor and one successor, cannot increase the Markowitz degree of other nodes but it may reduce them.

These two preparatory simplifications result in the structure displayed in Stage a of Figure 4 with no nontrivial multiplication having been expended so far. Now we face the alternative either to eliminate the central node near the bottom or one of the upper pair. The lower one seems a natural choice as it has the minimal Markowitz degree 4 and its elimination results in Stage b, where the degree of the remaining two vertices has gone up from 6 to 9. Their subsequent yields first Stage c and then the final bipartite Stage d at a total cost of 4+2*9 = 22 multiplications. If instead we eliminate both of the symmetric nodes before the lower one as shown in Stages b' and c', the total count is only 3*6=18. Interestingly, the mixed strategy of eliminating the lower central one after the first and before the second vertex in the upper pair yields the worst count, namely, 6+8+9 =23.

The last observation confirms that these kinds of combinatorial problems cannot be solved by local improvements, i.e., by exchanging one pair of nodes in the elimination order at a time. Here we have assumed that the goal is to minimize the number of multiplications, which represents also an upper bound on the number of fill-in arcs. One also sees on this small example that following the Markowitz route (a,b,c,d) costs 4 more multiplication than the alternative elimination order (a,b',c',d). In this particular case the Markowitz order happens to coincide with the forward mode and the alternative order with the reverse mode, which is generally expected to do better when there are fewer dependents than independents. However, this is certainly not a hard rule, especially if the problem is sparse. Overall we conclude that there is a need to investigate heuristics for the efficient calculation of Jacobians through the application of the chain rule to combine the elementary partials in a suitable order.

Another combinatorial problem arises if one wants to calculate a gradient, or more generally a vector-Jacobian product, without saving all intermediates on the forward sweep. Rather than establishing global checkpoints as suggested in Section 3.6, one could break the forward and reverse sweep into advances and retreats to and from individual nodes in the computational graph. John Gilbert has formulated the problem as a two color pebble game, which we have slightly modified as an edge elimination process for the following presentation.

Start:

Take a fixed number of pebbles.

Node Rule:

A pebble can be placed on any node whose predecessors all have pebbles.

Edge Rule:

An edge is eliminated when it connects pebbles and its destination has no more successor.

Goal:
Eliminate all edges using as few placements as possible.

Placing a pebble means (re-)evaluating an intermediate quantity or (re-) reading an independent variable from an external medium. It is always possible to remove a pebble from a node, which corresponds to erasing a particular value from internal memory. Eliminating an arc corresponds to incrementing the adjoint contribution from its destination node to its origin. If one has enough pebbles to cover all vertices the task can be solved trivially with each node being evaluated once. On the original graph in Figure 4 one can get by with 4 pebbles rather than the ten required to cover all. The downside of this saving in space is that the number of placements roughly doubles.

If the graph is merely a chain the "binomial" placement scheme suggested in [19] yields minimal time, i.e., number of placements, for given space, i.e., number of pebbles, and vise versa. To add realism one may attach nonuniform weights for the temporal complexity of placing a pebble on a particular node and the spatial complexity of leaving it there. Such heterogeneous models would then also allow the treatment of quotient graphs whose supernodes could represent for example subroutine calls or other composite calculations. The pebble game could be combined with the vertex elimination task to yield procedures for calculating Jacobians rather than just gradients with an acceptable temporal and spatial complexity. Obviously, a lot remains to be investigated theoretically and implemented practically. Nevertheless, it should not be overlooked that existing CD tools can solve many or even most current differentiation tasks at an acceptable and a priori predictable cost in code preparation and run-time.

Acknowledgments
There have been many developments and contributions in automatic differentiation that I was unable to mention or cite explicitly. My own research would not have been possible at all without the help of my co-authors and collaborators as well as the general support of my colleagues and superiors at Argonne National Laboratory. As many times before Gail Pieper greatly improved the readability of the manuscript at a moments notice. Figures 1 and 4 were generated by Dr. J. Benary and J. Utke, TU Dresden, using visualisation tools developed within the VISAMAD project funded by the German Ministry for Research and Technology.

References

[1] B. Averick, J. Moré, C. Bischof, A. Carle, and A. Griewank, "Computing large sparse Jacobian matrices using automatic differentiation," accepted for publication in *SIAM Journal on Scientific Computing*, Preprint MCS-P348-0193, Mathematics and Computer Science Division, Argonne National Laboratory, Argonne, Illinois, 1993.

[2] M. Bartholomew-Biggs, L. Bartholomew-Biggs, and B. Christianson, "Optimization & automatic differentiation in ada: Some practical experience," *Optimization Methods and Software*, 4 (1992), 47-73.

[3] C. H. Bennett, "Logical Reversability of Computation," *IBM Journal of Research and Development*, 17 (1973), 525-532.

[4] C. Bischof, A. Carle, G. Corliss, A. Griewank, and P. Hovland, "ADIFOR: Generating derivative codes from Fortran programs," *Scientific Programming*, **1** (1992), 11-29.

[5] C. Bischof, G. Corliss, L. Green, A. Griewank, K. Haigler, and P. Newman, "Automatic differentiation of advanced CFD codes for multidisciplinary design," *Journal on Computing Systems in Engineering*, **3** (1992), 625-638.

[6] C. Bischof and A. Griewank, "ADIFOR: A Fortran system for portable automatic differentiation," *Proceedings of the 4th Symposium on Multidisciplinary Analysis and Optimization*, AIAA Paper 92-4744, 1992, 433-441.

[7] C. Bischof, A. Carle, G. Corliss, A. Griewank, P. Hovland, and A. Mauer "ADIFOR Working Note #3: ADIFOR Exception Handling, Version 2.0", Technical Report ANL/MCS–TM–159, Mathematics and Computer Science Division, Argonne National Laboratory, Illinois, 1992.

[8] C. Bischof and P. Hovland, "Using ADIFOR to compute dense and sparse Jacobians," 1991.

[9] J. Burns, K. Ito, and G. Propst, "On Nonconvergence of Adjoint Semigroups for Control Systems with Delays" *SIAM J. Control and Optimization*, **26** (1988), 1442-1454.

[10] A. Brook, D. Kendrick, and A. Meeraus, *GAMS: A User's Guide*, The Scientific Press, Redwood City, CA, 1988.

[11] C. Broyden, "Block conjugate gradient methods," *Optimization Methods and Software*, **2** (1994), 1-18.

[12] B. Christianson, "Automatic Hessians by reverse accumulation," *IMA J. of Numerical Analysis*, **12** (1992), 135-150.

[13] T. F. Coleman and J. J. Moré, "Estimation of sparse Jacobian matrices and graph coloring problems," *SIAM Journal on Numerical Analysis*, **20** (1983), 187-209.

[14] A. R. Curtis, M. J. D. Powell, and J. K. Reid, "On the estimation of sparse Jacobian matrices," *J. Inst. Math. Appl.*, **13** (1974), 117-119.

[15] H. Elbanna and L. Carlson, "Determination of aerodynamic sensitivity coefficients in the transonic and supersonic regimes," *Proceedings of the 27th AIAA Aerospace Sciences Meeting*, AIAA Paper 89-0532, 1989.

[16] R. Giering, "Adjoint Model Compiler," Manual Version 0.2, AMC Version 2.04a, Max-Planck Institut für Meteorolgie, Hamburg, 1992.

[17] A. Griewank, "On automatic differentiation," *Mathematical Programming: Recent Developments and Applications*, M. Iri and K. Tanabe (editors), Kluwer Academic Publishers, Dordrecht, 1989, 83-108.

[18] A. Griewank, "Direct calculation of Newton steps without accumulating Jacobians," *Large-Scale Numerical Optimization*, T. F. Coleman and Y. Li (editors), Philadelphia, Pa., 1990, SIAM, 115-137.

[19] ———, "Automatic Evaluation of First- and Higher-Derivative Vectors," **97**, Birkhäuser Verlag, Basel, Switzerland, 1991, 135-148.

[20] ———, "Achieving logarithmic growth of temporal and spatial complexity in reverse automatic differentiation," *Optimization Methods & Software*, **1** (1992), 35-54.

[21] A. Griewank, "Some Bounds on the Complexity of Gradients, Jacobians, and Hessians," *Complexity in Numerical Optimization*, P. M. Pardalos (editor), World Scientific Publishers, 1993, 128-161.

[22] A. Griewank and G. Corliss, *Automatic Differentiation of Algorithms*, SIAM, Philadelphia, Penn., 1991.

[23] A. Griewank, D. Juedes, and J. Utke, "ADOL-C, a package for the automatic differentiation of algorithms written in C/C++," to appear in *ACM TOMS*, 1994.

[24] A. Griewank and S. Reese, "On the calculation of Jacobian matrices by the Markowitz rule," *Automatic Differentiation of Algorithms: Theory, Implementation, and Application*, A. Griewank and G. F. Corliss, eds., SIAM, Philadelphia, Penn., 1991, 126-135.

[25] A. Griewank, C. Bischof, G. Corliss, A. Carle, and K. Williamson "Derivative Convergence for Iterative Equation Solvers," *Optimization Methods & Software*, **2** (1993), 321-355.

[26] K. Herley, *On the NP-completeness of optimum accumulation by vertex elimination*, Unpublished manuscript, 1993.

[27] Jim E. Horwedel, Brian A. Worley, E. M. Oblow, and F. G. Pin. GRESS version 1.0 users manual. Technical Memorandum ORNL/TM 10835, Oak Ridge National Laboratory, Oak Ridge, Tenn., 1988.

[28] Masao Iri, "History of automatic differentiation and rounding estimation," *Automatic Differentiation of Algorithms: Theory, Implementation, and Application*, A. Griewank and G. Corliss (editors), SIAM, Philadelphia, 1991, 1-16 .

[29] D. Juedes, "A taxonomy of automatic differentiation tools," *Automatic Differentiation of Algorithms: Theory, Implementation, and Application*, A. Griewank and G. Corliss, eds., SIAM, Philadelphia, 1991,

[30] Koichi Kubota, "PADRE2, a FORTRAN precompiler yielding error estimates and second derivatives," *Automatic Differentiation of Algorithms: Theory, Implementation, and Application*, A. Griewank and G. Corliss, eds., SIAM, Philadelphia, 1991, 251-262.

[31] M. Liepelt, and K. Schittkowski, "PCOMP: A Fortran Code for Automatic Differentiation," Report No. 254, Mathematisches Institut Bayreuth, 1990. To appear in *TOMS*.

[32] N. M. Nachtigal, S. C. Reddy and L. N. Trefethen, "How fast are nonsymmetric matrix iterations," *SIAM J. Matrix Anal. Appl.*, **13** (1992), 778-795.

[33] G. N. Newsam and J. D. Ramsdell, "Estimation of sparse Jacobian matrices," *SIAM J. Alg. Disc. Meth.*, **4** (1983), 404-417 .

[34] L. B. Rall, "Automatic Differentiation: Techniques and Applications," Lecture Notes in Computer Science, 120, Springer-Verlag, Berlin, 1981.

[35] D. J. Rose and R. E. Tarjan, "Algorithmic aspects of vertex elimination on directed graphs," *SIAM Journal of Applied Mathematics*, **34** (1978), 176-197.

[36] N. Rostaing, S. Dalmas, and A. Galligo, "Automatic Differentiation in Odysee," Submitted to *Tellus Special Issue on Adjoint Methods in Dynamic Metereology*, 1993.

[37] O. Talagrand and P. Courtier, "Variational assimilation of meteorological observations with the adjoint vorticity equation. I: Theory," Q.J.R. Meteorological Society, **113** (1987), 1311-1328.

[38] T. Steihaug and A. K. M. Shahadat Hossain, "Graph coloring and the estimation of sparse Jacobian matrices using row and column partitioning," Report 72, Department of Informatics, University of Bergen, 1992.

Can There Be a Unified Theory of Complex Adaptive Systems?

John H. Holland

The University of Michigan

Ann Arbor, Michigan

1 Introduction

Many of our most troubling long-range problems—trade balances, sustainability, AIDS, genetic defects, mental health, computer viruses—center on certain systems of extraordinary complexity. The systems that host these problems—economies, ecologies, immune systems, embryos, nervous systems, computer networks—appear to be as diverse as the problems. Despite appearances, the systems share enough significant characteristics to make it possible, even probable, that common general principles explain their dynamics. For this reason, we group these systems under a single classification at the Santa Fe Institute, calling them COMPLEX ADAPTIVE SYSTEMS (CAS). This is more than terminology. It signals our intuition that there are general principles that govern all CAS behavior, principles that point to ways of solving the attendant problems. Much of our work is aimed at turning this intuition into fact.

It is an easy exercise to produce a list of significant characteristics common to all CAS:

1. All CAS consist of large numbers of components, agents, that incessantly interact with each other.

2. It is the concerted behavior of these agents, the aggregate behavior, that we must understand, be it an economy's aggregate productivity, or the immune system's aggregate ability to distinguish antigen from self.

3. The interactions that generate this aggregate behavior are nonlinear, so that the aggregate behavior cannot not be derived by simply summing up the behaviors of isolated agents.

4. The agents in CAS are not only numerous, they are also diverse. An ecosystem can contain millions of species melded into a complex web of interactions; the mammalian brain consists of a panoply of neuron morphologies organized into an elaborate hierarchy of modules and interconnections; and so on.

5. The diversity of CAS agents is not just a kaleidoscope of accidental patterns; remove one of the agent types and the system reorganizes itself with a cascade of changes, usually "filling in the hole" in the process.

6. The diversity evolves, with new niches for interaction emerging, and new kinds of agents filling them. As a result, the aggregate behavior, instead of settling down, exhibits a perpetual novelty, an aspect that bodes ill for standard mathematical approaches.

7. CAS agents employ internal models to direct their behavior, an almost diagnostic character. An internal model can be thought of, roughly, as a set of rules that enables an agent to anticipate the consequences of its actions. Even an agent as simple as a bacterium employs an "unconscious" internal model when it swims up a glucose gradient in the search for food, while humans make continual prosaic use of internal models, as in our unconscious expectation that room walls are unmoving structures.

The combination of internal models with a diversity of agents, along with the attendant nonlinearities, undercuts most traditional approaches to system dynamics. Anticipations based on internal models, even when they are incorrect, may substantially alter the aggregate behavior. And the evolving diversity of agents in a CAS produces a perpetual novelty in dynamics. CAS will certainly remain mysterious until we can take such effects into account.

2   Adaptive Agents

It is easier to produce a definition of an adaptive agent than it is to produce a general formal definition of CAS. The definition turns on the computation-based implementation of two processes (Holland et al. [1989]): (1) a performance system that specifies the agent's capabilities at a given point in time—its abilities in the absence of further learning—and (2) an inductive apparatus that modifies the performance system as experience accumulates—the learning mill.

The performance system is conveniently defined in terms of rules. Any given environmental situation is defined by the concurrent activation of a cluster of rules that describe, and act upon, parts of the overall situation. The rules are treated as hypotheses, rather than as facts, and are subject to progressive confirmation, or replacement, as the system accumulates experience.

Three mechanisms help the agent to balance exploration (acquisition of new information and capabilities) against exploitation (the efficient use of information and capabilities already available):

1. PARALLEL EXECUTION of rules allows transfer of experience to novel situations by the combined activation of relevant rules—building block rules—that describe aspects of the situation.

2. TAGS AND RULE COUPLING provide for directed sequential action, making a RbridgeS for credit assignment to stage-setting actions. Tags also provide for adaptive clustering of rules.

3. COMPETITION, based on rule specificity and strength, allows rules to be treated as hypotheses to be marshalled and progressively confirmed (by strength revision under credit assignment) as required by the changing environmental situation.

To provide new hypotheses, a genetic algorithm (Holland [1992]) treats strong rules as parents, recombining parts of the parents to provide offspring rules that replace weak rules (hypotheses). The resulting adaptive agent is well-defined in computational terms. Via these mechanisms it constructs increasingly sophisticated internal models (default hierarchies, for example) that enable it to anticipate its environment. Its predictions are continually tested against outcomes, with falsifications being used to improve the models, even in the absence of payoff. Though the agent readily improves it performance, it uses only computationally simple procedures to do so. As such, it conforms to reasonable notions of bounded rationality.

3   A Broader Perspective

In order to study these effects, I have defined the ECHO class of models. The prototype of Echo is a simulated closed world with almost trivial representations of geography, physics, chemistry, biochemistry, etc. Echo does provide for a distinction between genotype and phenotype, so that the fitness of a genotype depends upon interactions of the phenotype with other agents and the local environment. Despite Echo's simplicity and its completely endogenous character (it receives no inputs or control signals once it is started), this "world" exhibits perpetual novelty within the finite limits set by the host computer. Specifically, Echo exhibits counterparts of sophisticated ecological processes, such as biological arms races and speciation. More advanced versions of Echo, involving agents displaying tags, and simple grammars controlling the relation between genotype and phenotype, should exhibit counterparts of the generation and exploitation of niches (by parasitism, symbiosis, mimicry, etc.), selective mating and recombination, the evolution and spread of multi-functional co- adapted sets of alleles, and, most importantly, the counterparts of mechanisms (such as competence and induction in morphogenesis) that permit the evolution of sophisticated organizations.

On the basis of this perspective, I do believe a useful unified theory is possible. It would, I think, involve the following elements:

1. Interdisciplinary comparison.
   Different CAS show different characteristics of the class to advantage.

2. A "Correspondence Principle".
   Using Bohr's principle, mutatis mutandis, CAS should encompass standard models from prior studies of particular examples of CAS (such as the Prisoner's

Dilemma, Wicksell's Triangle, Overlapping Generations Models, Lotka-Volterra models, and the like).

3. Computer-based gedanken experiments.
   Computer-based experiments, by varying parameters under different "restarts" from known initial conditions, allow a systematic search for invariants and critical patterns. Such experiments provide existence proofs of the sufficiency of given mechanisms for generating observed CAS phenomena.

4. A mathematics of competition-based recombination.
   A mathematics so-oriented would emphasize invariant features of evolutionary, far-from-equilibrium, trajectories generated by recombination, such as time to first occurrence of certain kinds of "building blocks", rate of spread of such "bulding blocks", mechanisms that maintain diversity, and so on.

References

[1] J. H. Holland, *Adaption in Natural and Artificial Systems*, 2nd Edition, MIT Press, 1992.

[2] J. H. Holland, K. J. Holyoak, R. E. Nisbett, and P. R. Thagard, *Induction: Processes of Inference, Lear, and Discovery*, Paperback Edition. MIT Press, 1989.

[3] J. H. Holland, "Echoing emergence," *Integrative Themes*, G. Cowan et al. (editors), Addison-Wesley, 1993.

A New Parallel Architecture for Scientific Computation Based on Finite Projective Geometries

Narendra Karmarkar

AT&T Bell Laboratories

Murray Hill, New Jersey

1 Introduction

The parallel architecture described in this paper was motivated by certain types of problems arising in scientific computation such as linear programming, solution of partial differential equations, signal processing, simulation of non-linear electronic circuits, non-linear programming etc. A number of massively parallel architectures have been explored in the last two decades. Some key features of the architecture described here that differentiate it from many others are as follows.

In this architecture, each elementary instruction you can give to the system as a whole automatically results in coherent, conflict-free operation and uniform, balanced use of all the resources in the system such as processors, memories and network connecting them. Thus, conflicts are ruled out by architectural design, instead of relying on some higher level software to resolve them. To accomplish this, an instruction for the system as a whole is not just any combination of instructions for the individual elements of the system. The allowed combinations obey restrictions derived from mathematical properties of projective geometry. Although the collection of these allowed combinations is small, it is "complete" in the sense that any desired computation can be carried out by a proper sequence of these restricted combinations. Furthermore, their mathematical structure permits the process of transforming user programs into a sequence of system instructions to be carried out efficiently. The programmer is not required to come up with a decomposition of the computational problem into pieces suitable for individual processors and program the communication between the pieces. Instead, the architecture offers a built-in decomposition algorithm using a rule based on projective geometry. This decomposition is carried out at the level of individual atomic operations (such as add, multiply etc.) thus exploiting fine-grain parallelism. The architecture does not need a routing algorithm to control the movement of data through the interconnection network. Instead, each routing element makes its decision at each machine cycle by computing a simple mathematical function over a finite field. This computation is built into the hardware and does not require any programming. The interconnection network is also based on subspaces of the projective geometry.

The remainder of this section contains a simple introduction to finite projective geometry. Section 2 describes the architecture based on projective geometry. Section 3 describes applications of 4-dimensional projective geometry. The final sections describes the group theoretic structure and symmetries of projective geometry and their significance to the architecture. I have been working on this architectural concept since 1988, and some earlier reports on this work can be found in references at the end.

Consider a finite field $F = GF(s)$ having s elements, where s is a power of a prime number q, $s = q^k$, and k is a positive integer.

A projective space of dimension d over the finite field F, denoted by $P^d(F)$, consists of one dimensional subspaces of a $(d + 1)$ dimensional vector space F^{d+1} over the finite field, F. Elements of this vector space can be represented as $(d + 1)$-tuples $(x_1, x_2, x_3, \ldots, x_{d+1})$ where each $x_i \in F$. Clearly, the total number of such elements is $s^{d+1} = q^{k(d+1)}$. Two non-zero elements $\underline{x}, \underline{y} \neq 0$ of this vector space are said to be equivalent if there exists a $\lambda \in GF(s)$ such that $\underline{x} = \lambda \underline{y}$. Each equivalence class gives a point in the projective space. Hence the number of points in $P^d(F)$ is given by

$$P_d = \frac{s^{d+1} - 1}{s - 1}.$$

An m-dimensional projective subspace of $P^d(F)$ consists of all one dimensional subspaces of an $(m+1)$-dimensional subspace of the vector space F^{d+1}. Let $\underline{b_0}, \underline{b_1}, \ldots, \underline{b_m}$ be a basis of the latter vector subspace. The elements of the vector subspace are given by

$$\underline{x} = \sum_{i=0}^{m} \alpha_i \underline{b_i} \quad where \ \alpha_i \in F.$$

Hence the number of such elements is s^{m+1}, and the number of points in the corresponding projective subspace is

$$p_m = \frac{s^{m+1} - 1}{s - 1}.$$

Let $r = d - m$. Then an $(m + 1)$-dimensional vector subspace of F^{d+1} can also be described as the set of all solutions of a system of r independent linear equations $\underline{a_i}^T \underline{x} = 0$ where $\underline{a_i} \in (F^*)^{d+1}$, the dual of F^{d+1}, $i = 1, \ldots, r$; this vector subspace and the corresponding projective subspace are said to have codimension r.

Let Ω_l denote the collection of all projective subspaces of dimension l. Thus Ω_0 is the set of all points in the projective space, Ω_1 is the set of all lines, Ω_{d-1} is the set of all hyperplanes, etc. For $n \geq m$, define

$$\phi(n, m, s) = \frac{(s^{n+1} - 1)(s^n - 1) \ldots (s^{n-m+1} - 1)}{(s^{m+1} - 1)(s^m - 1) \ldots (s - 1)}.$$

Then the number of l-dimensional projective subspaces of $P^d(GF(s))$ is given by

$$\phi(d, l, s).$$

The number of points in an l-dimension projective subspace is given by

$$\phi(l, 0, s) .$$

The number of distinct l-dimensional subspaces through a given point is

$$\phi(d - 1, l - 1, s),$$

and the number of distinct l-dimensional subspaces, $l \geq 2$, containing a given line is

$$\phi(d - 2, l - 2, s) .$$

More generally, for $0 \leq l < m \leq d$, the number of m-dimensional subspaces of $P^d(GF(s))$ containing a given l-dimensional subspace is

$$\phi(d - l - 1, m - l - 1, s),$$

and the number of l-dimensional subspaces contained in a given m-dimensional subspace is

$$\phi(m, l, s) .$$

2 Architecture Based on Projective Geometry

2.1 Definition

Given a prime number q, two positive integers k and d, and two integers m and p such that $0 \leq m < p \leq d$, we can define an architecture as follows:

Let $P^d(GF(q^k))$ be the projective geometry of dimension d over $GF(q^k)$. As before, Ω_m and Ω_p denote the collection of projective subspaces of dimension m and p respectively. Let $M = \phi(d, m, q^k) = |\Omega_m|$ be the number of subspaces of dimension m, and $P = \phi(d, p, q^k) = |\Omega_p|$ be the number of subspaces of dimension p. The architecture has M memory modules numbered $1, 2, \ldots M$, put in one-to-one correspondence with the subspaces of dimension m. It has P processors numbered $1, 2, \ldots P$, put in one-to-one correspondence with subspaces of dimension p. A memory module and a processor are connected if and only if the subspace corresponding to the memory module is contained in the subspace corresponding to the processor. For a symmetric architecture having the same number of processors as memories, it is necessary that $m + p + 1 = d$. Asymetric architectures also have applications, since it may be desirable to have more memories than processors for the sake of bandwidth matching. This paper deals primarily with the symmetric case.

2.2 Example of the Two-Dimensional Scheme

In this section, we introduce the simplest scheme that is based on a two dimensional projective space $P^2(F)$, where $F = GF(s)$, also called the projective plane of order s.

The number of points in a projective plane of order s is

$$\phi(2, 0, s) = \frac{s^3 - 1}{s - 1} = s^2 + s + 1 .$$

The number of lines is given by

$$\phi(2,1,s) = \frac{(s^3-1)(s^2-1)}{(s^2-1)(s-1)} = s^2 + s + 1\,,$$

which is the same as the number of points.

Each line contains $(s+1)$ points and through any point there are $(s+1)$ lines. Every distinct pair of points determine a line and every distinct pair of lines intersect in a point. Note that there are no exceptions to the latter rule in a projective geometry since there are no parallel lines.

Let $n = s^2 + s + 1$. Given n processors and a memory system partitioned into n memory modules $M_1, M_2, \ldots M_n$, put the memory modules in one-to-one correspondence with points in the projective space, and processors in one-to-one correspondence with lines. A memory module and a processor are connected if the corresponding point belongs to the corresponding line. Thus each processor is connected to $(s+1)$ memory modules and vice versa.

2.3 Rule-Based Load Assignment

The assignment of computational load to processors is done at a fine-grain level, based solely on the location of the data. For example, consider a binary operation that takes two operands a and b as inputs and modifies one of them, say a, as output

$$a \leftarrow a \circ b.$$

Let A denote the projective subspace corresponding to the memory module that contains the operand 'a' and let B denote the projective subspace corresponding to the memory module that contains the operand 'b'. Then the processor responsible for carrying out the above operation is determined by computing the projective subspace C associated with the processor as follows:

$$C = f(\langle A, B \rangle)\,.$$

Here $\langle A, B \rangle$ denotes the projective subspace spanned by A and B and f is an assignment function that maps subspaces to subspaces.

As a simple example of this mechanism, consider an architecture based on the two-dimensional projective geometry and choose the assignment function as follows.

- f maps Ω_1 (lines) to Ω_1 acting like the identity function; i.e., $f(X) = X$ if $X \in \Omega_1$.

- f maps Ω_0 (points) to Ω_1 is a one-to-one manner such that $X \subseteq f(X)$ for $X \in \Omega_0$; i.e., it maps each point to one of the lines going through the point. It is easy to generate such mappings using the duality of projective geometry.

Thus if $A \neq B$, $C = \langle A, B \rangle$, and the processor responsible for the binary operation corresponds to the unique line determined by the two points A and B. If $A = B$, C is some line containing A. In either case, according to the interconnection rule described earlier, the processor corresponding to C has direct connections to the memory modules containing the operands it needs to access in order to carry out the operation. Therefore, the interconnection network in this architecture is as effective as a fully connected network, in spite of being much sparser, provided the load assignment is done by the rule described above. Additional advantage of the rule-based load assignment is that the system can infer the assignment by performing a simple computation thus relieving the programmer from the burden of specifying "intelligent" problem mapping. The rule described above also tends to distribute the load evenly among the processors. It is easy to show that if all sequences of binary operations of a given length are considered to be equally likely, the expected amount of work assigned to each processor is the same.

2.4 Perfect Patterns

The memory system of this architecture is partitioned into m modules denoted by $M_1, M_2, \ldots, M_m$. The type of memory access possible in one machine cycle in this architecture is between the two extremes of random access and sequential access and could be called "structured" access. Only certain combinations of words can be accessed in one cycle. These combinations are designed using certain symmetries present in the projective geometry so that no conflicts can arise in either accessing the memory or sending the accessed data through the interconnection network. The total set of allowed combinations is powerful enough so that a sequence of such structured accesses is as effective as random accesses. On the other hand, such a structured-access memory has a much higher bandwidth than a random-access memory implemented using comparable technology. The allowed access patterns are called "perfect patterns."

This section defines perfect access patterns for a two-dimensional geometry.

Let the number of points (and hence the number of lines) in the geometry be n. Hence $m = p = n$.

A **perfect access pattern** is a collection of n ordered pairs of points

$$P = \{(a_1, b_1), (a_2, b_2), \ldots (a_n, b_n) | a_i \neq b_i ,$$
$$a_i, b_i \in \Omega_0, i = 1, \ldots n\}$$

having the following properties:

1. First members $\{a_1, a_2, \ldots a_n\}$ of all the pairs form a permutation of all points of the geometry.

2. Second members $\{b_1, b_2, \ldots b_n\}$ of all pairs form a permutation of all points of the geometry.

3. Let l_i denote the line determined by the ith pair of points, i.e.,

$$l_i = \langle a_i, b_i \rangle .$$

Then the lines $\{l_1, l_2, \ldots l_n\}$ determined by these n pairs form a permutation of all lines of the geometry.

	Point Pairs	Corresponding Lines
1	(a_1, b_1)	$l_1 = \langle a_1, b_1 \rangle$
2	(a_2, b_2)	$l_2 = \langle a_2, b_2 \rangle$
$\vdots$	$\vdots$	
n	(a_n, b_n)	$l_n = \langle a_n, b_n \rangle$

Table 1: Perfect access patterns for 2-d geometry

Clearly, if one schedules a collection of binary operations corresponding to such a set of point-pairs for simultaneous parallel execution, then we have the following situation:

1. There are no read or write conflicts in memory accesses.

2. There is no conflict or waiting in processor usage.

3. All processors are fully utilized.

4. Memory bandwidth is fully utilized.

Hence the name "perfect pattern." A collection of perfect patterns is called **complete** if every pair occurs in exactly one pattern in the collection.

2.5 Perfect Sequences

The perfect patterns described above can be combined to form certain special sequences, again using the symmetries present in the projective geometry. Each such sequence, called a "perfect sequence" defines the operation of the hardware for several consecutive machine cycles at once, and has the effect of utilizing the communication bandwidth of the machine fully. As a result, it is possible to connect the memories and processors in the system so that the number of wires needed grows linearly with respect to the number of processors and memories in the system.

Recall that a memory module corresponding to a subspace A is connected to a processor corresponding to a subspace B if and only if $A \subseteq B$. This connection is denoted by the ordered pair (A, B). Let C denote the collection of all memory-processor connections. A perfect pattern is said to *exercise* a connection if execution of the pattern involves data movement over that connection.

A sequence of perfect access patterns is called a **perfect sequence** if each connection in C is exercised the same number of times collectively by the patterns contained in the sequence. If such a perfect sequence is packaged as a single instruction that defines the operation of the machine over several machine cycles, it leads to uniform utilization of the communication bandwidth of the wires connecting processors and memories.

Perfect access patterns and perfect sequences can be easily generated using the group-theoretic structure of projective geometries, as described later.

3 Applications of Four-Dimensional Projective Geometries
3.1 Introduction
Subspaces of the four-dimensional projective geometry $P^4(GF(s))$ are:

$$\Omega_0 : \text{points,}$$
$$\Omega_1 : \text{lines,}$$
$$\Omega_2 : \text{planes,}$$
$$\Omega_3 : \text{3 spaces, and}$$
$$\Omega_4 : \text{the entire space.}$$

A symmetric architecture can be constructed using Ω_0 and Ω_3, or Ω_1 and Ω_2. Here we will consider the latter case. For this case, parameters of the architecture defined in section 2.1 are chosen as follows: q: any prime number, k: any positive integer, $d = 4$, $m = 1$, and $p = 2$.

In the projective space $P^4(F)$ where $F = GF(s)$ the number of lines is

$$n = \phi(4,1,s) = \frac{(s^5 - 1)(s^4 - 1)}{(s^2 - 1)(s - 1)} \ .$$

Therefore, the number of memory modules is

$$n = (s^2 + 1)(s^4 + s^3 + s^2 + s + 1) \ .$$

The number of planes is
$$\phi(4,2,s) = \phi(4,1,s) = n \ .$$

Hence the number of processors is also n. The number of planes containing a given line is

$$\begin{aligned}
\phi(d - 2, 0, s) &= \phi(2, 0, s) = \frac{s^3 - 1}{s - 1} \\
&= s^2 + s + 1 \ .
\end{aligned}$$

Let $k = s^2 + s + 1$. Thus $k = 0(\sqrt[3]{n})$. Hence each memory module is connected to $k = 0(\sqrt[3]{n})$ processors.

Similarly the number of lines contained in a given plane is $\phi(2,1,s) = s^2 + s + 1 = k$. Hence each processor is connected to $k = 0(\sqrt[3]{n})$ memory modules.

As described earlier, the number of interconnections required between memories and processors can be further reduced if the communication between processors and memories is carried out in a disciplined and coordinated manner, based on the *perfect sequences* of *perfect patterns*.

3.2 Application Examples
As a first example, consider parallelization of join operation of two binary relations in a relational database. Structured access memory for this application consists of a collection of disks put in one-to-one correspondence with lines of a four dimensional geometry, and connected to processors corresponding to two dimensional planes.

Data for a binary relation R is partitioned into m components $R_1, R_2, \ldots, R_m$ according to the scheme described below and the component R_i is stored on the ith disk. Map individual components of each tuple to points of the projective geometry by means of a hash function. Thus a tuple (a, b) is mapped to a pair of points $(f(a), f(b))$. This pair of points determines a unique line $\langle f(a), f(b) \rangle$ provided $f(a) \neq f(b)$. (If $f(a) = f(b)$ there is a choice). The tuple (a, b) is stored on the disk corresponding to this line. Suppose we want to join two binary relations R and S on the second attribute. To accomplish this, each processor P_i carries out $R_j | X | S_k$ for all lines j and k belonging to the plane i, in parallel. Note that each tuple $(a, b) \in R$, needs to be brought together with each tuple $(c, b) \in S$. This is done by the processor corresponding to the plane $\langle f(a), f(b), f(c) \rangle$. In order to do the join, the processor needs to access (a, b) and (c, b) which are stored on the disks corresponding to lines $\langle f(a), f(b) \rangle$ and $\langle f(c), f(b) \rangle$. But the plane $\langle f(a), f(b), f(c) \rangle$ contains the lines $\langle f(a), f(b) \rangle$ and $\langle f(c), f(b) \rangle$. Therefore all the required connections between processors and disks exist, according to the interconnection rule based on the projective geometry. This is another example of how a scheme based on projective geometry makes a sparse network as effective as a fully connected network.

As a second example, consider sparse Gaussian elimination, an operation required in many scientific and engineering computations.

A typical operation in symmetric gaussian elimination applied to a matrix A is

$$A(i, k) \leftarrow A(i, k) - \frac{A(i, j) * A(j, k)}{A(j, j)}$$

where $A(j, j)$ is the pivot element. Such an operation needs to be carried out only if $A(i, j) \neq 0$ and $A(j, k) \neq 0$. When the non-zero elements in the matrix are not in consecutive locations, it is very difficult to obtain uniformly high efficiency on vector machines. On the projective geometry architecture the matrix is stored as follows.

1. Map the row and column indices $1, 2, \ldots$ to points of the projective space by means of an assignment function f, which can be a hash function.

$$\alpha = f(i), \quad \alpha \in \Omega_0 .$$

2. Let $A(i, j)$ be a non-zero element

$$let \quad \alpha \ = f(i)$$
$$and \quad \beta \ = f(j) \quad where \ \alpha, \beta \in \Omega_0 .$$

Then the pair of points α, β in the projective space determines a line $l \in \Omega_1$ (if $\alpha = \beta$ we have some freedom in determining the line). The element $A(i, j)$ is stored in the memory module corresponding to line l.

Again consider a typical operation in gaussian elimination

$$A(i, k) \leftarrow A(i, k) - \frac{A(i, j) * A(j, k)}{A(j, j)}$$

	Triplet of Pts	Triplet of Lines			Planes
1	(a_1, b_1, c_1)	$u_1 = \langle a_1, b_1 \rangle$	$v_1 = \langle b_1, c_1 \rangle$	$w_1 = \langle c_1, a_1 \rangle$	$h_1 = \langle a_1, b_1, c_1 \rangle$
2	(a_2, b_2, c_2)	$u_2 = \langle a_2, b_2 \rangle$	$v_2 = \langle b_2, c_2 \rangle$	$w_2 = \langle c_2, a_2 \rangle$	$h_2 = \langle a_2, b_2, c_2 \rangle$
$\vdots$	$\vdots$				
n	(a_n, b_n, c_n)	$u_n = \langle a_n, b_n \rangle$	$v_n = \langle b_n, c_n \rangle$	$w_n = \langle c_n, a_n \rangle$	$h_n = \langle a_n, b_n, c_n \rangle$

Table 2: Perfect patterns for four-dimensional geometry

and let $\alpha = f(i), \beta = f(j)$ and $\gamma = f(k)$. Then all the point-pairs involved in the above operation are subsets of the triplet (α, β, γ).

Assuming that the triplet of points (α, β, γ) are in general position, they determine a plane, say δ, of the projective space; i.e., $\delta \in \Omega_2$. The above operation is assigned to the processor corresponding to δ. In order to carry out this operation, the processor needs to be able to communicate with memory modules corresponding to the pairs $(\alpha, \beta), (\beta, \gamma)$ and (α, γ). Note that the lines determined by these pairs are contained in the plane determined by the triplet (α, β, γ), hence the necessary connections exist.

3.3 Perfect Access Patterns for 4-d Geometry

In $P^4(GF(s))$, let n denote the number of lines, which is also equal to the number of 2-dimensional planes.

A **perfect access pattern** is a collection n non-collinear triplets.

$$P = \{(a_i, b_i, c_i) | a_i, b_i, c_i \in \Omega_0, \dim\langle a_i, b_i, c_i \rangle = 2, i = 1, \ldots n\}$$

having the following properties.

1. Let $u_i, i = 1, \ldots n$, denote the lines generate by first two points from each triplet, i.e., $u_i = \langle a_i, b_i \rangle$. Then the collection of lines $\{u_1, u_2, \ldots u_n\}$ forms a permutation of all the lines of the geometry.

2. Let $v_i, i = 1, \ldots n$, denote the lines $v_i = \langle b_i, c_i \rangle$. Then the collection of lines $\{v_1, v_2, \ldots v_n\}$ forms a permutation of all the lines of the geometry.

3. Let $w_i, i = 1, \ldots n$, denote the lines $\langle c_i, a_i \rangle$, $w_i = \langle c_i, a_i \rangle$. Then the collection of lines $\{w_1, \ldots w_n\}$ forms a permutation of all the lines of the geometry.

4. Let $h_i, i = 1, \ldots n$, denote the planes generated by the triplets (a_i, b_i, c_i), $h_i = \langle a_i, b_i, c_i \rangle$. Then the collection of planes $\{h_1, h_2, \ldots h_n\}$ forms a permutation of all the planes of the geometry.

When an operation having (a_i, b_i, c_i) as the associated triplet is performed, the three memory modules accessed correspond to the three lines $\langle a_i, b_i \rangle$, $\langle b_i, c_i \rangle$ and $\langle c_i, a_i \rangle$, and the processor performing the operation corresponds the plane $\langle a_i, b_i, c_i \rangle$. Hence it is clear that if we schedule n operations having the property that the associated triplets form a perfect pattern, to execute in parallel in the same machine cycle then

1. there is no read or write conflicts in memory accesses

2. there is no conflict in the use of processors

3. all the processors are fully utilized

4. the memory bandwidth is fully utilized.

Hence such a collection of triplets is called **perfect pattern**. A collection of perfect patterns is called **complete** if each non-collinear triplet occurs in exactly one pattern in the collection.

4   Group-Theoretic Structure of Projective Spaces

4.1   Introduction

Recall that points in the projective space of dimension d over $GF(s)$ were defined as rays through the origin in the vector space of dimension $(d + 1)$ over $GF(s)$, which contains s^{d+1} elements. Since $s = q^k$, a power of a prime number, so is $s^{d+1} = q^{k(d+1)}$. Hence there is a unique finite field with s^{d+1} elements. One might suspect that there may be some relation between the finite field $GF(s^{d+1})$ and the projective space $P^d(GF(s))$, which would give the projective space $P^d(GF(s))$ additional structure based on the multiplication operation in $GF(s^{d+1})$, besides its geometric structure. Indeed, there is such a relation, and we want to elaborate it further.

If q is a prime number, then $GF(q^m)$ contains $GF(q^n)$ as a subfield if and only if $n|m$. Therefore $GF(s^{d+1})$ (where $s = q^k$) contains $GF(s)$ as a subfield and the degree of $GF(s^{d+1})$ over $GF(s)$ is $(d + 1)$. Hence $GF(s^{d+1})$ is a vector space of dimension $(d+1)$ over $GF(s)$. Each non-zero element of this vector space determines a ray through the origin and hence a point in the projective space $P^d(GF(s))$.

Let $G^* =$ the multiplicative group of non-zero elements in $GF(s^{d+1})$ and $H^* =$ the multiplicative group of non-zero elements in $GF(s)$. Clearly H^* is a subgroup of G^* and both groups are cyclic. Let $x \in G^*$. The ray through origin determined by x consists of the points

$$\{\lambda x | \lambda \in GF(s)\} \, .$$

This is precisely the coset of H^* in G^* determined by x, along with the origin.

This establishes a one-to-one correspondence between points of the projective space $P^d(GF(s))$ and elements of the quotient group G^*/H^*. This correspondence allows us to define a multiplication operation in the projective space.

Let g be a fixed point in the projective space $P^d(GF(s))$ and consider the mapping of $P^d(GF(s))$ onto itself defined by

$$L_g : x \to g \circ x,$$

where $g \circ x$ is the multiplication operation introduced above. It is easy to check that this operation maps lines in the projective space onto lines, planes onto planes and, in general any projective subspace of dimension k onto another projective subspace of the same dimension. Such a mapping is called an *automorphism* of the projective geometry.

Since G^* and H^* are cyclic, so is G^*/H^*. Let g be a generator of G^*/H^*. (By abuse of notation, we will not henceforth distinguish between elements of G^*/H^* and points of $P^d(GF(s))$). Thus we can denote points in $P^d(GF(s))$ as $g^i, i = 0, \ldots n-1$. The mapping L_g becomes

$$L_g : g^i \to g^{i+1} .$$

This will be called a *shift* operation.

Any power L_g^k of the shift operation is also an automorphism of the geometry and the collection of all powers of the shift operation forms an automorphism group denoted henceforth by L, which is a subgroup of the group of all automorphisms of the geometry. A subgroup G of the full automorphism group is said to act transitively on subspaces of dimension k if for any pair of subspaces, H_1, H_2 of dimension k, there is an element of G which maps H_1 to H_2.

For any projective space, the shift-operation subgroup L acts transitively on the points and on subspaces of co-dimension one (hyperplanes). This property is used for generating perfect patterns and sequences for the two dimensional geometries. Perfect patterns for 4-dimensional case requires other automorphisms described later.

4.2 Virtual to Physical Mapping Based on Symmetries

Number of memories and processors in the logical or virtual view of the system can be different from the actual physical number of memories and processors. There are many advantages to "folding" the virtual architecture along the symmetries of projective geometry.

To define this notion more precisely, let H be any subgroup of the automorphism group. It's action on Ω_k partitions Ω_k into orbits. All subspaces in an orbit can be thought of as forming a "virtual" group that corresponds to a single physical resource. This can be done for memories or processors or both. Such virtual to physical mapping makes it possible to have interleaved memories and multithreaded processors consistent with the projective geometry structure to obtain benefits of higher memory bandwidth and latency-hiding.

4.3 Generation of Perfect Patterns

In case of a two-dimensional geometry, the hyperplanes (co-dimension $= 1$) are the same as lines (dimension $= 1$). Hence the shift operation L_g and its powers L_g^k act transitively on lines.

To generate a perfect pattern using the shift operation, take any pair of points $a, b, a \neq b$, in the geometry. Let l denote the line generated by a and b, i.e., $l = \langle a, b \rangle$. Set $a_0 = a, b_0 = b$ and $l_0 = l$ and define a_k, b_k and $l_k, k = 1, \ldots n - 1$, by successive application of the shift operation L_g as follows:

$$
\begin{aligned}
a_k &= L_g \circ a_{k-1}, \\
b_k &= L_g \circ b_{k-1}, \\
\text{and } l_k &= L_g \circ l_{k-1} .
\end{aligned}
$$

Since L_g is an automorphism of the geometry, we have

$$l_{k-1} = \langle a_{k-1}, b_{k-1} \rangle \Leftrightarrow l_k = \langle a_k, b_k \rangle .$$

Then $P = \{(a_k, b_k), k = 0, \ldots n-1\}$ is a perfect pattern of the geometry.

Now in order to generate a complete collection of such patterns, take any line $l = \{a_1, a_2, \ldots a_k\}$ of the geometry. Form all pairs $(a_i, a_j), i \neq j$, from the points on l. Generate a perfect pattern from each of the pairs. It is easy to show that the collection of perfect patterns obtained this way is complete.

Automorphisms of the geometry based on cyclic shifts are not enough for generating perfect access patterns for the 4-d geometry. First we consider other types of automorphism.

In a finite field of characteristic p, the operation of raising to the pth power, i.e., $x \rightarrow x^p$ forms an automorphism of the field, i.e., $(x+y)^p = x^p + y^p$ and $(xy)^p = x^p y^p$. Since the points in the projective space $P^d(GF(s))$ correspond to elements of the multiplicative group G^*/H^*, the operation of raising to the pth power can also be defined on the points of $P^d(GF(s))$. It is easy to show that such operation is an automorphism of the projective space.

The operations cyclic shift and raising to the pth power together are adequate for generating the perfect access patterns for the smallest 4-dimensional geometry $P^4(GF(2))$ which has 155 lines and planes. For bigger 4-d geometries we need other automorphisms. The most general automorphism of $P^d(GF(s))$ is obtained by means of a non-singular $(d+1) \times (d+1)$ matrix over $GF(s)$. Particular choices of automorphisms are important from the point of view of efficient hardware or chip implementation.

In any $P^d(GF(s))$, the group of all automorphism acts **transitively** on **all subspaces**; i.e., given any pair of subspaces H_1 and H_2 of the same dimension, there is an automorphism that maps H_1 to H_2.

Perfect patterns for 4-d geometry are generated as follows:

- Take any non-collinear triplet (a, b, c) of points

- Apply the elements of automorphism group to generate the **orbit**

$$(a^{(k)}, b^{(k)}, c^{(k)}) \quad k = 0, 1, 2, \ldots, n-1 .$$

The orbit is a perfect pattern.

To generate a **complete set** of **perfect patterns**

- Take any plane $H = (x_1, x_2, \ldots, x_k)$

- Form all non-collinear triplets from points of H

- Generate a perfect pattern/triplet as its orbit

Together they give a **complete set**.

References

[1] G. C. Bell and A. Newell, *Computer Structures: Readings and Examples*, McGraw-Hill, 1971.

[2] I. S. Dhillon, *A Parallel Architecture for Sparse Matrix Computations*, B. Tech. Project Report, Indian Institute of Technology, Bombay, 1989.

[3] M. Hall Jr., " Combinatorial Theory," *Wiley-Interscience Series in Discrete Mathematics*, 1986.

[4] N. Karmarkar, "A new parallel architecture for sparse matrix computations," Proceedings of the Workshop on Parallel Computing, BARC, Bombay, February 1990, 1-18.

[5] N. Karmarkar, "A new parallel architecture for sparse matrix computations based on finite projective geometries," invited talk at the SIAM Conference on Discrete Mathematics, Atlanta, June 1990.

[6] I. Dhillon, N. Karmarkar, and K. G. Ramakrishnan, "Performance analysis of a proposed parallel architecture on matrix vector multiply like routines," Technical Report #11216-901004-13TM, AT&T Bell Laboratories, Murray Hill, N.J., October 1990.

[7] N. Karmarkar, "A new parallel architecture for sparse matrix computation based on finite projective geometries," Proceedings of the Supercomputing 91, Albuquerque, New Mexico, November 1991, 358-369.

[8] I. Dhillon, N. Karmarkar, and K. G. Ramakrishnan, "An overview of the compilation process for a new parallel architecture," Proceedings of the Fifth Canadian Supercomputing Conference, Fredericton, N. B., Canada, June 1991.

[9] Van Der Waerden, *Algebra*, **1**, Unger, 1970.

Mathematical Programming and the Algebra of Polynomials

Laszlo Lovász

Yale University
New Haven, Connecticut

The role of linear programming in optimization, especially discrete optimization, is well understood; we routinely use techniques like LP duality, complementary slackness, and LP algorithms in proving theorems and designing algorithms. Also, complexity questions in this field have been extensively studied.

Much less is known about the role of non-linear inequalities. It is of course discouraging that one almost immediately runs into NP-hard problems. For example, it is NP-hard to decide whether a given set of quadratic equations is solvable—even if we allow complex solutions. So we cannot appeal to non-linear algebra as freely as to linear programming.

On the other hand, quadratic (and, of course, even more higher-degree) polynomials provide very elegant an concise algebraic formulations for combinatorial optimization problems. As an example, consider the stable set problem. Let $G = (V, E)$ be a graph and consider the following system of equations:

$$x_i^2 = x_i$$

for every node $i \in V$, and

$$x_i x_j = 0$$

for every edge $ij \in E$. Trivially, the solutions of the first set of inequalities are precisely the 0-1 vectors, and so the solutions of the whole system are precisely the incidence vectors of stable sets.

One fact this formulation shows is that even the solvability of such a simple system of quadratic equations (together with a linear equation $\sum_i x_i = t$) is NP-hard (whether we want to find solutions in the rational, real, complex, or any finite field). But we can also use this in the positive direction.

First note that it *can* be decided in polynomial time whether a single quadratic inequality (in n variables) has a solution. This can be reduced to the positive definiteness of a matrix, which can be checked by Gaussian elimination. It is also known that a quadratic polynomial is identically non-negative iff it is a sum of squares of linear polynomials, and is identically positive if and only if it is a sum of a positive constant and squares of linear polynomials. The positive definiteness check gives such a decomposition.

Second, it *can* be decided in polynomial time whether a given system of quadratic inequalities has a linear combination with non-negative coefficients that is identically false. Basically, this can be reduced to checking whether some non-negative linear combination of matrices belongs to the convex cone of positive definite matrices. This problem is equivalent) to a *semidefinite program*. Much of the machinery of linear programming extends to semidefinite programs; in particular, duality theory, the ellipsoid method, and interior point methods.

Unfortunately, for quadratic inequalities there is no Farkas lemma: a given system may be unsolvable and yet have no identically false combination with non-negative coefficients. Let us say that the system is *strongly unsolvable* if it does. However, replacing unsolvability by strong unsolvability leads to relaxations that are often of interest. For example, the smallest t for which the system above, together with $\sum_i x_i \geq t$ is not unsolvable is the stability number $\alpha(G)$; the smallest t for which this same system is not strongly unsolvable is the polynomial time computable upper bound $\vartheta(G)$ on the stability number.

On the other hand, we gain from the fact that among quadratic inequalities, there are other natural rules of inference besides linear combination with non-negative coefficients. For example, we can multiply two linear inequalities to get a quadratic one. It turns out that these rules can be used to derive all valid linear inequalities, and in fact in a very simple and short way. This derivation also has algorithmic implications.

Specific problems where quadratic inequalities and/or semidefinite programming have been applied include perfect graphs, stability number, chromatic number, maximum cuts, discrepancy of hypergraphs, interactive proof systems, graph homomorphisms, and the satisfiablity problem.

One field of very interesting applications emerging recently is the design of approximation algorithms. The richer structure of the optimum solutions of quadratic relaxations (a matrix instead of a vector) gives more power to (usually randomized) rounding methods.

Financial Planning via Multi-Stage Stochastic Programs

John M. Mulvey

Princeton University

Princeton, New Jersey

1 Introduction

Stochastic programming provides an ideal framework for modeling financial decisions and investment strategies over time. The main drawbacks–computational efficiency and the ability to model stochastic parameters in a dynamic environment–are slowly receding as computers become cheaper and more powerful, algorithms become more efficient, and financial economists and others begin to understand the structure of markets.

The area of finance has numerous applications of stochastic programs. The need for an optimal solution is obvious. There are clear sources of uncertainty that can be represented by means of stochastic processes and related models. The problems are significant and complex enough to warrant a formal optimization model. In many cases, we are now able to solve the resulting large scale (non)linear program.

The proposed multi-stage stochastic program depends upon two discretizations. First, we divide the planning period into a finite number of stages in which investment decisions are rendered. In our model, decisions are made at the last instant of each period. Second, stochastic parameters are estimated by means of a modest number of sample paths over the planning period. Each path is called a **scenario**, which represents a single realization of all uncertainties for parameters in the optimization model. Problem size grows quickly as a function of the number of time stages and scenarios. Thus, from a practical standpoint, it is critical to keep both factors from becoming too large, while at the same time ensuring that the stochastic program represents the real-world problem to sufficient accuracy. Achieving this goal will be critical to the success of future stochastic programming related research and real-world implementations. Regarding applications, we devote much effort to constructing the key scenarios and their selection for use in the financial planning system.

The dynamic investment problem has attracted much interest in the literature, going back to the early papers by Samuelson (1969) and Merton (1969) Besides stochastic programming, there are several alternative modeling frameworks and solution strategies. For example, many economists employ continuous time models and optimal control theory. See Dixit (1990), Dixit and Pindyck (1994) for a sampling of these techniques. Davis (1993) suggests another approach based on piecewise-deterministic Markov processes (PDPs). These approaches often employ dynamic programming for finding the optimal dynamic stochastic control. Also, see Chow (1993) for a related solution strategy. Unfortunately, these methods place severe restrictions on the decision domain and hence have limited utility for many applications. On the other hand, the results are quite strong when they apply.

This paper focuses on discrete-time, scenario-based financial planning. The next section presents the basic machinery for calculating fair economic value of securities and portfolios, including a short discussion of interest rate models. Section 3 defines the multi-stage financial planning model as a stochastic program; several alternative objective functions are proposed. The integration of assets and liabilities is discussed in Section 4. Solution strategies for the large-scale stochastic programs are taken up in Section 5. The importance of developing efficient parallel/distributed algorithms is emphasized. Last, Section 6 suggests topics for future research in this growing area.

2   Preliminary Concepts in Financial Engineering

Calculating investment risks has become increasingly complex as markets have broadened to include international securities, as volatility has increased due to the deregulation of interest rates and other controls, as computerized trading has grown in use, and as real-time data has become available. Also, new exotic securities are being developed at a rapid pace. Common approaches for evaluating securities are briefly discussed in this section.

2.1   Fair Market Computations

The last decade has seen the introduction of securities with complex payouts and embedded options. Mortgage backed securities (MBS) are a prototypical example. A group of mortgages is collected in a large pool and sold in the marketplace as a single instrument. There are two major uncertainties for investors who purchase these securities. First, the mortgages can be paid off by the mortgagees. They hold an option which can be exercised at any time. Second, the value of the securities will depend upon changes in interest rates due to the discounting of future cashflows at different rates. In order to sort out these issues, models have been built for projecting prepayment activity (Zenios 1993b) based on changes in interest rates and other factors. These models generate cashflows conditional on the future course of interest rates. They are path dependent. Figure 1 shows the future paths of interest rates in a scenario tree; cashflows are shown at each node in this tree.

Once the cashflows are determined, the program adds an adjustment term so that the discounted cashflows are set equal to the current market price of the security. The relevant equation is

$$P_0 = \frac{1}{\mid S_0 \mid} \sum_{s=1}^{|S_0|} \sum_{t=0}^{T} \frac{c_t^s}{\prod_{i=1}^{t}(1 + \rho_i^s + OAS)} \tag{1}$$

where OAS (option adjusted spread) is determined by solving this nonlinear equation.

Here, P_0 reflects the current market price, S_0 denotes a set of scenarios that emanate from the root node of the scenario tree, i.e. the current state. The cashflows at each node are equal to c_t^s, and ρ_t^s is the discount rate at time t, scenario s.

The fair market value at any future time period can be computed by discounting the cashflows at the relevant discount rate, including the *OAS*. Many firms who own large numbers of fixed-income securities carry out this form of pricing calculation each evening for all securities in their portfolios (Holmer and Zenios 1993; Mulvey and Zenios 1993).

2.2 Interest Rate Models

Underlying the scenario tree is a stochastic process for generating the interest rates. This area has been the subject of much recent research (Hull 1993; Jarrow 1994; Marsh 1994) and there are numerous competing stochastic models. The models can be divided into single and multiple factors and according to the basic concepts such as ensuring arbitrage-free conditions. These assumptions have been criticized for long term projections. Mulvey (1994c) and Tilley (1992) discuss issues relating to the use of interest rate models.

An example is the Brennan-Schwartz interest rate model. First, the short and long spot rates are calculated based on a two-factor pair of diffusion equations. At its simplest, we assume that long and short interest rates are linked in their movements through a correlated white noise term. The requisite equations are:

$$\text{Short rate:} \quad dr_t = a(r_0 - r_t)dt + b\sqrt{r_t}dZ_1 \tag{2}$$

$$\text{Long rate:} \quad dl_t = c(l_0 - l_t)dt + e\sqrt{l_t}dZ_2 \tag{3}$$

where r_t and l_t represent the short and long interest rates at time t, respectively; a and c are drift coefficients; and b and e are instantaneous volatility coefficients. Mean reverting levels equal r_0 and l_0. The random coefficients–dZ_1 and dZ_2–depict correlated Wiener terms. These two diffusion equations provide the building blocks for the remaining spot interest rates, and then the full yield curve.

The remaining points on the yield curve are determined by making economic assumptions such as arbitrage free or equilibrium. See Brennan and Schwartz (1982) for more details, and Hull (1993) for a description of related interest rate models. These interest rate equations are discretized so that the branches in the scenario tree can be computed (Figure 1).

2.3 Portfolio Issues

Diversification is a fundamental investment concept. The idea is to spread one's investments across a variety of securities so that risks can be reduced. To diversify requires the computation of security co-movements. This calculation is relatively straightforward in some cases: Markowitz proposed the use of single-period variance as an indicator of risks–requiring a covariance matrix. However, covariances can be unstable over longer periods of time due to intertemporal dependencies and other causes. Co-movements are more complex when placed in a dynamic system such as

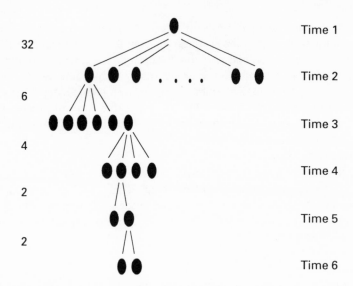

Figure 1: A Scenario Tree for Modeling Stochastic Parameters in Financial Planning Model (3072 Scenarios)

linking assets and liabilities within a multiple period setting, or when putting together a portfolio of instruments such as mortgage backed securities. It is more difficult to compute overall risks since we must employ a method for evaluating cashflows as they depend upon economic factors. Mulvey and Zenios (1993) describe a systematic approach for computing co-movements based on pricing models. The scenario tree provides a mechanism for generating the relative price changes and hence the correlations between securities. Then, one can link the predicted cashflows in a portfolio setting as described in the next section.

In the remainder of this paper, we discuss long term planning models and investment strategies. Hence, we focus on broad asset categories rather than individual securities. Typical asset categories include: U.S. large stocks (SP500), government long-term bonds, money market (cash), real estate, international stocks and bonds, and smaller stocks. The strategic planning problem is called asset allocation; it includes critical decisions pertaining to the investor's asset mix. Experience has shown that the asset mix plays a crucial role in establishing an investor's risk profile, especially for investors who are well diversified. Asset allocation decisions are the critical issues for long-term investors.

3 Multi-Stage Financial Planning

This section presents the asset allocation (AA) decision problem as a multi-stage stochastic program (Mulvey and Vladimirou 1992; Mulvey 1993). Many specific applications can be posed as special cases or approximations of this model. The problem is portrayed as a network graph as shown in Figure 2. Some real-world issues are difficult to accommodate within the network context and must be handled as general linear constraints. Nevertheless, the network provides a visual reference for the AA planning system.

The first critical issue is to establish the requisite planning period. We discretize time within two intervals: $t = \{0, 1, \ldots, \tau\}$ and $t = \{\tau + 1, \ldots, T\}$. The former corresponds to the period over which investment decisions are made. Period t defines the date of the planning horizon; we focus on the investor's position at the beginning of period t. We assume that decisions are made only at the last instant of each time stage. Much flexibility exists. An active trader might see his time interval as short as minutes, whereas a pension plan advisor will be more concerned with much longer planning periods such as the dates between the annual Board of Directors' meetings. It is possible for the time steps to vary–for instance, short time steps at the beginning of the planning period and longer time steps towards the end. The second interval $t = \{\tau + 1, \ldots, T\}$ is needed in order to properly handle the horizon at time τ. The model will calculate economic and other factors beyond period t up to period T and these may depend upon the resolution of previous uncertainties. However, the investor cannot render any active decisions after the end of period $\tau - 1$.

Asset investment categories are defined using the set $i = \{1, 2, \ldots, I\}$ representing broad investment groupings such as stocks, bonds, real estate, or cash. The categories should track a well-defined market segment. Also, whenever possible, the co-movements between pairs of asset returns should be relatively low so that diversification can be done across the asset categories. There should be an index which tracks the market segment for instance the S&P 500 index, the Russell 3000 index, or Morgan Stanley's EAFE. Ideally, the asset category should be available as marketable security via index funds or a futures contract. In the context of the global asset allocation, we differentiate hedged from unhedged investments as two asset categories.

The next critical issue involves the uncertainties. We employ the notion of **scenarios** for modeling the stochastic parameters. This corresponds to the robust optimization framework posed in Mulvey, Vanderbei and Zenios (1994). A scenario $s \in S$ depicts a single plausible set of outcomes for all of the random coefficients over the entire planning period $t = \{1, \ldots, \tau, \ldots, T\}$. General dynamic relationships are included in the discrete-time framework. We extend the economic projections beyond period t for the economic factors –interest rates, currencies, inflation. This allows for a concise procedure for modeling the end of the planning period. We assume that a representative set of scenarios has been constructed which "covers" the set of possibilities to a sufficient degree. Developing this representative set is a topic of current research; see Mulvey (1994a). Forecasting a single scenario, say for placing one-time investment bets, is much different than developing a spectrum of possibilities. See Section 2.4 for more details on the scenario generation process.

There are numerous alternatives to handling uncertainties via a finite set of discrete-time scenarios. See Chow (1994), Dantzig and Infanger (1993), Davis (1993), Dixit and

Pindyck (1994) and their references for competing techniques which address the modeling of the stochastic processes and the accompanying solution of dynamic optimization problems under uncertainty. Some of these papers employ random sampling. Others specialize the modeling conditions in order to find closed-form solutions–in continuous time or in discrete time. Still others employ dynamic programming to find solutions to the multi-stage investment problem. The relative pros/cons of these competing methods have not yet been fully explored.

In our approach, the primary decision variable $x_{i,t}^s$ denotes the amount of investment in asset category i at the beginning of time period t under scenario s. Units are consistent with the investor's home country. Thus, for instance, if an investor decides not to hedge currency risks, returns are always defined in terms of the investor's original currency. The basic model can be readily extended in order to deal with multicurrency variables, but model size will grow as a consequence. The x-vector depicts the state of the system after the rebalancing decisions have been made in the previous period. At that time the investor's total assets are equal to:

$$\sum_i x_{i,t}^s = assets_t^s, \ s \in S, \ t \in T. \tag{4}$$

The uncertain returns $r_{i,t}^s$ for the asset categories–for asset i, time t, and scenario s–are projected by the stochastic modeling subsystem. Each scenario is internally consistent. Thus , $\nu_{i,t}^s$ the wealth accumulated at the end of the t-th period before rebalancing in asset i, is

$$x_{i,t}^s(1 + r_{i,t}^s/100) = \nu_{i,t}^s, \ \forall i \in I, \ t \in T, \ s \in S. \tag{5}$$

Rebalancing decisions are rendered at the very end of each period. Purchases and sales of assets are accommodated by the variables $ybuys_{i,t}^s$ and $ysells_{i,t}^s$ with transaction costs defined via the coefficients t_s assuming symmetry in the transaction costs.

Using the terminology of robust optimization (Mulvey, Vanderbei, and Zenios 1994), we next construct the relationships of the various investment categories at each period as the structural constraints. The flow balance constraint for each asset category and time period is

$$x_{i,t+1}^s = \nu_{i,t}^s + ybuys_{i,t-1}^s(1 - t_i) - ysells_{i,t-1}^s, \ \forall i \in I, \ t \in T, \ s \in S. \tag{6}$$

This equation restricts the cashflows at each period to be consistent. Each node in Figure 1 corresponds to a flow balance equation of the type shown in (3) and (6) below. It is assumed that dividends and interest are forthcoming simultaneously with the rebalancing decisions. Thus, the ysell variables consist of two parts corresponding to the involuntary cash outflow–dividend or interest–and a voluntary component for the cashflow–the amount actively sold (sales). The requisite equation is

$$ysells_{i,t}^s = div_{i,t}^s + sales_{i,t}^s, \ \forall i \in I, \ t \in T, \ s \in S, \tag{7}$$

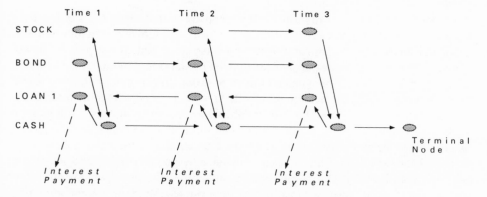

Figure 2: Network Model indicating Flows of Funds between Asset Categories and Cash during Planning Horizon.

where

$$div_{i,t}^s = x_{i,t}^s (divp)_i^s \tag{8}$$

and *divp* is the dividend payout percentage ratio for asset i under scenario s. The cash node at each period t also requires a flow balancing equation

$$
\begin{aligned}
cash_t^s \;=\;& cashin_{t-1}^s + \sum_i (sales)_{i,t-1}^s (1 - t_i) + div_{i,t-1}^s) \\
-\;& \sum_i (ybuys_{i,t}^s + bor_{i,t-1}^s) + cash_{t-1}^s - liab_{t-1}^s \\
+\;& bor_{i,t}^s - prin_{t-1}^s, \; \forall t \in T, \; s \in S
\end{aligned}
\tag{9}
$$

with two new decision variables: $bor_{i,t}^s$ corresponding to the amount of borrowing in each period t; and $liab_t^s$ corresponding to committed liabilities other than borrowing. The variable $prin_t^s$ represents the reduction in borrowed funds that occurs during period t under scenario s. The liability decisions may be dependent upon the state of the world, as depicted by scenario s. We assume that all borrowing is done on a single-period basis. (This assumption can be avoided by adding new decision variables for each category of multi- period borrowing.) Initial wealth at the end of period 0 equals $\nu_{i,0}$, for all scenarios s.

In practice, investors restrict their investments in asset categories for a diversity of purposes such as company policy, legal and historical rules and other considerations. These policy constraints may take any form, but we keep the structure to a set of linear restrictions as specified by

$$A^s x^s = b^s, \; \forall s \in S, \tag{10}$$

where A is $(m \times n)$ matrix of coefficients that depend upon scenario s.

For example, investors may set a lower limit–say 5%–on cash for liquidity considerations. Or investors may wish to restrict their foreign exposure to 10-20% of their portfolio's value. Or there might be a constraint that limits the stock-to-bond exposure to a ratio such as 0.7. Or there might be a minimum dividend inflow at each stage during the planning horizon. Allowing these constraints is valuable on several counts. First, they improve the chances that the model recommendations will be implemented. If an institutional investor must observe a legal constraint, there is little purpose in proposing an investment mix that cannot be made. Second, information can be gained by observing the dual variables on constraints that are binding at the proposed solution. These dual variables can guide the investor who is exploring alternative investment strategies. The use of constraints is the most direct way to approach the what-if analysis that is an inevitable byproduct of asset allocation studies. Third, an investor who places constraints on the feasible region becomes part of the solution algorithm. By this, the investor puts his mark on the recommendations and they become his recommendations, as opposed to a mathematical model simply making an independent decision. Any investor must be convinced that the proposed decisions are sensible. Otherwise, he will not place money in the suggested categories. See Mulvey (1994b) and Cariño et al. (1993) for further implementation details relating to asset-liability investment planning systems.

The constraints and variables have been posed using the split variable representation of the multi-stage stochastic program (Mulvey and Ruszczynski 1994). In this context, separate decision variables are defined for each scenario and asset category. To reflect reality, we must include a set of constraints to prevent the optimization model from anticipating the future. The constraints are known as non-anticipativity conditions (rules) and they have a particularly simple structure. The split-variable model is considerably larger than the traditional "compact" model, but it has the benefit that it accommodates a variety of alternative frameworks and decision policies. Also in some situations the split variable model is easier to solve than the compact model (See Lustig et al. 1991).

There are a number of possible ways to model the non-anticipativity conditions. For instance, in stochastic programming the path of uncertainties is typically revealed as a scenario tree. Non-anticipativity conditions stipulate that the decision variables $\{x, y, bor, \nu, etc.\}$ must be equal to each other as long as they have a common historical past until some time t in the planning horizon $\{0, 1, \ldots, t\}$. We can write this restriction as:

$$x_{i,t}^s = x_{i,t}^{s'} \tag{11}$$

for scenarios s and s' inheriting an identical past up to time t. Similar constraints are needed for selected pairs of the decision variables appearing in the scenario tree. While these constraints are extremely numerous, solution algorithms take advantage of their simple form–a pair of +1 and -1 for each row. Non-anticipativity equations (8) are commonplace in the field of stochastic programming.

An alternative method for handling non-anticipativity is to reduce the decision universe to a set of control policies that do not depend upon knowing the future. We call this approach "dynamic stochastic control" (DSC). For example, we could limit the investment decisions at each stage to a certain fixed percentage rule. Letting $\Lambda_{i,t}^s$ be the proportion of wealth at period t invested in asset i under scenario s, we define a set of constraints which prevent the model from using unavailable information in rendering its decisions. A set of constraints that enforce this policy is shown below

$$\lambda_{i,t}^{s_1} = \lambda_{i,t}^{s_2} \ \forall s_1, s_2 \in S, \ t \in T, \ i \in I. \tag{11'}$$

where

$$\lambda_{i,t}^s = \frac{x_{i,t}^s}{assets_t^s}$$

for asset i at time t. In each period, the investor rebalances to a target fixed investment mix. Thus, we can reduce the number of lambda variables to λ_i for $i \in I$. This restriction greatly shrinks the problem's size. The solution to the asset allocation decision provides the recommended fixed proportions. Adding constraint (11') turns the model into a nonconvex optimization problem. See Maranas et al. (1994) for details on a practical global algorithm for solving this nonconvex problem.

The fixed-mix investment policy may lead to sub-optimal performance since it restricts the decision universe at each step during the planning horizon. For alternatives to the fixed-mix rule, Perold and Sharpe (1988) describe control policies in the context of portfolio insurance. Fixed-mix is a form of selling portfolio insurance and performs best when markets are volatile and drifting sideways. There are many other dynamic investment policies to consider, but in most cases they lead to non-convex optimization problems due to the inclusion of non-linear dynamics when rebalancing the portfolio at the end of each period. The recent results in global optimization (Maranas et al. 1994) are significant since these highly efficient algorithms, which provide performance guarantees, can readily be applied to other classes of DSC policies.

Several issues arise when considering a control policy. These include the background and experience of the investor as well as her computer capabilities. Many investors are unable to set up and run multi-stage stochastic optimization models at each time period as proposed by the model implied by equations (11). On the other hand, once specified, the control strategy is generally very easy to understand and implement. In addition, the control approach can be readily tested with out-of-sample scenarios. The stochastic programming approach may provide improved performance, however, since it generates a larger feasible region. Further testing is needed in order to understand the pros/cons of these alternative modeling frameworks.

3.1 Objective Functions

A fundamental issue in carrying out a financial modeling effort is to settle on the choice of an objective function and underlying preference structure. There are numerous possibilities. In our basic model, the proposed objective maximizes the investor's wealth at the beginning of period τ, subject to the payout of intermediate cash outflows (liabilities) under each of the $s \in S$ scenarios. The investor's true wealth at the horizon τ equals the following

$$wealth_\tau^s = \sum_i x_{i,\tau}^s - PV(liab_{\tau,T}^s) - prin_\tau^s, \tag{12}$$

where $liab_{\tau,T}^s$ is the liability stream from period τ to period T, and $prin_\tau^s$ depicts the amount of loans outstanding at time period τ. This calculation may cause no particular difficulty. In many cases, however, the investor's liabilities are not readily marketable and must be projected along with the accompanying discount factors. Take the case of a pension plan.

Actuaries measure the amount of cashflows that are needed to pay pension beneficiaries as they are demanded. Discounting of the cashflows depends upon the current government interest rates, such as 30-Year Treasury Bonds, as dictated under scenario s. In this situation, we must estimate market value by calculating the present value (PV) of the liabilities under a variety of possible forecasts–scenarios. Part of the right hand side of equation (12) must therefore be estimated through actuarial simulation models. In fact, the actual cash liabilities will depend upon the scenario that occurs. Pension plans generally pay out greater amounts during times of high inflation. Hence, the liabilities must be conditioned on the scenario and may be partially under the control of the decision maker, e.g. for insurance companies when they price their life insurance products.

There are a number of alternative objective functions. One possibility is to employ the classical mean-variance function:

$$maxExp(Wealth_\tau) - \rho \; Variance(Wealth_\tau) \tag{13}$$

where ρ indicates the relative importance of variance as compared with the expected value. This objective leads to an efficient frontier of wealth at period τ by varying ρ. Numerous techniques exist for selecting a point on the efficient frontier, depending upon the risk tolerance of the investor and other factors. See Markowitz (1959) and Ankrim (1992) for discussions. The basic idea is to find a point on the efficient frontier in which the marginal rate of substitution for the investor equals the negative slope of the efficient frontier. While this issue is well understood for single-period models, the introduction of temporal factors greatly complicates the decision: Investors when approaching their goal (e.g. as they near retirement age) will often become more risk averse. Conversely, young investors should be relatively less risk averse with their retirement funds than someone in retirement. Mulvey and Atkins (1994) discuss these issues in fuller detail.

An obvious alternative to mean-variance is the von Neumann-Morgenstern expected utility (EU) of wealth at period τ. Here, the objective becomes

$$max \sum_s prob_s Utility(wealth_\tau^s) \tag{13'}$$

where $prob_s$ is the probability of scenario s, and Utility(wealth) is the VM utility function as derived via certainty equivalence and risk premium questions.

An investor is risk averse when his risk premium (RP) is positive for all lotteries; see Keeney and Raiffa (1976). This condition corresponds to a concave shape. But it is important to distinguish concave EU utility functions from those of a classical economist's utility function which represents decreasing marginal utility as measured in satisfaction (happiness, pleasure). The EU utility function has nothing to say about happiness as a deterministic value. The location of a local optimal solution is equivalent to the global solution when the feasible region for the constraints consists of a convex set and when the objective function is concave. The mean-variance model and the EU model are equivalent under certain conditions on the distribution of returns and the shape of the VM utility function; see Markowitz (1959).

A third objective is a direct extension of the previous EU model. Most investors are not only interested in their wealth at the end of some planning horizon: They prefer one set of trajectories over another, even when the results at the horizon are identical. An example is the rapid attainment of a certain level of wealth and the subsequent flattening of the wealth curve, as compared with a trajectory in which the growth in wealth comes at the end of the planning horizon. Given these two paths, most investors prefer the former. Thus, we ought to model intermediate preferences and the path to achieve a target wealth. In order to accomplish this goal, we employ a multi-objective formulation rather than the previous single EU attribute in which we focus on wealth at the beginning of period t. Several dimensions are added to the certainty equivalence EU questions, corresponding to the investor's wealth at several key junctures during the planning period. This extension is clearly a multi-objective problem that can be addressed using the efficient algorithms discussed in the literature. A general objective function for this problem is:

$$\max \sum_s prob_s Utility(Wealth_1^s, Wealth_2^s, \ldots, Wealth_\tau^s) \tag{13''}$$

The recommended course for implementing this temporal multicriteria problem is an issue for future research. King (1993) presents a penalty approach for handling time- dimensional preferences.

The complete financial planning model (FPM) is to maximize equation (13) subject to the restrictions implied by equations (4) to (12). In most cases, this model takes the form of a nonlinear program possessing linear constraints and multiple objectives.

3.2 Special Cases

The most famous special case of FPS is the mean-variance model [MV] proposed by Harry Markowitz in the 1950s (Markowitz 1959). This single-period quadratic program is among the most widely applied nonlinear programs (Perold 1984). The usual objective function possess two conflicting terms: the expected value and the variance of assets at the end of the planning period $\tau = 1$. Variance of assets provides a surrogate for risk in this framework.

The mean variance model has been extended in several directions. Konno and Yamazaki (1991), Konno, Pliska and Suzuki (1993), Markowitz et al. (1993) and others have proposed a modified objective in which risk is defined differentially between upside (profit) and downside (loss). Another idea is to penalize the absolute value deviations from a target wealth (Konno, Shirakawa, and Yamazaki, 1993).

Another special case of the FPS model is maximizing the single-period expected utility (EU) of assets. This myopic strategy presents an obvious special case of FPS in which $\tau = 1$ and there are no liabilities. Grauer and Hakansson (1985) and Mulvey (1993, 1994b) showed that the expected utility approach can generate superior returns when evaluated by means of historical evidence. This type of *ex post* backtesting analysis is commonly done when investment strategies are proposed.

3.3 Why A Multi-Period Framework?

The single period models possess several inherent drawbacks that can be eliminated by modeling the dynamic aspects of the investment problem within a multi-period context. These difficulties include: 1) handling of transaction and market impact costs; 2) the lack of temporal independence for many of the economic factors (e.g. interest rates); 3) a consistent specification of risk over time; and 4) the ability to make sensible tradeoffs between short-term risk and long-term gains.

The first factor–market impact costs–is generally ignored in investment studies– partially because of the difficulty in measuring the degree of this factor and the con- sequential increase in modeling complexity (Mulvey and Vladimirou 1992; Davis and Norman 1990). In many real world cases and especially in severe market downturns, however, the market impact costs can be substantial. For example, an investor in real estate might like to sell his building once the market begins dropping, before the price drops even further. Selling in these situations can be quite expensive in terms of causing continued erosion in prices. Anyone who has tried to sell a house in such a situation knows that buyers' markets can be deadly for sellers. A symmetric situ- ation occurs when buying during an inflationary spiral. Many institutional investors are faced with similar phenomena when they invest in thinly traded markets or when they are dealing with substantial positions that must be unwound.

The second issue–intertemporal dependencies–has gained in importance as evi- dence mounts that certain economic factors display mean reversion over time. For instance, most interest rate models are stochastic processes that mean revert, such as the Ornstein-Uhlenbeck process that forms the basis for the Cox, Ingersoll and Ross single factor model (Hull 1993). Research on equities has shown that the variance of long term returns is inconsistent with short term variance, if temporal independence exists. The historical variance ratio is too small to support the case of independence.

A third issue involves defining a preference function that can be employed suc- cessfully over a number of years. One of the advantages of a systematic approach to investing is consistency–year after year maintaining a unified plan of action. It does the investor no good to become conservative after a dramatic drop in price–selling all of his stocks! This behavior is easy to understand from a psychological stand- point, but makes poor sense for investors and especially for long term investors. A multi-year financial planning system provides an opportunity for the investor to look both at the long and the short term consequences of today's investment decisions. For example, investing in stocks over the long run has dominated most other forms of asset investments. Yet many individuals are hesitant to place a majority of their assets in stocks, even for long term investments such as pension plans (401s or IRAs). A multi-period planning system pinpoints the consequences of this form of severe risk aversion.

3.4 Modeling Stochastic Parameters

The stochastic parameters that are needed for the FPS model can be placed in three groupings: 1) a small set of economic factors; 2) projected returns for the asset categories as implied by the values of the economic factors in the prior group; and 3) projected liabilities based on the implied values of the same economic factors. Again, the notion of a scenario is critical to the scenario generation process. A scenario consists of a complete and consistent set of parameters across the extended planning horizon $t = \{1, \ldots, T\}$ as required by the constraints in FPS.

There are several goals to keep in mind when building a model for the stochastic parameters. First, the procedures must be based on sound economic principles. For instance, the interest rates must be totally consistent with the returns for the fixed-income asset categories. International investments should be designed so that the foreign currencies are a separate category–for the purposes of developing reasonable hedging strategies. The basic trends should be preserved whenever possible–such as mean reversion in interest rates over an extended horizon. And the projections should be evaluated with regard to their fit with historical data and trends.

A second goal is to design a stochastic process flexible so that the system can be tailored to individual investor's circumstances. An investor will not trust the recommendations of a planning system unless he or she is confident that the investor's general beliefs are properly portrayed in the stochastic models. In this regard, the model should be simple enough so that the investor can understand the basic philosophy and the key linkages among the modeling components. An understandable model will go a long way to gaining the confidence of the investor–thus increasing the chances that FPS will be employed in an active manner.

The primary aim of the scenario generation is to construct a number of scenarios that provide a reasonable representation of the universe of possible outcomes. This objective is much different than the generation of a single scenario, say for forecasting and trading strategies. Rather, we are interested in constructing a **representative** set of scenarios. In this regard, we must include scenarios that are both optimistic and pessimistic–of course, within a general modeling framework. Such an effort was undertaken by Towers Perrin (one of the largest actuarial firms in the world). Their scenario generation process is called CAP:Link for capital market projections (Mulvey 1994c). The process entails a cascading set of submodels, starting with the interest rate component. TP uses a version of the Brennan-Schwartz (1982) two-factor interest rate model as modified to avoid the problem cited by Hogan (1993). The other submodels are driven by the interest rates and other economic factors.

The TP scenario projections are long term–over 10 to 20 years or more. Thus, their objectives would be different than for someone who is interested in a short term planning horizon. For example, covariances between asset returns are unreliable over extended periods. TP could not use a simple multi-normal covariance structure for their scenario generation. This process would be inconsistent with historical evidence. However, an investor who has a short horizon and who is actively trading and rerunning her model might be able to use a covariance matrix to generate the requisite scenarios. It all depends upon the characteristics of the investor and his or her aims for using a financial planning system.

4 Integrating Assets and Liabilities

Many investors use the mean-variance model without reference to their liabilities or the intended purposes for funds. Focusing on assets alone misrepresents the risks and relative rewards for dynamic investment strategies, especially for investors who are risk averse (Berger and Mulvey 1994). Mulvey (1989) extended the mean-variance and the expected utility model to handle liabilities in the context of asset allocation strategies.

To accommodate the integration of asset allocation and liability projections requires a set of scenarios that bring together the two sides of the investment equation. Generally, historical data provides a benchmark for calibrating asset returns. However, forward looking simulations must be designed for projecting liabilities. These simulations must consider the organizational decisions regarding liabilities. For instance, an insurance company can adjust its payouts for many products such as annuities. These pricing decisions must be integrated with investment decisions to determine the overall joint risks to the entire enterprise. The dynamic stochastic control approach discussed above provides a perfect framework to conduct this form of strategic planning. There is simply no excuse for a financial entity to misunderstand the risks that it encounters, such as occurred in the 1980s for the U.S. Savings and Loan crisis. Once it becomes clear that an acceptable risk is present, say for one or more economic scenarios, the financial organization can hedge this risk in the marketplace through derivatives or futures, or by contracting with a Walls Street firm for a custom tailored product that fits the firm's needs. To this end, an integrative asset-liability system provides the ideal vehicle for identifying the risks to a financial organization.

There are several noteworthy asset-liability investment systems (Kusy and Ziemba 1986; Mulvey 1989; Mulvey 1994a). One of the largest is the Russell-Yasuda-Kasai effort (Cariño et al. 1993). This project received second prize in the 1993 Edelman competition for the best application in Management Science. A multi-stage stochastic program was built to assist the Yasuda company in making investment decisions and analyzing overall risks. The Japanese insurance industry must abide by restrictions on the dividends and other rules. These rules provide the basis for the linear constraints (10). The objective function combines expected profit with piecewise linearized penalties for violating specified targets with a specified probability. The stochastic programs are solved using IBM RS/6000 Model 530 workstations in conjunction with the OSL software package. The largest problem consists of 2048 scenarios. The delivered software system handles 256 scenarios.

The second example involves the world-wide benefit consulting company Towers Perrin. The objectives of the TP asset-liability investment system are to provide actuarial advice regarding the soundness of pension plans and to render careful recommendations as asset consultants. The TP system depends upon dynamic stochastic control, rather than the stochastic program employed in the Russell system. The TP staff devoted considerable effort in the scenario generation process (Mulvey 1994c). The scenario generation program assists both actuaries for setting return assumptions as well as asset consultants who make recommendations concerning the risk/rewards for investment strategies. The dynamic stochastic control model is solved using PCs in conjunction with Lasdon's GRG nonlinear programming software. The system is designed to handle 500 to 1000 scenarios.

5 Solution Strategies

This section reviews solution algorithms for the financial planning model and its specializations. We do not cover algorithms designed for alternative investment models, although these techniques can be clearly implemented. Higle and Sen's stochastic decomposition (1991) could be employed, for instance, in a setting with a finite number of scenarios. We focus on the solution of multi-stage stochastic programs possessing discrete-time decisions, a modest number of scenarios–typically under 1000 to 3000, and nonlinear objective functions for addressing risk aversion.

The model's size depends upon the number of decision variables and the form of the non- anticipativity rules. If the classical stochastic or robust program is selected, FPS becomes a linear or convex nonlinear program whose size hinges on the scenarios that are placed in $\{S\}$. If a dynamic stochastic control approach is selected, problem size can be much smaller since conditional decisions are greatly restricted. DSC models, however, can result in difficult non-convex optimization problems. Maranas et al. [34] have developed a practical algorithm for solving a popular version of DSC.

The simplest approach to solve FPS when the objective is linear is to employ one of the modern efficient linear programming solvers; Cplex and IBM's OSL are the two leading contenders. In some cases, the nonlinear terms in the objective function can be piecewise linearized. The Russell team, for example, linearized the penalty terms for missing target contributions so that OSL could be employed (Cariño et al. 1993). Of course, this transformation requires separable terms.

In the mean variance applications, another useful problem transformation is to factor the covariance matrix Q into two component parts:

$$Q = F^T F, \tag{14}$$

where F is a $(k \times n)$ matrix of factors and $k \ll n$. We convert the mean-variance objective function into a separable function by introducing *y-variables* such that

$$y = Fx. \tag{15}$$

The separable inner product $y^T y$ replaces the variance term $x^T Q x$ in the objective function. Adding constraints (15) to the model completes the problem transformation.

As long as k is relatively small, e.g. when the number of scenarios is modest, the expanded model can be solved more efficiently than the original dense nonseparable model. Konno and Suzuki (1992) addressed the transformed problem via piecewise linearization. Vanderbei and Carpenter (1993) proved that nonlinear interior methods can take advantage of the separable structure despite the increase in the number of constraints. Indeed, interior methods are less dependent on the number of constraints than the buildup density in the Cholesky factors. A similar transformation is possible with the expected utility objectives as discussed in Berger, Mulvey and Ruszczyński (1993). With this transformation, the solution of nonlinear objectives is not much harder than solving linear programs of the same size and structure. These computational breakthroughs will improve the chances that nonlinear risk aversion will be addressed directly.

Substantial progress has been made in the design of efficient algorithms for solving multi-stage stochastic programs since the original L-shape proposal by Van Slyke and Wets (1969). These algorithms take advantage of the stochastic program's structure. See Birge et al. (1994), Birge and Holmes (1992), Dantzig and Infanger (1993), Ermoliev and Wets (1988), Gassmann (1990), Mulvey and Vladimirou (1991), Rockafellar and Wets (1991), and Nielsen and Zenios (1993) for examples. In several cases, implementation has occurred in a parallel or distributed computing environment. The use of parallel computer will greatly accelerate over the next decade. Parallel implementations of stochastic programs presents a promising avenue for future research.

In the original stochastic programs (Dantzig and Madansky 1961), the model maximized expected value. As discussed, the inclusion of risk aversion requires nonlinear objectives. Recently, nonlinear algorithms have been developed that are well suited to solving convex stochastic programs. Carpenter et al. (1993) extended the OB1 code to include nonlinear objectives within an efficient interior-point algorithm. This research has been further expanded by Vanderbei and colleagues at Princeton University. Berger, Mulvey and Ruszczyński (1993) extended the diagonal approximation algorithm (DQA) method to handle convex objectives as specified by risk averse EU functions.

The size of stochastic programs quickly grows as a function of the number of scenarios. As an example, we generated a six-stage financial planning problem with 156 variables and 96 constraints per scenario. This model grows to 15,600 variables and 9,600 constraints with 100 scenarios–beyond the range of most NLP solvers! Nevertheless, we can take advantage of the partially separable structure of the expected utility function. The DQA algorithm, for instance, splits the stochastic program into pieces in an arbitrary fashion, depending upon the hardware configuration. The goal of the decomposition is to match the software algorithm to available hardware. DQA forms an augmented Lagrangian function by dualizing non-anticipativity constraints, but not just any of the non-anticipativity constraints. Instead, we keep most of the non- anticipativity constraints explicit or use the compact formulation; the choice depends upon the capability of the solver with respect to dense columns. Subproblems

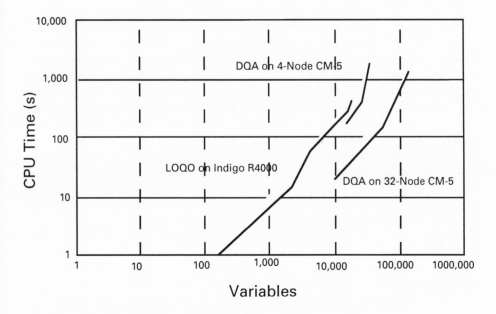

Figure 3: Solution Times for the Convex DQA Algorithm (6-Stage Financial Planning System)

are kept as large as possible in order to reduce total iteration count.

Suppose that two powerful processors are available and we wish to solve a problem with 1000 scenarios. We might form two subproblems of 500 scenarios–sending one to each processor. However, the processors may not be able to handle efficiently a NLP with 500 scenarios. Then we would further split the model into four subproblems of 250 scenarios. The objective is to carry out as little decomposition as possible while staying within the limits of efficient solution time via the solvers for the subproblems.

This flexible decomposition strategy can be highly effective. To give some idea of the efficiency of the convex-DQA algorithm, for example, we plot the solution time as a function of the number of scenarios for a large financial planning model (Figure 3). The largest problem in this domain that has been solved to date consists of 3072 scenarios–corresponding to 480,000 total decision variables and almost 295,000 linear constraints (not including nonanticipativity constraints). Although these problems remain difficult, integrative financial models are generally applied to long term planning–hence, execution time has less concern so long as a number of test examples can be solved. Investors must get an idea of the range of recommendations under differing sets of assumptions. Those investors interested in real time planning can employ massively parallel computing (Zenios, 1991). Likely, in the next 3-5 years, we will be able to solve substantial stochastic programs using desktop computers.

6   Conclusions and Future Directions

The proposed financial planning model provides a general framework for integrating all asset and liability decisions for a large financial entity–such as an insurance company, bank or pension plan–as well as for individuals. This comprehensive approach is needed to properly measure the risk and rewards of alternative investment strategies. Without a comprehensive and integrative model, the investor will be unable to properly measure *risks*. The usual asset-only model does not adequately evaluate the impact of the investment decisions on the investor's wealth, especially for risk averse decision makers or organizations. The main lesson is that investment models must be tailored to individual circumstances. The multi-stage stochastic program provides an ideal vehicle for developing a financial planning system that fits the investor's needs.

Although the FPS may result in a large stochastic optimization problem or a non-convex NLP, efficient algorithms are now available for solving these nonlinear programs. The continued sharp decline in computer costs and the commensurate increase in usability will improve prospects that FPS will become a common place in large financial organizations as well as a critical tool for individual investors (Mulvey and Atkins 1993).

Future research should continue along several dimensions. First, we must increase the size of solvable stochastic programs so that a larger number of scenarios can be handled in a practical fashion. I see no fundamental reason why over the next decade we cannot address 10,000 to 100,000 scenarios using parallel and distributed computers. Certainly the raw computing power will be available. Whether or not we can employ direct solvers or must revert to decomposition algorithms due to numerical limits is a matter for future analysis.

Another critical research issue involves the scenario selection process. In particular, out-of-sample testing will be critical in order to measure valid bounds on the optimal solution value. When it applies, the dynamic stochastic control approach can be useful. The control system can assist in the selection of the scenarios–for instance by generating importance estimates for adding (or deleting) scenarios as they affect the solution to the control problem. These scenarios should be linked to the stochastic program. Of course, the approach of embedding a stochastic program within a simulation system such as carried out by Worzel, Vassiadou-Zeniou, and Zenios (1994) to evaluate the precision of the recommendations is possible. This approach requires enormous computational resources and may not be practical in many situations. Linking simulation and optimization models, however, will be increasingly important as stochastic programs become more widespread in practice.

The area of temporal preferences will require considerable research, especially relating to individual investors. Once a long term planning system is available, an individual can observe the tradeoff between accepting short-term volatility versus achieving greater potential long-term gains. This tradeoff must be done in the context of an individual's circumstances, for example, based on his age or wealth. Many other uses of a long-term planning system are possible such as developing strategies for minimizing taxes over time after the time preference issue becomes more clearly specified.

In terms of algorithmic issues, there are a number of items to explore. One topic is to take advantage of the structure of the multi-stage stochastic program within the interior-point algorithm. For instance, we can conduct the Cholesky factorization using modern sparse matrix calculations on parallel or distributed computers. Rothberg's approach (1993) seems to be a potential winner. Our preliminary tests show that large matrices can be quickly factored using this procedure. Also, Jessup, Yang, and Zenios (1993) have conducted similar tests on the CM-5 computer at Thinking Machines. Solving the stochastic program as quickly as possible will increase the chances that the models will be used by investors. What can be done in the area of decomposition strategies? The sparse matrix calculations are key for methods such as DQA which use an interior point algorithm for solving the subproblems. Any substantial progress on this issue leads to immediate gains in the decomposition algorithms.

Another topic involves the simultaneous solution of hundreds or thousands of stochastic programs. If individual investors begin to require a long-term financial plan based on stochastic programming, there would be substantial interest in a computer that could handle this requirement. Parallel approaches could be applied since the problems can be constructed such that the non-zero matrix structure is identical across individuals with differing coefficient values. This problem will tax the largest computer, but will have enormous potential for improving the way that individuals manage their financial affairs.

References

[1] E. M. Ankrim, "Risk tolerance, Sharpe ratios, and implied investor preferences for risk," Research Commentary, Frank Russell Company, Tacoma, Washington, 1992.

[2] A. J. Berger and J.M. Mulvey, "Errors in investment risk management," SOR Report, Princeton University, 1994.

[3] A. J. Berger, J.M. Mulvey, and A. Ruszczyński, "An extension of the DQA algorithm to convex stochastic programs", SOR Report 93-16, Princeton University, to appear in SIAM Journal On Optimization, 1993.

[4] J. R. Birge, C. J. Donohue, D. F. Holmes, and O. G. Svintsitski, "A parallel implementation of the nested decomposition algorithm for multistage stochastic linear programs," Technical Report 94-1, Dept. of Industrial and Operations Engineering, University of Michigan, 1994.

[5] J. R. Birge and D. F. Holmes, "Efficient solution of two-stage stochastic linear programs using interior point methods," Computational Optimization and Applications, 1 (1992), 245-276.

[6] M. J. Brennan and E. S. Schwartz, 1982. "An equilibrium model of bond pricing and a test of market efficiency," Journal of Financial and Quantitative Analysis, 17 (1982), 75-100.

[7] T. J. Carpenter, I. J. Lustig, J. M. Mulvey, and D. F. Shanno, "Separable quadratic programming via a primal-dual interior point method and its use in a sequential procedure," ORSA Journal on Computing, 5 (1993), 182-191.

[8] D. R. Cariño, T. Kent, D. H. Myers, C. Stacy, M. Sylvanus, A. Turner, K. Watanabe, and W. T. Ziemba, "The Russell-Yasuda Kasai financial planning model," Technical Report 93-MSC-018, Frank Russell Company, Tacoma, Washington, 1993.

[9] G. C. Chow, "Optimal control without solving the Bellman equation," Journal of Economic Dynamics and Control, 17 (1993), 621-630.

[10] G. C. Chow, "The Lagrange method of optimization in finance," Econometric Research Program Memorandum 369, Princeton University, 1994.

[11] G. Dantzig and A. Madansky, "On the solution of two stage linear programs under uncertainty," *Proceedings Of the Fourth Berkeley Symposium On Mathematics And Probability*, **1** (1961), Univ. Of California Press, Berkeley, CA, 165-176.

[12] G. B. Dantzig and G. Infanger, "Multi-stage stochastic linear programs for portfolio optimization," *Annals of Operations Research*, **45** (1993), 59-76.

[13] M. H. A. Davis and A. R. Norman, "Portfolio selection with transaction costs," *Mathematics of Operations Research*, **15** (1990), 676-713.

[14] M. H.A. Davis, *Markov Models and Optimization*, Monographs on statistics and applied probability, **49**, Chapman and Hall, 1993.

[15] A. K. Dixit, *Optimization in Economic Theory*, Oxford University Press, 1990.

[16] A. K. Dixit, and R. S. Pindyck, *Investment Under Uncertainty*, Princeton University Press, Princeton, NJ, 1994.

[17] Y. Ermoliev and R. J. -B. Wets (editors), *Numerical Techniques for Stochastic Optimization Problems*, Springer, Berlin, 1988.

[18] H. I. Gassmann, "A computer code for the multi-stage stochastic linear programming problem," *Mathematical Programming*, **47** (1990), 407-423.

[19] R. R. Grauer, and N. H. Hakansson, "Returns on levered actively managed long-run portfolios of stocks, bonds and bills," *Financial Analysts Journal*, 1985, 24-43.

[20] J. L. Higle and S. Sen, "Stochastic decomposition: an algorithm for two-stage linear programs with recourse," *Mathematics of Operations Research*, **16** (1991), 650-669.

[21] M. Hogan, "Problems in certain two-factor term structure models," *The Annals of Applied Probability*, **3** (1993), 576-581.

[22] M. Holmer and S. Zenios, "The productivity of financial intermediation and the technology of financial product management," Wharton Report 93-02-01, University of Pennsylvania, 1993.

[23] J. C. Hull, *Options, Futures, and Other Derivative Securities*, Prentice Hall, 1993.

[24] R. Jarrow, "Pricing interest rate options," *Finance*, R. Jarrow, V. Madsimovic, and W.T. Ziemba (editors), North Holland, 1994.

[25] E. R. Jessup, D. Yang, and S. A. Zenios, "Parallel factorization of structured matrices arising in stochastic programming," Report 93-02, School of Economics and Management, University of Cyprus, 1993.

[26] R. L. Keeney and H. Raiffa, *Decisions with Multiple Objectives: Preferences and Value Tradeoffs*, John Wiley, New York, 1976.

[27] A. J. King, "Asymmetric risk measures and tracking models for portfolio optimization under uncertainty," *Annals of Operations Research*, **45** (1993), 165-178.

[28] H. Konno, and K. Suzuki, "A fast algorithm for solving large scale mean-variance models by compact factorization of convariance matrices," *Journal Operations Research Society of Japan*, **35** (1992), 93-104.

[29] H. Konno, and H. Yamazaki, "Mean-absolute deviation portfolio optimization model and its applications to Tokyo stock market," *Management Science*, **37** (5) 1991, 519-531.

[30] H. Konno, H. Shirakawa, and H. Yamazaki, "A mean-absolute deviation-skewness portfolio optimization model", *Annals of Operations Research*, **45** (1993), 205-220.

[31] H. Konno, S. Pliska and K. Suzuki, "Optimal portfolio with asymptotic criteria," *Annals of Operations Research*, **45** (1993), 205-220.

[32] M. I. Kusy and W. T. Ziemba, "A bank asset and liability management model," *Operations Research*, **34** (1986), 356-376.

[33] I. J. Lustig, J. M. Mulvey, and T. J. Carpenter, "Formulating stochastic programs for interior point methods," *Operations Research*, **39** (1991), 757-770.

[34] C. D. Maranas, I. P. Androulakis, C. A. Floudas, A. J. Berger, and J. M. Mulvey, "Solving Stochastic Control Problems in Finance via Global Optimization," SOR Report 94-01, Princeton University, 1994.

[35] H. M. Markowitz, *Portfolio Selection: Efficient Diversification of Investments*, John Wiley, New York, 1959.

[36] H. M. Markowitz, P. Todd, G. Xu, and Y. Yamane, "Computation of mean- semivariance efficient sets by the critical line algorithm," *Annals of Operations Research*, **45** (1993), 307-318.

[37] T. Marsh, "Term structure of interest rates and the pricing of fixed income claims and bonds," *Finance*, R. Jarrow, V. Madsimovic, and W.T. Ziemba (editors), North Holland, 1994.

[38] R. C. Merton, "Lifetime portfolio selection under uncertainty, the continuous- time case," *Review of Economics and Statistics*, **51** (1969), 247-257.

[39] J. M. Mulvey, "A surplus optimization perspective," *Investment Management Review*, 3 (1989), 31-39.

[40] J. M. Mulvey, "Incorporating transaction costs in models for asset allocation," *Financial Optimization*, 243-259, Cambridge University Press, 1993.

[41] J. M. Mulvey, "Integrating assets and liabilities for large financial organizations," *New Directions in Computational Economics*, W. W. Cooper and A. B. Whinston (editors), Kluwer Academic Publishers, 1994.

[42] J. M. Mulvey, "An asset-liability investment system," to appear *Interfaces*, 1994.

[43] J. M. Mulvey, "Generating scenarios for the Towers Perrin investment systems," SOR Report, Princeton University, 1994.

[44] J. M. Mulvey and C.A. Atkins, "The home account: Financial analysis, planning, and management system for individuals," Technical Report, Princeton University, September, 1993.

[45] J. M. Mulvey and A. Ruszczyński, "A new scenario decomposition method for large-scale stochastic optimization," to appear in *Operations Research*, 1994.

[46] J. M. Mulvey, R. Vanderbei, and S. Zenios, "Robust optimization of large-scale systems," Princeton University, Report SOR-91-13, 1991, to appear in *Operations Research*, 1994.

[47] J. M. Mulvey and H. Vladimirou, "Solving multistage stochastic networks: an application of scenario aggregation," *Networks*, **21** (1991), 619-643.

[48] J. M. Mulvey and H. Vladimirou, "Stochastic network programming for financial planning problems," *Management Science*, **38**(11) 1992, 1642-1664.

[49] J. M. Mulvey and S. A. Zenios, "Diversifying a portfolio of fixed-income securities: modeling dynamic effects," *Financial Analysts Journal*, September-October 1993.

[50] S. Nielsen and S. Zenios, "A massively parallel algorithm for nonlinear stochastic network problems," *Operations Research*, **41**(2) 1993, 319-337.

[51] A. F. Perold, "Large-scale portfolio optimization," *Management Science*, **30** (1984), 1143-1160.

[52] A. F. Perold and W. F. Sharpe, "Dynamic strategies for asset allocation," *Financial Analysts Journal*, 1988, 16-27.

[53] R. T. Rockafellar and R. J. -B. Wets, "Scenarios and policy aggregation in optimization under uncertainty," *Mathematics of Operations Research*, 16 (1991), 119-147.

[54] E. Rothberg, "Performance of Panel and Block Approaches to Sparse Cholesky Factorization on the iPSC/860 and Paragon Multicomputers," Intel Supercomputer Systems Division Technical Report, Beaverton, Oregon, 1993.

[55] P. A. Samuelson, "Lifetime portfolio selection by dynamic stochastic programming," *Review of Economics and Statistics*, 1969, 239-246.

[56] J. A. Tilley, "An actuarial layman's guide to building stochastic interest rate generators," Study Note for the Society of Actuaries, 1992.

[57] R. Van Slyke and R. J. -B. Wets, "L-shaped linear programs with applications to optimal control and stochastic programming," *SIAM Journal of Applied Mathematics*, 17 (1969), 638-663.

[58] R. J. Vanderbei and T. J. Carpenter, "Symmetric indefinite systems for interior- point methods," *Mathematical Programming*, **58** (1993), 1-32.

[59] K. Worzel, C. Vassiadou-Zeniou, and S. A. Zenios, "Integrated simulation and optimization models for tracking fixed-income indices," *Operations Research*, 1994.

[60] S. A. Zenios, "Massively parallel computations for financial planning under uncertainty," *Very Large Scale Computing in the 21-st Century*, J. Mesirov (editor), SIAM, Philadelphia, PA, 1991.

[61] S.A. Zenios, "Parallel Monte Carlo simulation of mortgage-backed securities," *Financial Optimization*, S. A. Zenios (editor), Cambridge University Press, 1993.

Algorithms for Large Nonlinear Problems

Walter Murray

Systems Optimization Laboratory
Department of Operations Research
Stanford University
Stanford, California

1 Introduction

The problem of interest is the following:

$$\begin{array}{ll} \underset{x \in \Re^n}{\text{minimize}} & F(x) \\ \text{s.t.} & c(x) \geq 0, \end{array} \qquad \text{NP}$$

where $F : \Re^n \to \Re$ and $c : \Re^n \to \Re^m$. If second derivatives are not known, computing x^*, a point satisfying the *first-order KKT conditions* for NP is the best that can be assured. Otherwise, it is possible to assure finding a point satisfying the second-order KKT conditions. Despite the low assurance when second derivatives are not known, we still have the expectation of finding a minimizer and not a saddlepoint. Suppose the problem is unconstrained, then a sequence could be generated that reduces the norm of the gradient of $F(x)$ (the first-order KKT condition is $\nabla F(x^*) = 0$). Alternatively a sequence that reduces $F(x)$, while converging to a stationary point of the gradient, may be generated. The second approach is much preferred even though both approaches are only assured of finding a stationary point.

When solving large problems the precise form of what are mathematically equivalent forms of NP is important. In practice the constraints may be a mixture of linear and nonlinear inequality constraints, simple bounds on the variables and equality constraints. Also there may be upper as well as lower bounds on some constraints. Such seemingly trivial considerations may assume quite large proportions in some cases. For example, an algorithm may require the problem to be in standard form. For NP this is relatively trivial and requires the introduction of m slack variables, say s. If m is much larger than n the new "n" is now much larger. The matrices that are required to be factorized are correspondingly larger. This may not be of great significance if the special structure induced by slack variables is catered for, but this may not be the case. Although s are linear variables the approach taken may use a nonlinear transformation such as a barrier or penalty function. Again this may not be

172

of significance, but it cannot be taken for granted. Note that in linear programming (LP) converting a problem to standard form may be done using duality, but this is rarely an option in nonlinear programming (NLP).

Success in solving large problems often depends on attention to a myriad of small details in the definition of the model and algorithm. Small changes in the formulation of the problem or an adjustment of an optional parameter can result in a huge difference in performance. Quite often users do not have a grasp of what is good and bad when they have a choice in formulation. For example, which of these two constraints sets is best: $l_i \leq c_i(x)/x_i \leq u_i$ or $0 \leq c_i(x) - x_i l_i$ and $c_i(x_i) - x_i u_i \leq 0$, where $x_i \geq 0$? Some users think there is nothing wrong with the first formulation.

2 A User's Perspective

Except for professional algorithm designers few users have an interest in solving a wide variety of problems. Indeed they usually have only one problem (or class of problems) of interest and in that their interest is intense. Few people solve large problems on a whim. What this implies is that people interested in solving large problems are usually willing to invest some effort, maybe even a considerable effort in obtaining a solution. Usually a problem is not a "one off" affair, but something that a user is interested in solving repeatedly and/or expanding into more complex problems. Interest in a specific problem may be sufficiently great as to warrant designing a tailor-made algorithm to solve it.

3 Computing Second Derivatives

The method used to solve NP depends on whether or not second derivatives are available. When solving small problems it is usually assumed second derivatives are not available and the field is dominated by quasi-Newton methods, which generate an approximation to second derivatives. At first sight getting second derivatives for large problems looks a daunting task. For example, if $n = 60,000$ and $m = 50,000$ that means generating $50,001$ matrices of dimension $60,000$. This would seem to imply that for large problems there ought to be even more emphasis on methods that do not require second derivatives. Fortunately, it is only a weighted sum of such matrices that is required and each individual matrix is usually "super" sparse.

Typically $c_i(x)$ will be a function of very few nonlinear variables, which implies the Jacobian matrix is sparse. Just as significant it implies the Hessian matrices are even sparser having only a few rows and columns that are not *all* zero. The total number of nonzero elements in all the Hessian matrices is often comparable to the number in the Jacobian matrix. Of course if just one constraint is a function of all the variables then this rosy picture is destroyed. However, such problems can usually be transformed to fit the rosy picture.

For example, suppose we have the constraint $\sum x_i^2 = 1$. Consider introducing the constraints $y_i - x_i^2 = 0$. The original constraint is now $\sum y_i = 1$, which is linear, and the Hessians of the new constraints have only one nonzero element. We have introduced n extra constraints, but any weighted sum of the $n + 1$ Hessians is a diagonal matrix.

For some algorithms it is not necessary for the *explicit* Hessian to be sparse only that it is easy to generate and to operate (e.g. form a matrix-vector product) efficiently with it. Having said all this in favor of second derivatives there are many problems for which methods requiring second derivatives are unsuitable.

When solving small problems the lack of interest in methods requiring second derivatives stems not so much from their unavailability, but from either satisfaction with quasi-Newton methods and/or an unwillingness to expend the effort to provide them. Usually when solving large problems a user is prepared to put much more effort (indeed it is often necessary) into formulating the problem and providing whatever is required to get a solution. Also the user may be less satisfied with the performance of quasi-Newton methods. One consequence of the view prevailing when solving small problems is that many who are solving large problems do not consider the provision of second derivatives.

In Section 9 we discuss the automatic provision of derivatives.

4 Measuring Efficiency

When comparing the efficiency of algorithms to solve small problems it is usual to compare the number of function evaluations required. This is because for small problems the effort to compute the user's functions is considered to be dominant. If this is not the case then such problems are trivial and are only of significance if very large numbers of them need to be solved. For large problems the housekeeping operations required to generate the iterates may be significant even if these operations are done efficiently. Indeed for some problems such operations may be dominant and for all problems will be of significance if n and m are large enough.

These two aspects of efficiency impact the nature of the "best" algorithm to solve a given problem and suggest that a wider variety of algorithms is likely to be required for large problems. It also brings into question what is meant by the term "algorithm". Algorithms for general nonlinear problems are often deemed the same if they generate the same iterates (a hang-over from small problems being the only ones of concern). However, for large problems just how to compute the iterates leaves a lot to be decided.

To give an example, consider MINOS (see [28] and [29]), which is a widely distributed package to solve large-scale nonlinear problems. MINOS is based on a method proposed by Robinson [32] and Rosen and Kreuser [33], who suggested solving a nonlinearly constrained problem by solving a sequence of linearly constrained problems. Although this is the *basis* of the method and as such defines the "major" iterates it is necessary to define *how* the linearly constrained subproblems are solved for the algorithm to be defined properly. In the case of MINOS a quasi-Newton method is used in which the Cholesky factor of an approximation to the *reduced* Hessian is recurred, which is a method proposed by Gill and Murray [15]. However, in the approach advocated by Gill and Murray a basis for the null-space was defined using an orthogonal matrix. In MINOS the null-space matrix is identical to that used by Wolfe [36]. However, in Wolfe's description an explicit inverse is used whereas in MINOS sparse LU factors are recurred and the null-space matrix is never stored explicitly.

Over the years the procedures of how to compute and update the LU factors have been continuously refined (see [19]. Also additional features such as an anti-cycling strategy have been added (see [21]). The point we are making here is that what would have been considered fairly irrelevant aspects of an algorithm when solving small problems assume significant proportions for large problems. As a consequence, algorithms which are very similar in a mathematical sense, may differ significantly in performance. To give an example of the significance of such differences, Conn, Gould and Toint report [11] that MINOS 5.3 took 2 hours 36 minutes to solve a particular problem while MINOS 5.4 took 1 hour 30 minutes. Most readers would have a hard time discerning the differences between MINOS 5.3 and 5.4. It is a mistake to assume that refinements made even to well established codes such as MINOS are simply nibbling at the edges.

5 Barrier Methods

Barrier and interior-point methods have proved efficient for linear programming. The issues are not quite the same for nonlinear programming, but barrier techniques do have an important although different role. Firstly, the basic system of equations that is required to be solved is the same whether one is using or not using a barrier method. In LP the simplex method requires solving non-symmetric systems whereas interior-point methods may be arranged to solve positive definite symmetric systems. Also in LP the only issue is the work involved in solving such systems whereas in nonlinear problems the functions (and possibly their derivatives) defining the problem need to be evaluated.

The useful feature of a barrier method in LP is it quickly identifies the correct active set. In NLP this is sometimes relatively easy (in process control problems one may have 200,000 variables, but 199,950 equality constraints) and is rarely a pressing issue. Some minimal number of iterations are usually required even if the correct active set is known. While performing these required iterations, adjustments can be made to the active set. Moreover, when constraints are nonlinear the active set can change significantly from one iteration to the next without any action to contrive such a change. This contrasts with linear constraints where unless one deletes a constraint from the active set it remains active. In NLP remaining at a vertex due to degeneracy for large numbers of iterations rarely (if ever) arises in practice. Difficulties do arise due to the Jacobian matrix being ill-conditioned, but this impacts barrier methods as well as alternative methods.

We expect barrier methods to be useful when the user's functions are relatively cheap to evaluate. They may also have a role if a problem has a significant number of linear constraints. Another plus in favor of barrier methods is that in some approaches it is necessary to update a symmetric indefinite factorization and this is relatively costly. Barrier methods are also useful for modeling some constraints such as ensuring the minimum eigenvalue of a matrix is nonnegative (see [31]). There is also some evidence (see [34] and [30]) that both barrier and penalty function methods may be useful for finding good approximations to global minimizers.

There are a variety of ways in which the barrier approach may be applied although some, such as the primal-dual, which is commonly used in LP, do not have the same validity for nonconvex problems. This is the same issue that we mentioned earlier that our interest is not in just any stationary point. However, it was shown in [16] that the primal-dual algorithm can be interpreted as a "modified" Newton algorithm applied to the barrier function, which gives some validity to applying primal-dual algorithms for nonconvex problems.

In general the use of barrier methods is unlikely to be critical to successfully solving a problem, but their use may result in a significant improvement in efficiency. If a problem is a slight modifications of an LP for which barrier methods have proved substantially better than the simplex algorithm then they are likely to be just as successful on the nonlinear problem. Barrier methods are likely to show a more significant improvement over active-set methods in the short term since for large problems they are inherently simpler to implement than active-set methods.

Since the Hessian of the barrier function may change significantly for small changes in the variables it seems likely that barrier methods are best when combined with second-derivatives methods as opposed to quasi-Newton methods. Also if a null-space method is being used the dimension of the null space for the transformed problem may be significantly larger than that of the untransformed problem. Ringertz [31] reports using a barrier method under such circumstances, but for the problems in the report only one or two active constraints arise in the transformed constraints.

Another difficulty with barrier methods is the need to provide an initial point not on any bound. In solving real problems it is essential to be able to use a good initial point if such is available and often the initial point is on many bounds. It is possible to make small adjustments to the initial point and this is easy if the constraints are simple bounds on the variables. However, it does raise the issue of what is "small". Another approach is to form the usual QP subproblem without the barrier transformation (see Section 6) at the given initial point and solve this first problem using a barrier method. In solving this subproblem *any* suitable initial point may be used. The solution will automatically be a strictly interior point. In solving the QP subproblem it is only necessary to drive the barrier parameter down to the initial choice for the nonlinear barrier method.

6 Sequential Quadratic Programming Methods (SQP)

We shall discuss some adaptations of SQP methods to solve large problems. Typically SQP methods generate a sequence of points $\{x_k\}$ converging to a solution, by solving at each point, x_k, a quadratic program (QP), which for problems in the form of NP will be of the form

$$
\boxed{
\begin{array}{ll}
\underset{p\in\Re^n}{\text{minimize}} & \nabla F(x_k)^T p + \frac{1}{2} p^T H_k p \\
\text{s.t.} & c(x_k) + \nabla c(x_k) p \geq 0
\end{array}
}
\qquad \text{QP}
$$

for some positive definite matrix H_k. Let p_k (referred to as the search direction) denote the unique solution to QP. We define $x_{k+1} \equiv x_k + \alpha_k p_k$, where the steplength α_k is chosen to achieve a reduction in a merit function.

SQP algorithms are viewed by many as the best approach to the solution of NP when n is small. However, for small problems the effort to solve the QP subproblems is rarely relevant. To adapt SQP methods to solve large problems requires being able to solve the QP subproblems efficiently. One approach taken in Murray and Prieto [26] is to show that a suitable search direction may be obtained without the need to solve the QP subproblem. They also prove convergence of SQP methods that incorporate a large degree of flexibility in the definition of the algorithm in order to accommodate the type of adaptation that is necessary when designing algorithms for large problems.

An often unappreciated feature of SQP methods is that they automatically take advantage of linearity in a linear constraint. For example, if the initial estimate satisfies a linear equality or inequality constraint then so do all subsequent iterates. Such constraints are also automatically excluded from the merit function. In contrast an augmented Lagrangian method would not have this property unless the linear constraint was identified as such and then treated differently from the nonlinear constraints. Treating linear constraints separately in an augmented Lagrangian method has a hugh impact on the resulting software. It is now necessary to solve a sequence of linearly constrained problems as opposed to a sequence of unconstrained problems.

The key to a successful SQP algorithm is a fast algorithm to solve (or partially solve) the QP subproblems. Solving a sequence of related QP problems is not quite the same as solving a single QP problem. Usually after a few nonlinear iterations, the active set of both the nonlinear problem and QP subproblem changes only slightly. The active set of one QP subproblem is then close to that of the following subproblem. It is important that full advantage is made of this information.

6.1 Solving large QP problems

There are a variety of methods to solve large QP problems, which is best will depend on the characteristics of the QP and its solution. We shall consider four approaches:

- A null-space method.

- A range-space method.

- A Schur-complement method.

- A barrier method.

The first three are active-set methods and differ only in how the relevant linear equations are solved. Which of the three active-set methods to use depends largely on the number of active constraints. Null-space methods are efficient when the number of active constraints is almost the same as the number of variables (note this is obviously true for some problems such as the process control problems referred to earlier). Range-space methods are efficient when there is only a few active constraints (obviously true if the problem only has a few constraints). If neither of these conditions holds then the Schur-complement approach is recommended. The use of a barrier algorithm to solve the subproblem as opposed to applying a barrier approach to the nonlinear problem is likely to be preferred when the user's functions are expensive to compute. Usually applying a barrier approach to the original nonlinear problem results in a substantial increase in the number of nonlinear iterations. In theory the number of nonlinear iterations is independent of *how* the QP subproblem is solved. In practice there are likely to be small but essentially negligible differences on most problems.

It is sometimes beneficial if the QP is in standard form although this is not always strictly necessary. It helps in the barrier approach since the ill-conditioning in the resulting subproblems is then benign (see [30]) and in the Schur-complement method it simplifies the updating procedures.

In all four methods we are required at each iteration to solve the KKT equations, which are of the form

$$K \begin{pmatrix} p \\ -\pi \end{pmatrix} = - \begin{pmatrix} g \\ 0 \end{pmatrix}, \quad \text{where} \quad K \equiv \begin{pmatrix} H_i & A_i^T \\ A_i & 0 \end{pmatrix}.$$

The application of the standard Schur-complement algorithm to the special case of solving the sequence of QP problems that arises in SQP methods is straightforward (see [20] and [22]). The Schur-complement method requires the KKT matrix to be sparse, hence H needs to be sparse. We would usually expect H to be the exact Hessian or approximated directly by finite differences. The range-space method may also be adapted in a reasonably straightforward manner. For this method to be efficient it is necessary to be able to solve systems with H efficiently. In the case of a barrier algorithm the method needs to be able to take advantage of a good initial point. The null-space method needs considerable adaptation if a direct approximation is made to a reduced Hessian. If the Hessian is known then again everything is straightforward. However, approximating the reduced Hessian has two benefits, it requires storing only a small matrix and it facilitates the solution of the QP subproblem, which is solved very efficiently if the predicted active set is correct. However, when the reduced Hessian that is known is not that at the solution of the QP subproblem then more information about H is required. In the approach taken by Gill et al. [17] and Eldersveld [13], a matrix $Q^T H Q$ is recurred, where Q is a nonsingular matrix. The matrix Q is never required explicitly and the null-space algorithm may be efficiently implemented provided equations of the form $Qy = b$ and $Q^T H Q y = b$ can be solved efficiently.

7 Some Real Applications

Solving large or even small arbitrary nonlinear problems may be extremely hard. It is not difficult to construct problems in two or three variables that defy solution. Fortunately, the real world is often (but not always) much less perverse. When modeling physical systems it is often easy to scale a problem sensibly and for large problems scaling is extremely important. Also obtaining a reasonable initial estimate is usually straightforward. Often a user is looking only for a modest improvement in current practice. When dealing with large sums even a 5% reduction may be very significant.

7.1 The Optimal Power Flow Problem

There are many interesting and important application of optimization in the planning and operation of utilities such as gas and water distribution. A problem that has received considerable attention is the optimal power flow (OPF) problem, which addresses the issue of how best to generate and distribute electrical power. Electric systems are relatively easy to model accurately. The objective may be to minimize transmission losses or cost of generation. The constraints are physical laws such as Kirchoff's law holding at each node in the distribution network, bounds on the operation of equipment, desirable operating conditions (the voltage in a transmission line needs to be within some prescribed limits) etc. For a typical large OPF problem, $n \approx 5000$ and $m \approx 4500$. However, if an operator is concerned about the system being secure in the event of an unexpected contingency (e.g., a transmission line being struck by lightning), the problems may be an order of magnitude larger. One may also be interested in stochastic constraints, which again would increase the problem by an order of magnitude. Of course until the base OPF problem was solved, there was no question of being able to address more elaborate models.

Before there was the "optimal" power flow problem there was the power flow problem, which could be interpreted as that of finding a feasible point. In other words how to operate the system while obeying the laws of physics and constraints on the use of equipment. It is typical of many applications of optimization that part of the problem was being solved prior to the use of optimization. (Of course solving nonlinear equations to get a feasible point may also be viewed as an optimization problem.) It gives some insight into the thinking of practitioners to know they have been addressing part of the problem for some time. It sometimes leads them to a fatal approach when solving the optimization problem, namely casting the problem only in what they see as the "optimization" variables. Sometimes, such as in structural problems, this is a valid approach, but all too often it is not.

In this case it leads to the need to solve the power flow equations in order to evaluate the problem functions. Another flawed viewpoint is that unless constraints are always satisfied there is no meaning to the model. While there is some truth to this view, usually the mathematical model behaves in a smooth way at least for some degree of infeasibility. From a user's point of view there is considerable comfort from knowing that premature termination of an algorithm still leaves a usable solution. However, such comfort may come at the high cost of frequently failing to solve the problem.

The problem functions consist primarily of products of the variables and simple trigonometric functions, and hence both first and second derivatives can easily be obtained and are cheap to compute. In OPF problems 95% of the work is in performing the housekeeping operations. This might suggest using a method that gave rise to simple subproblems, but was not too efficient in terms of the number of iterations. However, it turned out that SQP methods, which have expensive subproblems, work incredibly well on these problems and often require fewer than a dozen iterations.

Although the problems are not convex there are few difficulties that arise as a result of nonconvexity even though the reduced Hessian is not always positive definite. Although the subproblems are complex the remarkably small number that are required to be solved makes the SQP method efficient. To improve on efficiency two approaches were tried (see [7]). A barrier approach was used to replace the active-set QP algorithm. Despite some massaging that was necessary to be able to use the usually good initial estimate and hot starts for subsequent subproblems, the approach did not prove effective. Applying a barrier approach to the original nonlinear problem proved much more effective. Although the number of nonlinear iterations increased by a factor that typically was in the range from 3 to 10 the effort to solve the increased number of subproblems was more than off-set by the ease with which they could now be solved (these were now equality quadratic programming problems).

Efficiency did not depend much on the initial choice of barrier parameter or the rate at which it was reduced. A consistent and significant improvement was achieved by switching to the primal-dual system. Interestingly the KKT systems of the primal-dual had the correct inertia more frequently and it may be this feature that contributed to the improvement since if the inertia was not correct no corrections were made. Since the problem was in standard form before the barrier transformation the resulting systems solved were not impacted by ill-conditioning.

There is not a "single" OPF problem, but a range of problems with different time horizons. The greater the time horizon the greater the degrees of freedom since more actions become possible. Planning models are also possible in this framework.

Many theoretical and practical difficulties remain for the development of a large-scale SQP algorithm based on second derivatives. Fortunately, it is somewhat easier to develop a successful method for a specific application, such as the OPF problem. For example, the QP subproblem may have an unbounded solution for a general problem, since the Hessian of the Lagrangian can be indefinite. This cannot happen in the OPF problem because of the presence of bounds on all the variables. Indeed, since the upper and lower bounds tend to be quite close, the presence of indefiniteness is unlikely to lead to inefficiency.

The OPF example illustrates that new approaches can be successfully applied to a particular problem even when they are not yet fully developed for general problems. The implication is that practitioners can exploit the latest developments in optimization and need not wait until all issues have been resolved. In fact, certain methods unsuitable for general use may still have a useful role in particular applications.

The OPF problem also highlights the interaction between the development of models and algorithms. Success in solving a given model inevitably leads to higher expectations and the formulation of more sophisticated models.

The OPF problem also illustrates a common occurrence when solving real problems. If the model does not have a solution, the user still wants an answer. Of course the question is now different. In this case "not having a solution" means there is no feasible point. What the user now wants is to obtain a solution that satisfies physical laws and minimizes the degree to which the equipment is being abused and desirable values are degraded (e.g. a brown out). In terms of the model this means some constraints are "hard" and some are "soft". However, there are hard limits on abuse and the degree to which desired values can be degraded may have contractual limits. Consequently, at some point the system may still fail.

What this means in terms of this problem is the system can no longer meet demand. It is now necessary to reduce demand (a black out) and what the user now wishes is to minimize the reduction required. Ideally the algorithm needs to recognize each of these situations. Not having a feasible point to the original problem has implications for the QP subproblems, which are unlikely to have feasible points. Demand, which may have been parameters in the original formulation, has to be modeled as "sleeping" variables. A good analogy is the cruise control of a car. The speed is a parameter and should be fixed. However, if a hill is steep enough once the pedal is pushed to the floor (on its upper bound) it is now necessary for the speed to fall. Many models do not quite fit the form of NP, but often an algorithm can be adapted to accommodate the variation from the ideal form.

7.2 Trajectory Problems

The optimal trajectory problems of interest may be loosely described as the optimal control of a vehicle such as a rocket or plane that is required to perform some maneuver (see [2], [35] and [17]). For example, we may have a plane flying at one altitude and speed and it is required to get to some other altitude and speed in minimum time or using the minimum amount of fuel. This is a continuous control problem. To convert it into the form of NP requires some discretization process.

The Boeing package OTIS (Optimal Trajectories by Implicit Simulation, see [25]) uses Hermite collocation to discretize the problem. A comparison with partial or nondiscretizing methods showed the approach adopted in OTIS to be efficient and more flexible than alternatives. Originally OTIS incorporated an early version of NPSOL (see [18]), a subroutine designed to solve dense problems based on an SQP algorithm. NPSOL proved extremely efficient on small problems, but inevitably proved to be inefficient for large problems. Nonetheless the number of iterations required by NPSOL was still small even for large problems, which showed the basic SQP approach would prove effective provided the QP subproblems were solved efficiently. Betts and Huffman in [2] used an SQP method (see [23]) coupled with a Schur-complement QP solver (see [22]) and used a finite-difference approximation to the Hessian. Gill et al [17] also used a similar SQP method, but one based on a quasi-Newton approximation to the reduced Hessian and a null-space method to solve the QP subproblem.

Trajectory problems contrast starkly with the OPF problems since even first derivatives of the objective are not available and evaluation of the user supplied functions is extremely time consuming. However, as we have already mentioned unless the linear algebra operations are done reasonably efficiently eventually, they will be damning. Also the number of controls dictate the maximum number of degrees of freedom in the problem, which for many problems is relatively small, implying a reduced Hessian approach should be efficient.

It is a testimony to the versatility of SQP methods that they have been successful on two such contrasting problems as OPF and trajectory problems.

8 Software

There are now several packages available to solve large general nonlinear problems, notably MINOS [29], LANCELOT [9], CONOPT [12] and a recent addition to the fold SNOPT (see [17]). MINOS and CONOPT have been integrated into two more general systems (see Section 10). A comprehensive guide of available software for optimization in general is given in Moré and Wright [27]. LANCELOT is based on an augmented Lagrangian algorithm (see [8]) and solves a sequence of bound constrained problems (possibly by a barrier algorithm). CONOPT is based on the generalized reduced gradient algorithm. SNOPT is a SQP algorithm based on a quasi-Newton approximation to a reduced Hessian matrix. LANCELOT has an elaborate user interface that allows a user to try and take advantage of the structure of their problem (see [10]). For many users a general package will be a first step rather than the final one.

9 Automatic Generation of Derivatives

In the last few years considerable strides have been taken in the automatic generation of derivatives (see [1], [3] and [4]). By this we are not referring to finite-difference approximations or methods based on symbolic manipulation, but rather the direct provision of the code that generates what may be viewed as the coding for an analytical form of the derivative. The basic technique used is the chain rule. The input to the program is the code for the function and the output is the code for the derivatives. Clearly derivatives to any degree may be produced. In general the error in the derivative is similar to that in the evaluation of the function although there are functions for which this is not true. For example, the derivative of the function $\det(A(x))$ is unlikely to be correct when $A(x)$ has a near zero eigenvalue. If one has differentiated this function manually there will be an awareness of this danger, but when the code to evaluate the derivative is generated automatically then potential danger may be hidden.

To date there has not been wide use of software to automatically generate derivatives to solve real problems. A much more rapidly growing use of software that also incorporates automatic differentiation is the use of modeling languages, which are described in the next section.

10   Modeling Languages

GAMS (General Algebraic Modeling System, see [6]), AMPL (see [14]) and AIMMS (Advanced Interactive Mathematical Modeling System, see [5]) are three modeling languages that enable NP to be specified in a relatively simple manner. All three systems include procedures to generate the derivatives of the nonlinear functions $f(x)$ and $c(x)$. Such systems are particularly useful for specifying large problems. To see why consider the need to specify the matrix A when using a solver for NP. If the solver is efficient on large problems A will need to be specified in compact form. For example, only the nonzero elements of A are given together with their location in A. There are several ways to do this, but a particular solver will do it in a particular manner. It is unlikely that a user has chosen to store A in the precise manner required for a given solver. Indeed it may be that A is not stored or known explicitly but is generated perhaps from other data. These systems provide tools for a user to be flexible in how they specify their problem and automatically generate the problem in the required format for some specified solver. All three systems come integrated with certain solvers. AMPL and GAMS are integrated with MINOS and generate A in the required format. AIMMS is integrated with CONOPT. In theory it is possible for a user to link these systems to other solvers, but that is not an exercise for the faint of heart.

In addition to formulating the problem and integrating a solver these packages also have some post-solution analysis. AIMMS differs from the other two systems in that it has a graphical interface that includes a full screen editor. For example, data may be in the form of a bar chart, which can be modified using the mouse. The current version is written for a PC running Windows. A version for a workstation running motif is under development.

11   Summary

Much of the software in current use is based on algorithms that were rejected or have become obsolete for solving small problems. This is due in part to the ease with which simpler methods can be adapted to solve large problems. It mirrors the development of software for small problems when the best software available used to be based on augmented Lagrangians long after the more efficient (and in my experience more robust) SQP methods were discovered. It is also due to the fact that for large problems efficiency is no longer measured by the same yardstick. It remains to be seen whether or not this state of affairs is transient. More sophisticated methods are being adapted to solve large problems. However, it seems likely such development will augment rather than replace existing software, which will continue to be developed and refined. It is also possible that software based on earlier and simpler algorithms will also emerge.

Although there is still considerable research on what might be viewed as the basic methodology, much research today is directed at the nuts and bolts of algorithms. A particular difficult nut to crack is to discover an efficient direct method to solve the KKT equations. In solving the OPF problem it was necessary to "fool" the analysis phase of the Harwell code MA27 to be able to obtain a good ordering. To date solvers for indefinite systems have not done well on KKT systems.

Rather than a "higher" level change, a class of problems may require changes at a "lower" level. For example, the structure of the Jacobian may suggest using a special LU factorization rather than the one in current use, which may cater for *general* sparse matrices. One can develop systems that give the user such choices or enable a user to construct an algorithm by piecing together a number of modules for each of which there is a choice.

It is usually the nature of a particular application that all the problems are essentially of the same type. For such people versatility is irrelevant what they want is something that solves their problem. General purpose software is likely to establish the feasibility of an approach, but if the application is one that is solved repeatedly it is likely to be worth the effort to tailor an algorithm to the problem.

References

[1] B. M. Averick, J. J. Moré, C. H. Bischof, A. Carle and A. O. Griewank, "Computing large sparse Jacobian matrices using automatic differentiation," *SIAM J. on Scientific Computing*, 15(2) 1994, 285-294.

[2] J. T. Betts and W. P. Huffman, "Path Constrained Trajectory Optimization Using Sparse Sequential Quadratic Programming," *J. of Guidance, Control, and Dynamics*, 16(1) 1993, 59-68.

[3] C. H. Bischof, A. Carle, G. Corliss, P. Hovland, and A. O. Griewank, "ADIFOR: Generating derivative codes from Fortran programs," Technical Report MCS-P263-0991, Argonne National Laboratory, Argonne, USA, 1991.

[4] C. H. Bischof and A. O. Griewank, " ADIFOR: A FORTRAN system for portable automatic differentiation," Technical Report MCS-P317-0792, Argonne National Laboratory, Argonne, USA, 1992.

[5] J. Bischop and R. Entriken, *AIMMS: The Modeling System*, Paragon Decision Technology, Holland, 1993.

[6] A. Brooke, D. Kendrick and A. Meeraus, *GAMS: A User's Guide*, The Scientific Press, South San Francisco, 1988.

[7] R. Burchett (private communications).

[8] A. R. Conn, N. I. M. Gould and Ph. L. Toint, "A globally convergent augmented Lagrangian algorithm for optimization with general constraints and simple bounds," *SIAM J. on Numerical Analysis* 28, 1991, 545-572.

[9] A. R. Conn, N. I .M. Gould, and Ph. L. Toint, "LANCELOT: A Fortran package for large-scale nonlinear optimization (Release A)," *Springer Series in Computational Mathematics*, 17, Springer Verlag, Heidelberg, Berlin, New York, 1992.

[10] A. R. Conn, N. I. M. Gould, and Ph. L. Toint, "Improving the decomposition of partially separable functions in the context of large-scale optimization: a first approach," *Large Scale Optimization: State of the Art*, W. W. Hager, D. W. Hearn, and P. M. Pardalos (editors), Kluwer Academic Publishers, 1993.

[11] A. R. Conn, N. I .M. Gould, and Ph. L. Toint, "Large-scale nonlinear constrained optimization: A current survey," to appear in *Algorithms for Continuous Optimization*, L. Dixon, D. F. Shanno, and Spedicato (editors), Kluwer Academic Publishers, 1993.

[12] A. S. Drud , "CONOPT-A GRG code for large-scale nonlinear optimization-Reference manual," ARKI consulting and development, Denmark 1992.

[13] S. K. Eldersveld, "Large-scale sequential quadratic programming algorithms", Report SOL 92-4, Department of Operations Research, Stanford University, Stanford, 1992.

[14] R. Fourer, D. M. Gay and B. W. Kernighan, *AMPL: A Modeling Language for Mathematical Programming*, The Scientific Press, South San Francisco, 1993.

[15] P. E. Gill and W. Murray, "Quasi-Newton methods for linearly constrained optimization," in *Numerical Methods for Constrained Optimization*, P. E. Gill and W. Murray (editors), Academic Press, London and New York, 1974, 67-92.

[16] P. E. Gill, W. Murray, D. B. Ponceleón and M. A. Saunders, "Primal-dual methods for linear programming," Report SOL 91-03, Department of Operations Research, Stanford University, Stanford, 1991.

[17] P. E. Gill, W. Murray and M. A. Saunders, "Large-scale SQP methods and their application in trajectory optimization", to appear in *Proceedings of the 9th IFAC Workshop on Control Applications in Optimization*.

[18] P. E. Gill, W. Murray, M. A. Saunders and M. H. Wright, "User's Guide for NPSOL (Version 4.0): a Fortran package for nonlinear programming," Report SOL 2 (1986), Department of Operations Research, Stanford University.

[19] P. E. Gill, W. Murray, M. A. Saunders and M. H. Wright, "Maintaining LU factors of a general sparse matrix," *Linear Algebra and its Applications* 88/89, 239-270.

[20] P. E. Gill, W. Murray, M. A. Saunders and M. H. Wright, "Inertia-controlling methods for quadratic programming," *SIAM Review*, **33** (1988), 1-33.

[21] P. E. Gill, W. Murray, M. A. Saunders and M. H. Wright, "A practical anti-cycling procedure for linear and nonlinear programming," *Mathematical Programming* **45** (1989), 437-474.

[22] P. E. Gill, W. Murray, M. A. Saunders and M. H. Wright, "A Schur-complement method for sparse quadratic programming," in *Reliable Numerical Computation*, M. G. Cox and S. Hammarling (editors), Oxford University Press, 1990, 113-138.

[23] P. E. Gill, W. Murray, M. A. Saunders, and M. H. Wright, "Some theoretical properties of an augmented Lagrangian merit function," in *Advances in Optimization and Parallel Computing*, P. M. Pardalos (ed.), North Holland, 1992, 101-128.

[24] A. O. Griewank, D. Juedes, J. Srinivasan, and C. Tyner, "ADOL-C, a package for the automatic differentiation of algorithms written in C/C++," to appear in *ACM Transactions on Mathematical Software*.

[25] C. R. Hargraves and S. W. Paris, "Direct trajectory optimization using nonlinear programming and collocation," *J. of Guidance, Control and Dynamics*, **10** (1987), 338-348.

[26] W. Murray and F. J. Prieto, "A sequential quadratic programming algorithm using an incomplete solution of the subproblem," to appear in *SIAM J. on Optimization*, Report SOL 93-3, Department of Operations Research, Stanford University, Stanford, 1993.

[27] J. J. Moré and S. J. Wright, *Optimization Software Guide*, SIAM, 1993.

[28] B. A. Murtagh and M. A. Saunders, "A projected Lagrangian algorithm and its implementation for sparse nonlinear constraints," *Mathematical Programming Study* **16** (1982), 84-117.

[29] B. A. Murtagh and M. A. Saunders, *MINOS 5.4 User's Guide*, Report SOL 83-20R, Department of Operations Research, Stanford University, 1993.

[30] D. B. Ponceleón, "Barrier Methods for Large-Scale Quadratic Programming," Report SOL 91-2, Stanford University, Stanford, 1991.

[31] U. T. Ringertz, "An algorithm for optimization of nonlinear shell structures," Report 93-7, Dept of Lightweight Structures, KTH, Stockholm, 1993.

[32] S. M. Robinson "A quadratically convergent algorithm for general nonlinear programming problems," *Mathematical Programming*, **3** (1972), 145-156.

[33] J. B. Rosen and J. Kreuser, "A gradient projection algorithm for nonlinear constraints," in *Numerical Methods for Non-Linear Optimization*, F. A. Lootsma (editor), Academic Press, London and New York, 1972, 297-300.

[34] L. Sha, "A macrocell placement algorithm using mathematical programming techniques," Ph.D. dissertation, EE Dept, Stanford University, Stanford, 1989.

[35] Y. Y. Shi, R. Nelson, D. H. Young, P. E. Gill, W. Murray, and M. A. Saunders, "The application of Nonlinear Programming and Collocation to Optimal Aeroassisted Orbital Transfer Technology," *proceedings of 30th AIAA Conference on Aerospace Sciences*, Reno, Nevada, 1992.

[36] P. Wolfe *The reduced gradient method*, unpublished manuscript, Rand Corporation, 1962.

Branch-and-Price: Column Generation for Solving Huge Integer Programs

Cynthia Barnhart
Ellis L. Johnson
George L. Nemhauser
Martin W. P. Savelsbergh
Pamela H. Vance

Georgia Institute of Technology
School of Industrial and Systems Engineering
Atlanta, Georgia

1 Introduction

The successful solution of large-scale mixed integer programming (MIP) problems requires formulations whose linear programming (LP) relaxations give a good approximation to the convex hull of feasible solutions. In the last decade, a great deal of attention has been given to the "branch-and-cut" approach to solving MIPs. Hoffman and Padberg [1985], and Nemhauser and Wolsey [1988] give general expositions of this methodology.

The basic idea of branch-and-cut is simple. Classes of valid inequalities, preferably facets of the convex hull of feasible solutions, are left out of the LP relaxation because there are too many constraints to handle efficiently and most of them will not be binding in an optimal solution anyway. Then, if an optimal solution to an LP relaxation is infeasible, a subproblem, called the separation problem, is solved to try to identify violated inequalities in a class. If one or more violated inequalities are found, some are added to the LP to cut off the infeasible solution. Then the LP is reoptimized. Branching occurs when no violated inequalities are found to cut off an infeasible solution. Branch-and-cut, which is a generalization of branch-and-bound with LP relaxations, allows separation and cutting to be applied throughout the branch-and-bound tree.

The philosophy of branch-and-price is similar to that of branch-and-cut except that the procedure focuses on column generation rather than row generation. In fact, pricing and cutting are complementary procedures for tightening an LP relaxation.

In branch-and-price, sets of columns are left out of the LP relaxation because there are too many columns to handle efficiently and most of them will have their associated variable equal to zero in an optimal solution anyway. Then to check the optimality of an LP solution, a subproblem, called the pricing problem, which is a separation problem for the dual LP, is solved to try to identify columns to enter the basis. If such columns are found, the LP is reoptimized. Branching occurs when no columns price out to enter the basis and the LP solution does not satisfy the integrality conditions. Branch-and-price, which is a generalization of branch-and-bound with LP relaxations, allows column generation to be applied throughout the branch-and-bound tree.

We have several reasons for considering formulations with a huge number of variables.

- A compact formulation of a MIP may have a weak LP relaxation. Frequently the relaxation can be tightened by a reformulation that involves a huge number of variables.

- A compact formulation of a MIP may have a symmetric structure that causes branch-and-bound to perform poorly because the problem barely changes after branching. A reformulation with a huge number of variables can eliminate this symmetry.

- Column generation provides a decomposition of the problem into master and sub problems. This decomposition may have a natural interpretation in the contextual setting allowing for the incorporation of additional important constraints.

- A formulation with a huge number of variables may be the only choice.

At first glance, it may seem that branch-and-price involves nothing more than combining well-known ideas for solving linear programs by column generation with branch-and-bound. However, there are fundamental difficulties in applying column generation techniques for linear programming in integer programming solution methods [Johnson 1989]. These include:

- Conventional integer programming branching on variables may not be effective because fixing variables can destroy the structure of the pricing problem.

- Solving these LPs to optimality may not be efficient, in which case different rules will apply for managing the branch-and-price tree.

Recently, several specialized branch-and-price algorithms have appeared in the literature. Our paper attempts to unify this literature by presenting a general methodology for branch-and-price and describing applications. It is by no means an extensive survey, but does develop some general ideas that have only appeared in very special contexts. Routing and scheduling has been a particularly fruitful application area of branch-and-price, see Desrosiers et al. [1994] for a survey of these results.

Section 2 presents the types of MIPs for which branch-and-price can be advantageous. Section 3 presents the special types of branching that are required for branch-and-price to be effective. Section 4 considers the implementation of branch-and-price in mixed-integer programming codes. Section 5 summarizes computational experience with branch-and-price algorithms for binary cutting stock problems, generalized assignment problems, urban transit crew scheduling problems and bandwidth packing problems.

2 Suitable Models for Column Generation
2.1 General Models
The general problem P we consider is of the form

$$
\begin{aligned}
\text{maximize } & cx \\
Ax & \leq b, \\
x & \in S, \\
x & \quad \text{integer},
\end{aligned}
\tag{1}
$$

where S is a bounded polyhedron. The boundedness assumption is not necessary and is made purely for simplicity of exposition.

The fundamental construct of column generation is that the set

$$
S^* = \{x \in S : x \text{ integer}\}
$$

is represented by the extreme points $y_1, ..., y_p$ of its convex hull. Note that if x is binary, then S^* coincides with the extreme points of $conv(S^*)$.

Any point y in $conv(S^*)$ can be represented as

$$
y = \sum_{1 \leq k \leq p} y_k \lambda_k,
$$

subject to the convexity constraint

$$
\begin{aligned}
\sum_{1 \leq k \leq p} \lambda_k & = 1, \\
\lambda_k & \geq 0 \quad k = 1, ..., p.
\end{aligned}
$$

This yields the column generation form of P given by

$$
\begin{aligned}
\text{maximize } & \sum_{1 \leq k \leq p} (cy_k)\lambda_k \\
& \sum_{1 \leq k \leq p} (Ay_k)\lambda_k \leq b, \\
& \sum_{1 \leq k \leq p} y_k \lambda_k \quad \text{integer}, \\
& \sum_{1 \leq k \leq p} \lambda_k = 1, \\
& \lambda_k \geq 0 \quad k = 1, ..., p.
\end{aligned}
\tag{2}
$$

When x is binary, the condition $\sum_{1 \leq k \leq p} y_k \lambda_k$ integer is equivalent to $\lambda_k \in \{0, 1\}$ for $k = 1, ..., p$. If the null vector is an extreme point of $conv(S^*)$, it may not be explicitly included in the formulation, in which case the convexity constraint can be written as an inequality, i.e., $\sum_{1 \leq k \leq p} \lambda_k \leq 1$.

If S can be decomposed, i.e., $S = \cup_{1 \leq j \leq n} S_j$, we can represent each set

$$S_j^* = \{x_j \in S_j : x_j \text{ integer}\}$$

by the extreme points $y_1^j, ..., y_{p_j}^j$ of its convex hull, i.e., any point y^j in $conv(S_j^*)$ can be represented as

$$y^j = \sum_{1 \leq k \leq p_j} y_k^j \lambda_k^j,$$

subject to the convexity constraint

$$\sum_{1 \leq k \leq p_j} \lambda_k^j = 1,$$

$$\lambda_k^j \geq 0 \quad k = 1, ..., p_j.$$

This yields a column generation form of P with separate convexity constraints for each S_j given by

$$\text{maximize} \sum_{1 \leq j \leq n} \sum_{1 \leq k \leq p_j} (cy_k^j) \lambda_k^j$$

$$\sum_{1 \leq j \leq n} \sum_{1 \leq k \leq p_j} (Ay_k^j) \lambda_k^j \leq b,$$

$$\sum_{1 \leq k \leq p_j} y_k^j \lambda_k^j \quad \text{integer} \quad j = 1, ..., n, \tag{3}$$

$$\sum_{1 \leq k \leq p_j} \lambda_k^j = 1 \quad j = 1, ..., n,$$

$$\lambda_k^j \geq 0 \quad j = 1, ..., n, \ k = 1, ..., p_j.$$

If the subsets in the decomposition are identical, i.e., $S_j = \overline{S}$ for $j = 1, ..., n$, then they can be combined into one subset $\overline{S}$ with the convexity constraints

$$\sum_{1 \leq k \leq p} \lambda_k^j = 1, \quad j = 1, ..., n$$

replaced by an aggregated convexity constraint

$$\sum_{1 \leq k \leq p} \lambda_k = n.$$

This results in the column generation form

$$\text{maximize} \sum_{1 \le k \le p} (cy_k)\lambda_k$$

$$\sum_{1 \le k \le p} (Ay_k)\lambda_k \;\le\; b,$$

$$\sum_{1 \le k \le p} y_k\lambda_k \qquad \text{integer,} \qquad\qquad (4)$$

$$\sum_{1 \le k \le p} \lambda_k \;\le\; n,$$

$$\lambda_k \ge 0 \qquad k = 1, ..., p,$$

where $y_1, ..., y_p$ are the extreme points of $conv(\overline{S}^*)$. Here we have chosen the inequality form of the aggregated convexity constraint because in most applications no elements are assigned to some subsets. Moreover, if n is not fixed as part of the input, then the aggregated convexity constraint can be omitted altogether.

The essential difference between P and its column generation form is that S has been replaced by the extreme point representation of its convex hull. We see that any fractional solution to the linear programming relaxation of P is a feasible solution to the linear programming relaxation of its column generation form if and only if it can be represented by a convex combination of extreme points of $conv(S^*)$. In particular, Geoffrion [1974] has shown that if the polyhedron S does not have all integral extreme points, then the linear programming relaxation of the column generation form of P will be tighter than that of P for some objective functions. Geoffrion has also shown that a Langrangian dual with respect to the constraints $Ax \le b$ gives the same bound as the column generation form of P.

However, since the column generation form frequently contains a huge number of columns, it may be necessary to work with restricted versions that contain only a subset of its columns, and to generate additional columns only as they are needed. The column generation form is called the master problem (MP) and when it does not contain all of its columns it is called a restricted master problem (RMP). Column generation is done by solving pricing problems of the form

$$\text{maximize}\{dx : x \in S^*\} \text{ or maximize}\{dx : x \in conv(S^*)\}$$

where d is determined from optimal dual variables of an RMP.

2.2   Partitioning Models

Since most of the branch-and-price algorithms we are aware of have been developed for set partitioning based formulations, they will be emphasized. In the general set partitioning problem, we have a ground set of elements and rules for generating feasible subsets and their costs, and we wish to find the minimum cost partitioning of the ground set into feasible subsets. Let $z_{ij} = 1$ if element i is in subset j, and 0 otherwise, and let z_j denote the characteristic vector of subset j, i.e., a vector with entries z_{ij} for each element i. Similarly, let c_{ij} denote the profit associated with having element i in subset j and let c_j denote the corresponding profit vector. The general partitioning problem is of the form

$$\text{maximize} \sum_{1 \le j \le n} c_j z_j$$

$$\sum_{1 \le j \le n} z_{ij} = 1 \quad i = 1, ..., m,$$

$$z_j \in S, \tag{5}$$

$$z_j \quad \text{binary},$$

where m is the number of elements in the ground set, n is the number of subsets, and S is the set of feasible subsets.

2.2.1 Enumerated Subsets

One important class of partitioning problems for which column generation is desirable occurs when we do not know a description of S by linear inequalities, but we do know a way of enumerating S. Hopefully, the enumeration can be done cleverly, but even a brute force approach may suffice.

This structure occurs, for example, in crew pairing problems, where a sequence of flights, called a pairing, has to be constructed and assigned to a crew. The first flight in the sequence must depart from the crew's base, each subsequent flight departs from the station where the previous flight arrived and the last flight must return to the base. The sequence can represent several days of flying. Pairings are subject to a number of constraints resulting from safety regulations and labor contract terms. These constraints dictate restrictions such as the maximum number of hours a pilot can fly in a day, the maximum number of days before returning to the base and minimum overnight rest times. The main point is that these restrictions are not efficiently described by linear inequalities. In addition, the cost of pairings is a messy function of several attributes of the sequence.

Although enumerating pairings is complex because of all the rules that must be checked, it can be accomplished by first enumerating all feasible possibilities for one day of flying and then combining the one-day schedules to form pairings. The major difficulty is the total number of pairings, which grows exponentially with the number of flights. For example, in a typical problem with 253 flights, there are 5,833,004 pairings [Vance 1993]. However, it is possible to represent pairings as paths in a graph, and to evaluate their costs with a multilabel shortest path or dynamic programming algorithm, see Desrochers and Soumis [1989], Barnhart et al. [1994], and Vance [1993].

The enumeration yields the following column generation form

$$\text{maximize} \sum_{1 \le k \le p} (c_k y^k) \lambda_k$$

$$\sum_{1 \le k \le p} y_i^k \lambda_k = 1 \quad i = 1, ..., m,$$

$$\lambda_k \in \{0,1\}, \quad k = 1, ..., p,$$

where each y^k is an element of S. This column generation form corresponds to the 'standard' formulation of the set partitioning problem.

2.2.2 Linearly Constrained Subsets

Now we suppose that the rules on feasible subsets in a set partitioning problem can be described by linear inequalities.

Different restrictions on subsets
First, we assume that the feasible subsets have different requirements. Assume the requirements are given by

$$S_j = \{z_j : D_j z_j \le d_j \quad j = 1, ..., n, \quad z_j \text{ binary}\}. \tag{6}$$

More explicitly, problem P is given by

$$\text{maximize} \sum_{1 \le j \le n} c_j z_j$$

$$\sum_{1 \le j \le n} z_{ij} = 1 \quad i = 1, ..., m, \tag{7}$$

$$D_j z_j \le d_j \quad j = 1, ..., n,$$

$$z \quad \text{binary},$$

and its column generation form by

$$\text{maximize} \sum_{1 \le j \le n} \sum_{1 \le k \le p_j} (c_j y_k^j) \lambda_k^j$$

$$\sum_{1 \le j \le n} \sum_{1 \le k \le p_j} y_{ik}^j \lambda_k^j = 1 \quad i = 1, ..., m, \tag{8}$$

$$\sum_{1 \le k \le p_j} \lambda_k^j \le 1 \quad j = 1, ..., n,$$

$$\lambda_k^j \in \{0,1\} \quad j = 1, ..., n, \quad k = 1, ..., p_j,$$

where the $\{y_k^j\}$, $1 \le k \le p_j$ are the extreme points of $conv(S_j^*)$ with elements y_{ik}^j for $i = 1, ..., m$. We have chosen to write the convexity constraint as an inequality, since in many of these applications we may not assign any elements to a given subset.

To illustrate, consider the generalized assignment problem (GAP). In the GAP the objective is to find a maximum profit assignment of m tasks to n machines such that each task is assigned to precisely one machine subject to capacity restrictions on the machines.

The standard integer programming formulation of GAP is

$$\text{maximize} \sum_{1\leq i\leq m} \sum_{1\leq j\leq n} p_{ij}z_{ij}$$

$$\sum_{1\leq j\leq n} z_{ij} = 1 \quad i=1,...,m,$$

$$\sum_{1\leq i\leq m} w_{ij}z_{ij} \leq d_j \quad j=1,...,n,$$

$$z_{ij} \in \{0,1\} \quad i=1,...,m, \; j=1,...,n,$$

where p_{ij} is the profit associated with assigning task i to machine j, w_{ij} is the claim on the capacity of machine j by task i, d_j is the capacity of machine j, and z_{ij} is a 0-1 variable indicating whether task i is assigned to machine j.

The column generation form is

$$\text{maximize} \sum_{1\leq j\leq n} \sum_{1\leq k\leq K_j} \left(\sum_{1\leq i\leq m} p_{ij}y_{ik}^j \right)\lambda_k^j$$

$$\sum_{1\leq j\leq n} \sum_{1\leq k\leq K_j} y_{ik}^j\lambda_k^j = 1 \quad i=1,...,m,$$

$$\sum_{1\leq k\leq K_j} \lambda_k^j \leq 1 \quad j=1,...,n,$$

$$\lambda_k^j \in \{0,1\} \quad j=1,...,n, \; k=1,...,K_j,$$

where the first m entries of a column, given by $y_k^j = (y_{1k}^j, y_{2k}^j, ..., y_{mk}^j)$, form a feasible solution to the knapsack problem

$$\sum_{1\leq i\leq m} w_{ij}y_i^j \leq d_j,$$

$$y_i^j \in \{0,1\} \quad i=1,...,m.$$

In other words, a column represents a feasible assignment of tasks to a machine. Note that by replacing the knapsack constraint by its feasible solutions, we have improved the quality of the linear programming relaxation.

Identical restrictions on subsets
Now, we assume that the feasible subsets have identical requirements. Then (6) is replaced by the single set of inequalities

$$\overline{S} = \{z_j : Dz_j \leq d \quad j=1,...,n, \quad z_j \text{ binary}\}. \tag{9}$$

More explicitly, problem P is given by

$$\text{maximize} \sum_{1\leq j\leq n} c_j z_j$$

$$\sum_{1 \le j \le n} z_{ij} = 1 \quad i = 1, ..., m, \tag{10}$$

$$Dz_j \le d \quad j = 1, ..., n,$$

$$z \quad \text{binary},$$

and its column generation form by

$$\text{maximize} \sum_{1 \le k \le p} (cy_k)\lambda_k$$

$$\sum_{1 \le k \le p} y_{ik}\lambda_k = 1 \quad i = 1, ..., m, \tag{11}$$

$$\lambda_k \in \{0, 1\} \quad k = 1, ..., p.$$

Here we have chosen to omit the convexity constraint because it is common in these applications for n not to be fixed.

Consider the 0-1 cutting stock problem where item i has length d_i, the demand for each item is 1, the length of each stock roll is d and the objective is to meet demand using the minimum number of stock rolls. An integer programming formulation is

$$\min \sum_{1 \le j \le n} w_j$$

$$\sum_{1 \le j \le n} z_{ij} = 1 \quad i = 1, ..., m,$$

$$\sum_{1 \le i \le m} d_i z_{ij} \le dw_j \quad j = 1, ..., n,$$

$$z_{ij}, w_j \in \{0, 1\} \quad i = 1, ..., m, \ j = 1, ..., n,$$

where $w_j = 1$ if roll j is selected and $z_{ij} = 1$ if item i is assigned to roll j.

The column generation formulation is

$$\min \sum_{1 \le k \le p} \lambda_k$$

$$\sum_{1 \le k \le p} y_{ik}\lambda_k = 1, \quad i = 1, ..., m,$$

$$\lambda_k \in \{0, 1\} \quad k = 1, ..., p,$$

where each y_k is a binary solution to the knapsack inequality

$$\sum_{1 \le i \le m} d_i y_{ik} \le d.$$

That is, each y_k represents a feasible pattern for cutting one of the stock rolls into some subset of the items. MP replaces the knapsack inequality by all of its 0-1 solutions. The resulting problem has an LP relaxation which very frequently provides a bound whose round up equals the value of an optimal solution, Marcotte [1985]. On the other hand, the LP relaxation of the formulation P yields the trivial bound of $\lceil(\sum_{1\le i\le m} d_i)/d\rceil$.

Another major advantage of MP for these problems with identical subset rules is that it eliminates some of the inherent symmetry of P that causes branch-and-bound to perform very poorly. By this we mean that any solution to P or its LP relaxation has an exponential number of representations as a function of the number of subsets. Therefore branching on a variable z_{ij} to remove a fractional solution will likely produce the same fractional solution with z_{ik} equal to the old value of z_{ij} and vice-versa, unless z_{ij} is fractional for all j. Formulation MP eliminates this symmetry and is therefore much more amenable to branching rules in which meaningful progress in improving the LP bound can be made as we go deeper in the tree.

Although the discussion above has focused on set partitioning type master problems, in many applications the problem structure allows the master problem to be formulated either as a set partitioning problem or as a set covering problem. Consider, for example, vehicle routing and scheduling problems, where several vehicles are located at one or more depots and must serve geographically dispersed customers. Each vehicle has a given capacity and is available in a specified time interval. Each customer has a given demand and must be served within a specified time window. The objective is to minimize the total cost of travel. A solution to a vehicle routing and scheduling problem partitions the set of customers into a set of routes for vehicles. This naturally leads to a set partitioning formulation in which the columns correspond to feasible routes and the rows correspond to the requirement that each customer is visited *precisely* once. Alternatively, the problem can be formulated as a set covering problem in which the columns correspond to feasible routes and the rows correspond to the requirement that each customer is visited *at least* once. Since deleting a customer from a route, i.e., not visiting that customer, results in another shorter less costly feasible route, it is easy to verify that an optimal set covering solution will be an optimal set partitioning.

In general, if any subcolumn of a feasible column defines another feasible column with lower cost, an optimal solution to the set covering problem will be an optimal set partitioning and we can work with either one of the formulations. When there is a choice, the set covering formulation is preferred since

- Its linear programming relaxation is numerically far more stable and thus easier to solve.

- It is trivial to construct a feasible integer solution from a solution to the linear programming relaxation.

3   Branching

An LP relaxation solved by column generation is not necessarily integral and applying a standard branch-and-bound procedure to the restricted master problem with its existing columns will not guarantee an optimal (or feasible) solution. The reason is that after branching, it may be the case that there exists a column that would price out favorably, but is not present in the master problem. However, this approach has been used to obtain good results, but to find an optimal solution we must generate columns after branching. Nonetheless, many problems have been solved successfully, but not to proven optimality, by the heuristic of limiting the column generation to the root node of the branch-and-bound tree.

Consider a branch-and-bound algorithm that has the possibility of generating columns at any node of the tree. In particular, suppose using the conventional branching rule based on variable dichotomy, we branch on fractional variable λ_k, and we are in the branch in which λ_k is fixed to zero. In the column generation phase, it is possible (and quite likely) that the optimal solution to the subproblem will be the set represented by λ_k. In that case, it becomes necessary to generate the column with the 2nd highest reduced cost. At depth l in the branch-and-bound tree we may need to find the column with l^{th} highest reduced cost. In order to prevent columns that have been branched on from being regenerated, we must choose a branching rule that is *compatible* with the pricing problem. By compatible, we mean that we must be able to modify the pricing problem so that columns that are infeasible due to the branching constraints will not be generated and the pricing problem will remain tractable.

So the challenge in formulating a branching strategy is to find one that excludes the current solution, validly partitions the solution space of the problem, and provides a pricing problem that is still tractable. We need a guarantee that a feasible integer solution will be found (or infeasibility proved) after a finite number of branches and we need to be able to encode the branching information into the pricing problem. In addition, a branch-and-bound algorithm is more likely to be effective if the branching scheme divides the feasible set of solutions to the problem evenly, i.e. each new subproblem created has approximately the same number of feasible solutions.

3.1 Set Partitioning Master Problems

Ryan and Foster [1981] suggested a branching strategy for set partitioning problems based on the following proposition.

Proposition 1 *If Y is a 0–1 matrix, and a basic solution to $Y\lambda = 1$ is fractional, then there exist two rows r and s of the master problem such that*

$$0 < \sum_{k:y_{rk}=1, y_{sk}=1} \lambda_k < 1.$$

The pair r, s gives the pair of branching constraints

$$\sum_{k:y_{rk}=1, y_{sk}=1} \lambda_k = 1 \quad \text{and} \quad \sum_{k:y_{rk}=1, y_{sk}=1} \lambda_k = 0.$$

This branching is analogous to requiring that rows r and s be covered by the same column on the first (left) branch and by different columns on the second (right) branch, i.e., elements r and s belong to the same subset on the left branch and to different subsets on the right branch. Thus on the left branch, all feasible columns must have $y_{rk} = y_{sk} = 0$ or $y_{rk} = y_{sk} = 1$, while on the right branch all feasible columns must have $y_{rk} = y_{sk} = 0$ or $y_{rk} = 0, y_{sk} = 1$ or $y_{rk} = 1, y_{sk} = 0$. Rather than adding the branching constraints to the master problem explicitly, the infeasible columns in the master problem can be eliminated. On the left branch, this is identical to combining rows r and s in the master problem giving a smaller set partitioning problem. On the right branch, rows r and s are restricted to be disjoint, which may yield an easier master problem since set partitioning problems with disjoint rows (sets) are more likely to be integral. Not adding the branching constraints explicitly has the advantage of not introducing new dual variables that have to be dealt with in the pricing problem.

Usually, enforcing the branching constraints in the pricing problem, i.e., forcing two elements to be in the same subset on one branch and forcing two elements to be in different subsets on the other branch, is fairly easy to accomplish. However, the pricing problem on one branch may be more complicated than on the other branch.

Proposition 1 implies that if no branching pair can be identified, then the solution to the master problem must be integer. The branch and bound algorithm must terminate after a finite number of branches since there are only a finite number of pairs of rows. The number of branches necessary will generally be considerably fewer than would be necessary if individual variables were branched on since there are in general many fewer pairs of rows than variables. In addition, each branching decision eliminates a large number of variables from consideration.

A theoretical justification for this branching rule is that the submatrix it excludes is forbidden in totally balanced matrices, see Hoffman, Kolen, and Sakarovitch [1985]. Total balancedness of the coefficient matrix is a sufficient condition for the LP relaxation of a set partitioning problem to have only integral extreme points and the branching rule eventually gives totally balanced matrices.

Applications of this branching rule can be found for urban transit crew scheduling in Desrochers and Soumis [1989]; for airline crew scheduling in Anbil, Tanga and Johnson [1992], Barnhart et al. [1994] and Vance [1993]; for vehicle routing in Dumas, Desrosiers and Soumis [1989], for graph coloring in Mehrotra and Trick [1993]; and for the binary cutting stock problem in Vance et al. [1994].

Different requirements on subsets
Now consider the situation where the rules on feasible subsets can be described by linear inequalities and where different subsets may have different requirements, i.e., the formulation for P has the block diagonal structure given by (7) and the associated explicit column generation form, with separate convexity constraints for each subset, is given by (8).

In this situation, if we apply the branching rule discussed above but always select one partitioning row, say row r, and one convexity row, say s, we obtain a special branching scheme that has a natural interpretation in the original formulation and some nice computational properties. The pair of branching constraints that results is given by

$$\sum_{1\leq k\leq p_s:y^s_{rk}=1} \lambda^s_k = 1 \quad \text{and} \quad \sum_{1\leq k\leq p_s:y^s_{rk}=1} \lambda^s_k = 0. \tag{12}$$

This branching rule corresponds to requiring element r to be in subset s on the left branch and requiring element r to be in any subset but s on the right branch. This branching strategy has a very natural interpretation based on the following proposition.

Proposition 2 *Let λ be a feasible solution to the LP-relaxation of (8) and let $z_{ij} = \sum_{1\leq k\leq p_j} y^j_{ik}\lambda^j_k$, then z constitutes a feasible solution to the LP-relaxation of (7) and z is integral if and only if λ is integral.*

Consequently, the branching strategy given by (12) corresponds precisely to performing standard branching in (7), since

$$\sum_{1\leq k\leq p_s:y^s_{rk}=1} \lambda^s_k = 1 \Rightarrow \sum_{1\leq k\leq p_s} y^s_{rk}\lambda^s_k = 1 \Rightarrow z_{rs} = 1$$

and

$$\sum_{1\leq k\leq p_s:y^s_{rk}=1} \lambda^s_k = 0 \Rightarrow \sum_{1\leq k\leq p_s} y^s_{rk}\lambda^s_k = 0 \Rightarrow z_{rs} = 0.$$

Furthermore, this branching strategy does not increase the difficulty of solving the pricing problem. In fact, Sol and Savelsbergh [1994] show that any algorithm for the pricing problem used in the root node can also be used in subsequent nodes. To prevent an element from being in a generated solution, we just ignore it altogether. To force an element to be in the solution, we modify the dual variables such that every solution that does not use the element has a nonpositive reduced cost.

Applications of this branching strategy are presented for crew scheduling in Vance [1993]; for generalized assignment in Savelsbergh [1993]; for multi-commodity flow in Barnhart, Hane, Johnson and Sigismondi [1991] and Parker and Ryan [1993]; for vehicle routing in Desrosiers, Soumis and Desrochers [1984] and Desrochers, Desrosiers and Solomon [1992]; and for pickup and delivery problems in Sol and Savelsbergh [1994].

3.2 General Mixed Integer Master Problems

So far, we have discussed branching strategies for set partitioning master problems. A branching strategy for general mixed integer master problems with different requirements on subsets can be derived directly from (3) as follows [Johnson 1989]. The optimal solution to the linear programming relaxation is infeasible if and only if

$$\sum_{1\leq k\leq p_j} y^j_k\lambda^j_k$$

has a fractional component r for some j, say with value α. This suggests the following branching rule: on one branch we require

$$\sum_{1 \leq k \leq p_j} y_{rk}^j \lambda_k^j \leq \lfloor \alpha \rfloor$$

and on the other branch we require

$$\sum_{1 \leq k \leq p_j} y_{rk}^j \lambda_k^j \geq \lceil \alpha \rceil.$$

This implies that when a new extreme point is generated, an upper bound of $\lfloor \alpha \rfloor$ on component r has to be enforced on the first branch, and a lower bound of $\lceil \alpha \rceil$ on the second branch. Note that each pricing problem differs only in the lower and upper bounds on the components.

General models with identical restrictions on subsets
Developing a branching strategy for general mixed integer master problems with identical restrictions on subsets is more complex. If the solution to (2) is fractional, we may be able to identify a single row r and an integer α_r such that

$$\sum_{k:(Ay_k)_r \geq \alpha_r} \lambda_k = \beta_r$$

and β_r is fractional. We can then branch on the constraints

$$\sum_{k:(Ay_k)_r \geq \alpha_r} \lambda_k \leq \lfloor \beta_r \rfloor \text{ and } \sum_{k:(Ay_k)_r \geq \alpha_r} \lambda_k \geq \lceil \beta_r \rceil.$$

These constraints place upper and lower bounds on the number of columns with $(Ay_k)_r \geq \alpha_r$ that can be present in the solution. In general, these constraints will not eliminate variables and have to be added to the formulation explicitly. Each branching constraint will contribute an additional dual variable to the reduced cost of any new column with $(Ay_k)_r \geq \alpha_r$. This may complicate the pricing problem.

It is easy to see that a single row may not be sufficient to define a branching rule. Consider a set partitioning master problem that has a fractional solution. The only possible value for α_r is 1. However, $\sum_{k:y_{kr} \geq 1} \lambda_k = 1$ for every row. Thus we may have to branch on multiple rows.

Assume that

$$\sum_{k:(Ay_k)_r \geq \alpha_r} \lambda_k = \beta_r$$

and β_r is integer for every row r and integer α_r. Pick an arbitrary row, say r, and branch on the constraints

$$\sum_{k:(Ay_k)_r \geq \alpha_r} \lambda_k \leq \beta_r - 1 \text{ and } \sum_{k:(Ay_k)_r \geq \alpha_r} \lambda_k \geq \beta_r.$$

Because, the current fractional solution is still feasible in the latter branch, we look for a row s such that

$$\sum_{k:(Ay_k)_r \geq \alpha_r \wedge (Ay_k)_s \geq \alpha_s} \lambda_k = \beta_s$$

and β_s is fractional. If such a row exists, we branch on the constraints

$$\sum_{k:(Ay_k)_r \geq \alpha_r \wedge (Ay_k)_s \geq \alpha_s} \lambda_k \leq \lfloor \beta_s \rfloor \text{ and } \sum_{k:(Ay_k)_r \geq \alpha_r \wedge (Ay_k)_s \geq \alpha_s} \lambda_k \geq \lceil \beta_s \rceil.$$

Otherwise we seek a third row. We note that if the solution is fractional it is always possible to find a set of rows to branch on and that a set of l rows gives rise to $l + 1$ branches.

The branching scheme presented above applied to set partitioning master problems gives precisely the branching scheme of Ryan and Foster [1981] discussed in Section 4.1. To see this, note that by Proposition 1 we can always branch on two rows, say r and s, that the first branch defined by

$$\sum_{k:y_{rk} \geq 1} \lambda_k \leq 0$$

is empty, and that the other two branches defined by

$$\sum_{k:y_{rk} \geq 1 \wedge y_{sk} \geq 1} \lambda_k \leq 0 \text{ and } \sum_{k:y_{rk} \geq 1 \wedge y_{sk} \geq 1} \lambda_k \geq 1$$

are exactly those defined by the branching scheme of Ryan and Foster.

4 Implementation

Although implementing branch-and-price algorithms (or branch-and-cut algorithms) is still a nontrivial activity, the availability of flexible linear and integer programming systems has made it a less formidable task than it would have been three years ago.

4.1 Software

Modern simplex codes, such as CPLEX [CPLEX Optimization, 1990] and OSL [IBM Corporation, 1990] not only permit column generation while solving an LP but also allow the embedding of column generation LP solving into a general branch-and-bound structure for solving MIPs.

The use of MINTO [Nemhauser, Savelsbergh, and Sigismondi 1994, Savelsbergh and Nemhauser 1993] may reduce the implementation efforts even further. MINTO (Mixed INTeger Optimizer) is based on the belief that to solve large mixed-integer programs efficiently, without having to develop a full-blown special purpose code in each case, you need an effective general purpose mixed integer optimizer that can be customized through the incorporation of application functions. Its strength is that it allows users to concentrate on problem specific aspects rather than data structures and implementation details such as linear programming and branch-and-bound.

4.2 Column Management

In a maximization linear program, any column with positive reduced cost is a candidate to enter the basis. The pricing problem finds the column with highest reduced cost. Therefore, if a column with positive reduced cost exists the pricing problem will always identify it. This guarantees that the optimal solution to the linear program will be found.

However, it is not necessary to select the column with the highest reduced cost; any column with a positive reduced cost will do. Using this observation can improve the overall efficiency when the pricing problem is computationally intensive.

Various column generation schemes can be developed based on using approximation algorithms to solve the pricing problem. To guarantee optimality, a two-phase approach is applied. As long as an approximation algorithm for the pricing problem produces a column with positive reduced cost, that column will be added to the restricted master. If the approximation algorithm fails to produce a column with positive reduced cost, an optimization algorithm for the pricing problem is invoked to prove optimality or produce a column with positive reduced cost. Such a scheme reduces the computation time per iteration. However, the number of iterations may increase, and it is not certain that the overall effect is positive. Depending on the pricing problem, it may even be possible to generate more than one column with positive reduced cost per iteration without a large increase in computation time. Such a scheme increases the time per iteration, since a larger restricted master has to be solved, but it may decrease the number of iterations.

During the column generation process, the restricted master problem keeps growing. It may be advantageous to delete nonbasic columns with highly negative reduced costs from the restricted master problem in order to reduce the time per iteration.

These ideas can be combined into the following general column generation scheme:

1. Determine an initial feasible restricted master problem.

2. Initialize the column pool to be empty.

3. Solve the current restricted master problem.

4. Delete nonbasic columns with high negative reduced costs from the restricted master problem.

5. If the column pool still contains columns with positive reduced costs, select a subset of them, add them to the restricted master, and go to 3.

6. Empty the column pool.

7. Invoke an approximation algorithm for the pricing problem to generate one or more columns with positive reduced cost. If columns are generated, add them to the column pool and go to 5.

8. Invoke an optimization algorithm for the pricing problem to prove optimality or generate one or more columns with positive reduced costs. If columns are generated, add them to the column pool and go to 5.

9. Stop.

A very fast and promising approach to generate columns with positive reduced costs is to use improvement algorithms that take existing columns with reduced cost equal to zero (at least all basic columns satisfy this requirement) and try to construct columns with a positive reduced cost by performing some simple changes [Sol and Savelsbergh 1994]. Notice the similarity between the column management functions performed in branch-and-price algorithms and the row management functions performed in branch-and-cut algorithms.

4.3   Combining Column Generation and Row Generation

Combining column and row generation can yield very strong LP relaxations. However, synthesizing the two generation processes is nontrivial. The principle difficulty is their incompatibility. That is, the pricing (separation) problem can be much harder after additional rows (columns) are added, because the new rows (columns) can destroy the structure of the pricing (separation) problem.

One remedy is to dualize the additional constraints using Lagrangian relaxation. Another is to do the pricing only over the original rows, i.e., assuming that the new columns have 0 coefficients in the additional rows. But then it may be necessary to update the columns coefficients over the additional rows in order to maintain validity, see Mehrotra [1992], or it may be desirable to lift the coefficients to increase the strength of the valid inequality. Then after the lifting is done, it may be the case that the column no longer prices out favorably.

Despite these difficulties, there have been some successful applications of combined row and column generation. In problem situations where the objective is to partition the ground set into a minimum number of feasible subsets, such as minimizing the number of vehicles required to satisfy customer demands in routing and scheduling problems, an LP solution with fractional objective function value v can be cut off by adding a constraint that bounds the LP solution from above by $\lfloor v \rfloor$. Because every column has a coefficient 1 in this additional constraint, the constraint does not complicate the pricing problem and can easily be handled.

The most successful optimization algorithms for the traveling salesman problem use branch-and-cut, see Junger, Reinelt and Rinaldi [1994] for a recent survey. However, these algorithms only maintain columns for a small subset of the edges. Consequently, when the LP is solved for this reduced edge set, it is necessary to price out all the edges not in this set to verify that a true lower bound has been found. If edges with favorable reduced costs are identified, they are added to the reduced edge set and the process is repeated.

Nemhauser and Park [1991] combine column and row generation in an LP based algorithm for the edge coloring problem. No branching in the master problem is required on the instances they solve. The edge coloring problem requires a partitioning of the edges of a graph into a minimum cardinality set of matchings. Therefore, it can naturally be formulated as a set partitioning problem in which the columns correspond to matchings of the graph. Consequently, the pricing problem is a weighted matching problem. However, to strengthen the linear programming relaxation, they add odd-circuit constraints to the restricted master, which destroys the pure matching structure of the pricing problem. The pricing problem now becomes a matching problem with an additional variable for each odd circuit constraint, and an additional constraint for each odd circuit variable which relates the odd circuit variable to the edge variables in the circuit. This problem is solved by branch-and-cut. The approach points out the need for recursive calling of integer programming systems for the solution of complex problems.

5 Computational Experience
5.1 The Binary Cutting Stock Problem
Vance et al. [1994] solve binary cutting stock problems using a branch-and-price algorithm. Columns are generated by solving a binary knapsack problem with the optimal dual prices from the rows of the master problem as the prices on the items. The branching rule is identical to the Ryan and Foster rule presented earlier. In cutting stock terms, this rule requires two items to be contained in the same pattern on one branch and different patterns on the other branch. On the branch where the two items must be in the same pattern, the resulting knapsack pricing problem has a new super item replacing those two items. Thus, on this branch, the column generation problem is a knapsack problem with one fewer item. On the other branch a constraint is added to the knapsack problem that allows at most one of the items to be chosen. While this pricing problem is somewhat more difficult than a knapsack problem, it can still be solved quickly if there are not too many additional constraints.

Computational results are reported for randomly generated test problems. The branch-and-price algorithm was able to solve in seconds problems that could not be solved using a standard branch-and-bound procedure on an explicit formulation. Standard branch-and-bound procedures were unable to solve problems with more than 70 items. The largest problems solved using the branch-and-price algorithm had 500 items and they were solved in less than an hour on an IBM RS/6000.

5.2 The Generalized Assignment Problem
Savelsbergh [1993] develops a branch-and-price algorithm for the generalized assignment problem discussed in Section 2. The pricing problem is given by

$$\text{maximize}_{1 \leq j \leq n}\{z(KP_j) - v_j\},$$

where v_j is the optimal dual price from the solution to the restricted master problem associated with the convexity constraint of machine j and $z(KP_j)$ is the value of the optimal solution to the knapsack problem

$$\text{maximize} \sum_{1 \leq i \leq m} (p_{ij} - u_i)y_i^j$$

subject to

$$\sum_{1 \le i \le m} w_{ij} y_i^j \le d_j,$$

$$y_i^j \in \{0, 1\}, \quad j = 1, ..., n,$$

with u_i being the optimal dual price from the solution to the restricted master problem associated with the partitioning constraint of task i. A column prices out to enter the basis if its reduced cost is positive. Consequently, if the objective value of the pricing problem is less than or equal to zero, then the current optimal solution for the restricted master problem is also optimal for the (unrestricted) master problem. The branching rule described in Section 4 for master problems with several convexity rows is used. This rule assigns task i to machine j on one branch and forbids machine j to perform task i on the other. In both cases, the size of the pricing problem for machine j is reduced by one task. Furthermore, on the branch where task i must be performed by machine j, task i may also be deleted from the pricing problem for each of the other machines.

Computational results indicate that the branch-and-price algorithm clearly out-performs existing algorithms and is able to solve much larger instances. In one of the computational experiments the average number of nodes required by the branch-and-price algorithm was compared with the average number of nodes required by the dual ascent algorithm of Guignard and Rosenwein [1989] for ten randomly generated instances in four different problem classes. The results are given in Table 1. Although in theory both algorithms use the same bounds, the branch-and-price algorithm clearly does better.

Table 1: Dual ascent versus branch-and-price

problem class	Dual ascent #nodes	branch-and-price #nodes
A,5,50	7.	1.0
B,10,50	101.	1.6
C,10,50	156.	5.7
D,5,30	102.	40.0

5.3 The Urban Transit Crew Scheduling Problem

Desrochers and Soumis [1989] use a branch-and-price algorithm to solve the urban transit crew scheduling problem (UTCS). An instance of UTCS is defined by a bus schedule and the collective agreement between the drivers and management. The schedule defines a set of tasks that must be performed and the collective agreement places restrictions on feasible workdays for the drivers and dictates the cost of those workdays. The agreement may also place global restrictions on the types of workdays included in the schedule. The master problem is a set covering problem with additional constraints. There is a set covering constraint for each task to be performed, and columns representing feasible workdays for a bus driver. The set covering constraints ensure that at least one driver is assigned to each task. The additional constraints enforce many global restrictions on the characteristics of the final solution. For example, the number of workdays in the solution whose total elapsed time is less than a given threshold may be limited to a certain percentage of the total number of workdays. Columns are generated by solving a constrained shortest path problem on a specially constructed network. The branching rule is similar to the Ryan and Foster rule presented earlier except that instead of branching on whether two tasks are executed in the same workday, they branch on whether two tasks are executed *consecutively* in the same workday. This rule is more easily enforced in the constrained shortest path procedure than the more general rule.

The authors present computational results for two real-world problems. In both cases, the branch-and-price algorithm constructed solutions with lower cost than the best known solutions.

5.4 The Bandwidth Packing Problem

The bandwidth packing problem is to decide which calls on a list of requests should be chosen to route on a capacitated network. An example is the routing of video data for teleconferencing within a private network. The objective is to minimize the costs of routing the selected calls plus the revenue lost from unrouted calls. Parker and Ryan [1994] formulate this problem as an integer program as follows.

Let P_i denote a set of feasible paths for call i. Let r_i denote the revenue of call i, d_i the demand of call i in units of bandwidth, u_e denote the capacity of link e in bandwidth, and c_e denote the usage cost of transmitting one bandwidth on link e. Let L be the set of all links. There are two types of variables

$$x_{ij} = \begin{cases} 1 & \text{if call } i \text{ uses path } j \\ 0 & \text{otherwise} \end{cases}$$

and

$$y_i = \begin{cases} 1 & \text{if call } i \text{ is not routed} \\ 0 & \text{otherwise} \end{cases}$$

Let δ_{ej} be the indicator that is 1 if link e is in path j and 0 otherwise and assume that there are n calls to be routed. The problem formulation is

$$\min \sum_{e \in L} c_e \sum_{i=1}^{n} \sum_{j \in P_i} \delta_{ej} d_i x_{ij} + \sum_{i=1}^{n} r_i y_i$$

subject to

$$\sum_{j \in P_i} x_{ij} + y_i = 1 \text{ for each call } i$$

$$\sum_{i=1}^{n} \sum_{j \in P_i} \delta_{ej} d_i x_{ij} \leq u_e \text{ for each link } e.$$

The first set of constraints ensures that each call is either routed or not, and the second set ensures the satisfaction of link capacities. The solution method uses column generation to solve the LP relaxations within a branch and bound scheme. The LP relaxations are similar to multi-commodity network flow problems and are solved using standard column generation solution techniques. To obtain integer solutions, Parker and Ryan use a hybrid branching strategy. Specifically, they first create one branch setting x_{ij} to 1. This rule is easy to enforce in the pricing subproblem by deleting call i from the problem, and removing d_i from the capacity of all links on path j. Since forcing x_{ij} to 0 is difficult to enforce in the pricing subproblem, $x_{ij} = 0$ is satisfied by creating several branches, one for each link of path j. If the links of path j are $e_1, e_2, \ldots, e_k$, k new branches are created. At the ℓth branch, they delete the column corresponding to x_{ij}, and any other column in which call i uses link e_ℓ. The pricing subproblem is prevented from generating any of the deleted paths by removing link e_ℓ from the network.

Parker and Ryan tested their algorithm on 14 problems ranging in size from 14 to 30 nodes and 23 to 93 calls. They report running times to find optimal solutions from 8 seconds to over 8 hours on a VAX 8800, and conclude that the algorithm is a practical procedure for solving a class of real world problem. Further investigations are proposed to use cutting planes at selected nodes in the branch and bound tree to reduce the computational effort.

References

[1] C. Barnhart, C. A. Hane, E.L. Johnson, and G. Sigismondi, "An Alternative Formulation and Solution Strategy for Multi-Commodity Network Flow Problems," Report COC-9102, Georgia Institute of Technology, Atlanta, Georgia, 1991.

[2] C. Barnhart, E. L. Johnson, R. Anbil, and L. Hatay, "A column generation technique for the long-haul crew assignment problem," T. Ciriano and R. Leachman (editors). *Optimization in Industry*, II, John Wiley and Son, 1994, to appear.

[3] CPLEX Optimization, Inc., *Using the CPLEX Linear Optimizer*, 1990

[4] M. Desrochers, J. Desrosiers, and M. Solomon, "A new optimization algorithm for the vehicle routing problem with time windows," *Operations Research*, **40** (1992), 342-354.

[5] M. Desrochers and F. Soumis, "A column generation approach to the urban transit crew scheduling problem," *Transportation Science*, **23** (1989), 1-13.

[6] J. Desrosiers, Y. Dumas, M. M. Solomon, and F. Soumis, "Time constrained routing and scheduling," in *Handbooks in Operations Research and Management Science, Volume on Networks*, M.E. Ball, T.L Magnanti, C. Monma, and G.L. Nemhauser (editors), 1994, to appear.

[7] J. Desrosiers, F. Soumis, and M. Desrochers, "Routing with time windows by column generation," *Networks*, **14** (1984), 545-565.

[8] Y. Dumas, J. Desrosiers, and F. Soumis, "The pickup and delivery problem with time windows," *European Journal of Operations Research*, **54** (1991), 7-22.

[9] A.A. Farley, "A note on bounding a class of linear programming problems, including cutting stock problems," *Operations Research*, **38** (1990), 922-924.

[10] A.M. Geoffrion, "Lagrangean relaxation for integer programming," *Mathematical Programming Studies*, **2** (1974), 82-114.

[11] M. Guignard and M. Rosenwein, "An improved dual-based algorithm for the generalized assignment problem," *Operations Research*, **37** (1989), 658-663.

[12] A.J. Hoffman, A. Kolen, and M. Sakarovitch, "Totally balanced and greedy matrices," *SIAM Journal on Algebraic and Discrete Methods*, **6** (1985), 721-730.

[13] K. Hoffman and M. Padberg, "LP-based combinatorial problem solving. *Annals of Operations Research,* **4** (1985), 145-194.

[14] IBM Corporation, *Optimization Subroutine Library, Guide and Reference.,* 1990

[15] E. L. Johnson, "Modeling and strong linear programs for mixed integer programming," *Algorithms and Model Formulations in Mathematical Programming,* S. W. Wallace (editor), NATO ASI Series 51, 1989, 1-41.

[16] E. L. Johnson, A. Mehrotra, and G.L. Nemhauser, "Min-cut clustering," *Mathematical Programming,* **62** (1993), 133-152.

[17] M. Junger, G. Reinelt, and G. Rinaldi, "The traveling salesman problem," *Handbooks in Operations Research and Management Science, Volume on Networks,* M. E. Ball, T. L Magnanti, C. Monma, and G. L. Nemhauser (editors), 1994, to appear.

[18] L.S. Lasdon, *Optimization Theory for Large Systems.* MacMillan, New York, 1970.

[19] O. Marcotte, "The cutting stock problem and integer rounding," *Mathematical Programming,* **33** (1985), 82-92.

[20] A. Mehrotra, *Constrained Graph Partitioning: Decomposition, Polyhedral Structure and Algorithms,* Ph.D. Thesis, Georgia Institute of Technology, Atlanta, GA, 1992.

[21] A. Mehrotra and M. A. Trick, *A Column Generation Approach to Graph Coloring,* Graduate School of Industrial Administration, Carnegie Mellon University, Pittsburgh, PA, 1993.

[22] G. L. Nemhauser and S. Park, "A polyhedral approach to edge coloring," *Operations Research Letters,* **10** (1991), 315-322.

[23] G.L. Nemhauser and L.A. Wolsey, *Integer and Combinatorial Optimization.* Wiley, Chichester, 1988.

[24] G. L. Nemhauser, M. W. P. Savelsbergh, and G. C. Sigismondi, "MINTO, a Mixed INTeger Optimizer," *Operations Research Letters,* 1994, to appear.

[25] M. Parker and J. Ryan, "A column generation algorithm for bandwidth packing," *Telecommunications Systems,* 1994, to appear.

[26] C. Ribeiro, M. Minoux, and M. Penna, "An optimal column generation with ranking algorithm for very large set partitioning problems in traffic assignment," *European Journal of Operations Research,* **41** (1989), 232-239.

[27] D. M. Ryan and B. A. Foster, "An integer programming approach to scheduling," in *Computer Scheduling of Public Transport Urban Passenger Vehicle and Crew Scheduling,* A. Wren (editor), North-Holland, Amsterdam, 1981, 269-280.

[28] M. W. P. Savelsbergh, "A Branch-and-Price Algorithm for the Generalized Assignment Problem," Report COC-9302, Georgia Institute of Technology, Atlanta, Georgia, 1993.

[29] M. W. P. Savelsbergh and G. L. Nemhauser, "Functional description of MINTO, a Mixed INTeger Optimizer," Report COC-91-03A, Georgia Institute of Technology, 1993.

[30] M. Sol and M. W. P. Savelsbergh, "A Branch-and-Price Algorithm for the Pickup and Delivery Problem," 1994, in preparation.

[31] P. H. Vance, *Crew Scheduling, Cutting Stock, and Column Generation: Solving Huge Integer Programs,* Ph.D. dissertation, School of Industrial and Systems Engineering, Georgia Institute of Technology, Atlanta, Georgia, 1993.

[32] P. H. Vance, C. Barnhart, E. L. Johnson, and G. L. Nemhauser, "Solving binary cutting stock problems by column generation and branch-and-bound," *Computational Optimization and Applications,* 1994, to appear.

Recent Advances in Large-Scale Nonlinear Optimization

Jorge Nocedal

Northwestern University

Evanston, Illinois

1 Introduction

At the European Center for Medium Range Weather Forecasts, nonlinear optimization problems in one million variables are solved routinely. These problems are not only large – they are also hard: the objective function is very expensive to evaluate (30 minutes on a CRAY 90), and the Hessian matrix, which is not sparse, cannot be computed in a realistic time. Meteorologists have taken great care in the computation of the gradient and, by using the adjoint method, can evaluate it at the same cost as the function. A gradient-based optimization method can therefore be applied (they use a limited memory BFGS method), and the problems are solved – albeit in a much longer time than they would hope for.

These weather forecasting problems are among the most difficult unconstrained problems solved at present, and are indicative of the types of applications that nonlinear optimization will face in the years to come. As in these problems, where the objective function requires the solution of a system of partial differential equations, many complex design and modeling problems will use nonlinear optimization techniques to perform sophisticated simulations. The efficient solution of problems of this kind requires good use of the computational resources, appropriate data structures and effective optimization algorithms.

In this talk I will describe some important recent advances in nonlinear optimization, paying particular attention to the large-scale case. We will see that considerable progress has been made in algorithms for large unconstrained problems, and that a variety of efficient codes is beginning to emerge. I will also describe advances in trust region algorithms for constrained optimization, and show how they are leading to large-scale implementations. I will briefly touch on the topic of interior-point methods for nonlinear programming – a research area that is likely to bring a fresh point of view to nonlinear programming. Finally I will describe important advances in software, such as the LANCELOT package and the CUTE testing system, as well as the Minpack-2 library which will be the first nonlinear optimization package that addresses issues arising in high-performance computation.

2 Unconstrained Optimization

As stated earlier, we will only consider methods for large problems, even though some interesting results have recently been obtained for quasi-Newton methods for small dense problems [13],[15],[1],[58],[75],[66]. We will first focus on the unconstrained optimization problem

$$\min f(x),$$

where f is a smooth function of n variables, and where n is assumed to be large, say $n > 500$. It is unlikely that large problems can be solved in an acceptable time if the gradient g of f is not available; for example, approximating the gradient at one point by finite differences requires n function evaluations, which is an unacceptable expense when n is large. On the other hand, the Hessian of f may or may not be available. Large unconstrained problems come in all sorts of forms, and since we cannot expect one method to be the best for all these cases, distinct classes of algorithms and software have been devised.

Limited memory methods are becoming quite popular. They represent one of the two contributions of quasi-Newton methods, which were extensively researched in the last twenty years, to large-scale optimization (the other contribution is the partially separable quasi-Newton approach described below). Limited memory methods are appropriate for large problems in which the Hessian matrix cannot be computed at reasonable cost, and for problems that do not possess a clearly partially separable objective function. In this talk we describe two problems of this kind – seismic inversion and weather forecasting – and discuss the difficulties they pose.

There are two distinct approaches to limited memory methods. The first [12] alternates cycles of BFGS steps (in which the quasi-Newton matrices are not formed, but are represented by a set of vectors) and cycles of preconditioned conjugate gradient steps, using the last BFGS matrix as a fixed preconditioner. At the end of a conjugate gradient cycle the BFGS information is discarded and the method is restarted. In this hybrid BFGS-CG method some steps are truly quasi-Newton steps and some are conjugate gradient steps – and the details of implementation, such as the line search, are different in each case. The second approach to limited memory methods [57], [65],[41] is simpler and takes a pure quasi-Newton point of view. This so-called L-BFGS method is identical to the standard BFGS method in all its implementation details, with one exception. The BFGS matrix is not formed, but (as in the previous approach) is implicitly defined by several pairs of correction vectors measuring the curvature observed during the last few steps, but unlike the hybrid limited memory approach, the information defining the BFGS matrix is continually updated, by discarding at each iteration the oldest pair of correction vectors and replacing it by the newest one. Thus the search direction is always of the form

$$d_k = -H_k g_k$$

where g is the gradient of f, and H_k is obtained by (implicitly) updating a scaling matrix of the form $\gamma_k I$ several times using the BFGS formula and the last few correction pairs $s_i = x_{i+1} - x_i, y_i = g_{i+1} - g_i$.

The approach taken in L-BFGS is both simpler and more versatile than that taken in the hybrid BFGS-CG method, and also appears to perform better in practice. L-BFGS can easily be generalized to constrained problems, as we will discuss below, but the hybrid approach may not have an efficient generalization due to the conjugate gradient cycle. Recently several authors have considered ways of improving the L-BFGS method by reducing the storage in half and by devising alternative implementations. As we explain in this talk, this is possible because a quasi-Newton update can be represented by only one correction vector. Fletcher [35] describes a method that cuts storage in half, but that does not allow for a continuous recycling of the correction vectors. Siegel [74] is able to achieve both goals – the information can be refreshed at each step and the storage is only half of that of L-BFGS – but the algorithm is different from L-BFGS and it is not clear at present if it performs as well. We discuss these new methods as well as two recent proposals by Reed [69], Kelly [54] and Wright [84].

A recent advance in limited memory methods has been the derivation of compact (or outer-product) representations of limited memory matrices [17]. Prior to this work, efficient procedures for handling limited memory matrices existed only for the inverse BFGS formula. The new compact representations allow us to represent and manipulate a variety of matrices (BFGS, SR1, Broyden) both in their inverse and direct forms. Due to this, limited memory methods can now be used in trust region settings, and more importantly, can be effectively incorporated into algorithms for solving constrained problems. In this talk we describe an extension of the L-BFGS algorithm to bound constrained problems

$$\min f(x) \tag{1}$$

$$\text{subject to } l \leq x \leq u. \tag{2}$$

The algorithm [16] is based on the gradient projection approach that has received much attention in the last years [59], [60], [23] and has been implemented in the L-BFGS-B code [88]. We present numerical results comparing this code with LANCELOT on a large collection of test problems from the CUTE testing environment [8].

It is now well established [27],[13] that the Symmetric Rank-One (SR1) method performs very well when solving small problems, and it is thus natural to investigate if it can give rise to a superior limited memory method. SR1 is appealing since it is not required to generate positive definite matrices, which is important in constrained problems and in the context of partially separable quasi-Newton methods. Byrd and Lu [14] propose an elegant limited memory SR1 method using a trust region approach, but the numerical results appear to be disappointing. This highlights the fact that, even though limited memory methods are conceptually very simple, little is understood about their behavior. For example it is not known if the strategy used in L-BFGS of keeping the most recent corrections can be improved by some other criterion. Research is also needed to explain, and possibly correct, the inability of limited memory methods to provide high accuracy in the answer, in some problems.

An important recent advance in software for limited memory methods for unconstrained problems is the routine LMVM of the Minpack-2 library. Even though the algorithm implemented in this routine is almost identical to the L-BFGS method used in [65] and [41], LMVM is more versatile, robust and portable than those two codes. In this talk we describe some of the results obtained by Averick and Moré when applying the LMVM code to problems from the Minpack-2 collection [5].

Newton's method is undoubtedly the best approach for solving most large problems. State-of-the-art implementations of Newton's method for large unconstrained optimization make use of sophisticated linear algebra techniques to control the cost of the iteration. The three main questions that need to be addressed are how to efficiently solve the Newton equations

$$\nabla^2 f(x_k)d = -g_k, \tag{3}$$

how to ensure convergence from remote starting points, and how to achieve a fast rate of convergence. These questions are related; for example the globalization technique may indicate how accurately (3) needs to be solved, and what type of iterative method for solving (3) can be terminated early. The rate of convergence is also dependent upon the accuracy with which (3) is solved [30].

The globalization of Newton's method is achieved by either line searches or trust region techniques. Since the paper by Murray in these proceedings (see also [38], [39]) discusses line search implementations of Newton's method and their corresponding linear algebra techniques, we will focus only on trust region implementations. For this reason our discussion of Newton methods will be very brief.

An attractive idea is to view the Newton step as the solution (or approximate solution) of the quadratic problem

$$\min g_k^T d + \frac{1}{2} d^T \nabla^2 f(x_k)d \tag{4}$$

$$\text{subject to} \quad \|d\|_2 \leq \Delta_k, \tag{5}$$

where Δ_k is a trust region radius that is updated at every iteration according to how well the model (4) approximates the function f. An approach that is becoming quite popular was proposed and analyzed by Steihaug [78], [79] (see also Toint [80]), and consists of applying the conjugate gradient method to the linear system (3) – even though this system may not be positive definite. The initial guess $d^{(0)}$ is zero and the conjugate gradient iteration generates a sequence of guesses $\{d^{(i)}\}$ until one of the following conditions is satisfied:

1. A convergence test for the inner conjugate gradient iteration is satisfied. This test is normally based on the residual of (3) and ensures a superlinear or quadratic rate of convergence of the (outer) Newton iteration.

2. A guess $d^{(l)}$ is larger in norm than the trust region Δ_k. In this case $d^{(l)}$ is scaled back so that it takes us to the border of the spherical trust region (5), and this step is accepted as the approximate solution to (3).

3. A direction of negative or zero curvature is encountered, i.e. the inner conjugate gradient iteration generates a direction $d^{(i)}$ such that $(d^{(i)})^T \nabla^2 f(x_k) d^{(i)} \leq 0$. In this case we follow that direction to the trust region, and define it as the approximate Newton step.

We call this *Steihaug's method*. In this talk we discuss the theoretical and computational strengths, as well as some potential drawbacks, of this method. An interesting variation [4] in which the Hessian matrix is implicitly perturbed during the conjugate gradient iteration, when necessary, will also be discussed. This new approach may overcome some of the potential weaknesses of Steihaug's method, and may also provide more freedom in the implementation of Newton's method.

The inner conjugate gradient iteration will often be inefficient unless properly preconditioned. Averick and Moré use the incomplete Cholesky factorization described by Jones and Plassman [51] as preconditioner, and obtain very good results on a few large problems. LANCELOT provides a variety of preconditioners, some naive and some sophisticated. These two recent codes – the Newton method of of Averick and Moré (which will be part of the Minpack-2 package), and the Newton method in LANCELOT – have therefore three layers: the outer Newton step is computed by a cycle of inner conjugate gradient iterations, each of which requires the solution of a sparse linear system of equations defining the preconditioner. Both codes are among the most sophisticated optimization software written to date, and are capable of handling very large problems.

Partially separable quasi-Newton methods were first introduced by Griewank and Toint [46], [47] and have given rise to some important advances in large-scale optimization, including the development of the LANCELOT code [26] which exploits the idea of partial separability. This idea is still not widely understood, and LANCELOT is often not used due to the mistaken impression that that problem under consideration does not possess this structure. Another obstacle is the difficulty of specifying partially separable objective functions in a way that can be exploited by the algorithms.

The simplest form of partial separability occurs when a function can be written as the sum

$$f(x) = \sum_{i=1}^{n_e} f_i(x^{[i]}), \tag{6}$$

where each element function f_i depends only on a small number of variables, indicated here as $x^{[i]}$. Objective functions arising from the discretization of differential equations often have this form. But this definition is too restrictive and can be relaxed to allow any element functions f_i that have a large invariant subspace. A simple example is [26]

$$f_i(x) = \left(\sum_{j=1}^{n} x_j \right)^4$$

which depends on all the variables, but which has an invariant subspace of dimension $n - 1$. The important point is that the Hessian of f_i has a simple structure, and can therefore be easily approximated. To do this it is best to introduce the change of variables $w = \sum_{j=1}^{n} x_j$, so that this function becomes simply w^4. LANCELOT exploits the more general concept of *group* partial separability which is a natural extension of the two cases just described.

LANCELOT provides both quasi-Newton and Newton methods and uses a trust region strategy to enforce convergence. The step is computed, as in Steihaug's implementation of Newton's method, by solving the system

$$B_k d_k = -g_k \qquad (7)$$

using the conjugate gradient method inside a trust region. The matrix B_k, which can be the Hessian or a quasi-Newton approximation, is represented in partially separable form. In particular, the best quasi-Newton option of LANCELOT uses the SR1 formula to update approximations to the Hessians of each of the element functions f_i, and these Hessian approximations can be assembled to define B_k. This partially separable quasi-Newton method is very powerful and our experience indicates that it is usually substantially better than the L-BFGS method; the latter only becomes the winner when the problem is not "sufficiently partially separable". We will present some numerical results illustrating this.

The Newton option of LANCELOT (which uses exact second derivatives) usually outperforms the quasi-Newton option and should be used whenever possible. But since in many large-scale applications the user is unwilling or unable to compute the Hessian, the quasi-Newton version of LANCELOT should be seriously considered. In this talk we argue that two important areas of future research are the automatic determination of partially separable structures and the development of interfaces that allow an easy formulation of the problem.

Nonlinear conjugate gradient methods are still widely used. In fact, some optimization researchers claim that a good implementation of the nonlinear conjugate gradient method performs as well, or perhaps even better, than limited memory methods. My experience does not support this claim, and I agree with Murray [62] that the performance of nonlinear conjugate gradient methods tends to be very uneven, and that it is hard to predict when they will perform poorly. In contrast, the limited memory L-BFGS method exhibits fairly uniform performance on a large variety of problems.

But the appeal of conjugate gradient methods is strong: the iteration is very simple, the storage requirements are minimal, and the method shares some of the remarkable properties of the linear conjugate gradient method. For this reason interest in the nonlinear CG method will continue for the foreseeable future. In this talk we briefly describe several algorithms that attempt to improve the performance of the Polak-Ribière method [42], [50].

3 Nonlinearly Constrained Problems

The most important open area of research in nonlinear optimization concerns the solution of large problems with general constraints,

$$\min f(x) \tag{8}$$

subject to

$$c_i(x) = 0 \quad i = 1, ..., m \tag{9}$$

$$c_i(x) \geq 0 \quad i = m + 1, ..., t. \tag{10}$$

Even though a large number of papers have been written during the last ten years on the theory and algorithms of nonlinear programming [61], many questions remain unanswered – even for the small dimensional case. Most of the research has focused on Sequential Quadratic Programming (SQP) approaches, but multiplier and interior-point methods have also received considerable attention. There are various competing approaches at present, and we can expect to see many new proposals in the near future. I will focus here only on trust region methods for solving (8)-(10) since the paper by Murray in these proceedings (see also [45]) addresses issues concerning line search methods.

During the 1980s important advances were made towards the development of trust region methods, based on sequential quadratic programming, for solving *equality constrained* optimization problems. This research focused on how to formulate the subproblems [18], [19], [68], [81], [83] and how to solve them [49], [86], [87]. A useful approach is to solve a sequence of quadratic subproblems of the form

$$\min g_k^T d + \frac{1}{2} d^T \nabla_{xx}^2 L_k d$$

subject to

$$A(x_k)^T d + c(x_k) = b_k,$$

$$\|d\| \leq \Delta_k,$$

where L denotes the Lagrangian, $A = [\nabla c_1, ..., \nabla c_m]$ is the matrix of constraint gradients corresponding to the equality constraints (we assume that inequalities are not present), and where b_k is a vector chosen to make the constraints of the subproblem compatible. As we discuss, the choice of b_k is crucial to the robustness and efficiency of the method.

In this talk I will describe an algorithm of this type proposed by Byrd and Omojokun [67] and argue that it is both economical and robust. I will then describe a software implementation of this method [55], and show that the linear algebra techniques used in the subproblems are of paramount importance. I will then discuss several novel convergence results [33],[31], [3] obtained for a class of trust region methods for equality constrained optimization that include the algorithm of Byrd and Omojokun as a special case. We will pay attention to the choice of merit function and the effects of linearly dependent constraint gradients.

Another important trust region method for solving (8)-(10) is implemented in LANCELOT. We will describe how the algorithm uses an augmented Lagrangian to transform the problem into a sequence of bound constrained problems. The resulting code is very robust, but has some weaknesses – particularly when linear constraints are present – and recent work has been devoted to overcoming them. We will discuss these improvements [29], together with some supporting theory [28], [70], and also present results comparing LANCELOT and the code developed at Northwestern [55] which implements the Byrd-Omojokun method.

We conclude by briefly discussing several primal-dual interior point methods and barrier methods for solving nonlinear programming problems [85], [10], [64], [25], [44]. We note, in particular, that important advances have been made in the solution of nonlinear optimization problems with simple bounds; some of the new algorithms are of the interior-point type [20], [21] [22], and others are based on the gradient projection method [59], [60], [52], [23], [24], [77].

The CUTE testing system will allow comparisons between the well established codes for large-scale nonlinear optimization [63], [76], [32], and the more recently developed codes. These numerical studies should clarify some open question and will undoubtedly stimulate new research.

4 Nonlinear Systems of Equations

Finally we consider the problem of finding a root x_* of a nonlinear system of equations

$$F(x) = 0, \qquad F : R^n \to R^n.$$

An interesting recent development is the introduction of tensor methods [73], [9], [72], and the analysis of some of their properties on problems where the Jacobian J of F is singular at the solution. Tensor methods are based on a quadratic model of F,

$$M(x_k + d) = F(x_k) + J(x_k)d + \frac{1}{2}T_k dd,$$

where T_k is an $n \times n \times n$ matrix (or tensor) chosen so that the model interpolates previous function values. The new iterate x_{k+1} is chosen so as to minimize $\|M(x_k + d)\|_2$. To make this approach tractable, T must be very simple, and in [34] it is chosen to have rank one. Numerical tests indicate that this tensor method performs better than Newton's method, particularly on problems where the Jacobian J has a small rank deficiency. In this talk we will outline a practical implementation of the tensor method, and discuss the analysis of Feng and Schnabel [34] which establishes a superlinear rate of convergence for problems where $J(x_*)$ has a null-space of dimension at most one.

We conclude by discussing some recent results on inexact Newton methods [82], [11].

References

[1] M. Al-Baali, "Convergence properties of self-scaling quasi-Newton methods with inexact line searches," Technical Report, Department of Electronics, Informatics and Systems, Calabria University, 1992.

[2] M. Al-Baali, "Analysis of a family of self-scaling quasi-Newton methods," Department of Mathematics, Faculty of Science, University of Damascus, Damascus, Syria, 1993.

[3] N. Alexandrov, "Multilevel algorithms for nonlinear equations and equality constrained optimization," Technical Report TR93-20, Department of Computational and Applied Mathematics, Rice University, Houston, TX, 1993.

[4] M. Arioli, T. F. Chan, I. S. Duff, N. I. M. Gould, and J. K. Reid, "Computing a search direction for large-scale linearly constrained nonlinear optimization calculations," Technical Report TR/PA/93/94, CERFACS Toulouse, France, 1993.

[5] B. M. Averick, R. G. Carter and J. J. Moré, "The MINPACK-2 test problem collection (Preliminary version)," Technical Memorandum No. 150, Mathematics and Computer Science Division, Argonne National Laboratory, 1991.

[6] D. P. Bertsekas, "Projected Newton methods for optimization problems with simple constraints," *SIAM J. Control and Optimization*, **20** (1982), 221-246.

[7] P. T. Boggs and J. W. Tolle, "A strategy for global convergence in a sequential quadratic programming algorithm," *SIAM Journal on Numerical Analysis*, **26**(3) 1989, 600-623.

[8] I. Bongartz, A. R. Conn, N. I. M. Gould, and Ph. L. Toint, "CUTE: Constrained and Unconstrained Testing Environment," Research Report RC 18860, IBM T. J. Watson Research Center, Yorktown Heights, NY, 1993.

[9] Ali Bouaricha and R. B. Schnabel, "A software package for large sparse nonlinear least squares using tensor methods," Technical report (in preparation), CERFACS, Toulouse, France, 1994.

[10] M. G. Breitfeld and D. F. Shanno, "Computational experience with modified log-barrier methods for nonlinear programming," Research Report RRR 17-93, Rutgers Center for Operations Research, New Brunswick, NJ, 1993.

[11] P. N. Brown and Y. Saad, "Convergence theory of nonlinear Newton-Krylov algorithms", to appear in *SIOPT*, 1992.

[12] A. Buckley and A. LeNir, "QN-like variable storage conjugate gradients," *Mathematical Programming*, **27** (1983), 155-175.

[13] R. H. Byrd, H. F. Khalfan, and R. B. Schnabel, "Analysis of a symmetric rank-one trust region method," Technical Report CU-CS-657-93, Department of Computer Science, University of Colorado at Boulder, Boulder, CO, 1993.

[14] R. H. Byrd and X. Lu, "A limited memory SR1 method for optimization," in preparation.

[15] R. H. Byrd, D.C. Liu and J. Nocedal, "On the Behavior of Broyden's Class of Quasi-Newton Methods," *SIAM Journal on Optimization*, **2**(4) (1992), 533-557.

[16] R. H. Byrd, P. Lu, and J. Nocedal, "A limited memory algorithm for bound constrained optimization," Technical Report NAM-07, Department of Electrical Engineering and Computer Science, Northwestern University, Evanston, IL, 1993.

[17] R. H. Byrd, J. Nocedal and R. Schnabel, "Representations of Quasi-Newton Matrices and their use in Limited Memory Methods," *Mathematical Programming*, **63**(4) 1994, 129-156.

[18] R. H. Byrd, R. B. Schnabel, G. A. Schultz, "A trust region algorithm for nonlinearly constrained optimization," *SIAM J. Numer. Anal.*, **24** (1987), 1152-1170.

[19] M. R. Celis, J. E. Dennis, R. A. Tapia, "A trust region strategy for nonlinear equality constrained optimization," in *Numerical Optimization 1984*, P. T. Boggs, R. H. Byrd, R. B. Schnabel (editors), SIAM, Philadelphia, 1985, 71-82.

[20] T. F. Coleman and Y. Li, "On the convergence of reflective Newton methods for large scale nonlinear minimization subject to bounds," Technical Report CTC 92TR110, Cornell Theory Center, Ithaca, NY, 1992.

[21] T. F. Coleman and Y. Li, "A reflective Newton method for minimizing a quadratic function subject to bounds on the variables," Technical Report CTC 92TR111, Cornell Theory Center, Ithaca, NY, 1992.

[22] T. F. Coleman and Y. Li, "An interior trust region approach for nonlinear minimization subject to bounds," Technical Report CTC, Cornell Theory Center, Ithaca, NY, 1993.

[23] A. R. Conn, N. I. M. Gould, and Ph. L. Toint, "Global convergence of a class of trust region algorithms for optimization with simple bounds," _SIAM J. Numer. Anal._, **25** (1988), 433-460.

[24] A. R. Conn, N.I.M. Gould, and Ph. L. Toint, "Testing a class of methods for solving minimization problems with simple bounds on the variables," _Mathematics of Computation_, **50** (1988), 399-430.

[25] A. R. Conn, N. I. M. Gould, and Ph. L. Toint, "A globally convergent Lagrangian barrier algorithm for optimization with general inequality constraints and simple bounds," Research Report RC 18049, IBM. T. J. Watson Research Center, Yorktown Heights, NY, 1992.

[26] A. R. Conn, N. I. M. Gould, and Ph. L. Toint, "LANCELOT: A Fortran package for large-scale nonlinear optimization (Release A)," _Sprinter Series in Computational Mathematics_, **17** (1992), Sprinter Verlag, Heidelberg, Berlin, New York.

[27] A. R. Conn, N. I. M. Gould, and Ph. L. Toint, "Numerical experiments with the LANCELOT package (Release A) for large-scale nonlinear optimization," Research Report RC 18434, IBM T. J. Watson Research Center, Yorktown Heights, NY, 1992.

[28] A. R. Conn, N. I. M. Gould, and Ph. L. Toint, "On the number of inner iterations per outer iteration of a globally convergent algorithm for optimization with general nonlinear equality constraints and simple bounds," _Proceedings of the 14th Biennial Numerical Analysis Conference_, Dundee, 1991, D. F. Griffiths and G. A. Watson (editors), 49-68.

[29] A. R. Conn, N. I. M. Gould, and Ph. L. Toint, "Convergence properties of minimization algorithms for convex constraints using a structured trust region," Research Report RC 18274, IBM T. J. Watson Research Center, Yorktown Heights, NY, 1992.

[30] R. S. Dembo, S. C. Eisenstat, and T. Steihaug, "Inexact Newton methods," _SIAM J. Numer. Anal._, **19** (1982), 400-408.

[31] J. E. Dennis, M. El-Alem and M. C. Maciel, "A global convergence theory for general trust-region-based algorithms for equality constrained optimization," Tech. Report, Department of Computational and Applied Mathematics, Rice University, 1993.

[32] A. Drud, "CONOPT: a large-scale GRG code," Technical report, ARKI Consulting and Developing, Bagsvaerd, Denmark, 1993.

[33] M. El-Alem, "A robust trust-region algorithm with a non-monotonic penalty parameter scheme for constrained optimization," Department of Mathematics, Faculty of Science, Alexandria University, Alexandria, Egypt, 1993.

[34] D. Feng, P. D. Frank, and R. B. Schnabel, "Local convergence analysis of tensor methods for nonlinear equations," _Mathematical Programming_, **62** (1992), 427-459.

[35] R. Fletcher, "Low storage methods for unconstrained optimization," _Computational Solutions of Nonlinear Systems of Equations_, E. L. Allgower and K. Georg, eds., Lectures in Applied Mathematics **26** (1990), AMS Publications, Providence, RI.

[36] R. Fletcher, "An optimal positive definite update for sparse Hessian matrices," Report NA/145, University of Dundee, 1992, to appear in _SIAM J. Optimization_.

[37] R. Fletcher, "Algorithms for unconstrained optimization," _Algorithms for continuous optimization: the state of the art_, L. Dixon, D. F. Shanno, and E. Spedicato, editors, Kluwer Academic Publishers, Dordrecht, Netherlands, 1994.

[38] A. L. Forsgren and W. Murray, "Newton methods for large-scale linear equality-constrained minimization," _SIAM Journal on Matrix Analysis and Applications_, **14** (1993), 560-587.

[39] A. L. Forsgren, P. E. Gill, and W. Murray, "Computing modified Newton directions using a partial Cholesky factorization," Report TRITA/MAT-93-02, Department of Mathematics, Royal Institute of Technology, Sweden, 1993.

[40] F. Facchinei and S. Lucidi, "Quadratically and superlinearly convergent algorithms for the solution of inequality constrained minimization problems," Submitted to _J.O.T.A._, 1990.

[41] J. C. Gilbert and and C. Lemaréchal, "Some numerical experiments with variable storage quasi-Newton algorithms," _Mathematical programming_, **45** (1989), 407-436.

[42] J. C. Gilbert and J. Nocedal, "Global convergence properties of conjugate gradient methods for optimization," _SIAM Journal of Optimization_, **2**(1) 1992.

[43] P. E. Gill, W. Murray, M. A. Saunders, and M. H. Wright, "User's guide for NPSOL (version 4.0): A Fortran package for nonlinear programming," Technical Report SOL9862, Department of Operations Research, Stanford University, Stanford, CA, 1986.

[44] P. E. Gill, W. Murray, M. A. Saunders, and M. H. Wright, "Shifted barrier methods for linear programming," Technical Report SOL88-9, Department of Operations Research, Stanford University, Stanford, CA, 1988.

[45] P. E. Gill, W. Murray, D. B. Ponceléon, and M. A. Saunders, " Preconditioners for indefinite systems arising in optimization," SIAM Journal on Matrix Analysis and Applications, 13 (1992), 292-311.

[46] A. Griewank and Ph. L. Toint, "On the unconstrained optimization of partially separable objective functions," Nonlinear Optimization 1981, M. J. D. Powell (editor), Academic Press, London, 1982, 301-312.

[47] A. Griewank and Ph. L. Toint, "Partitioned variable metric updates for large structured optimization problems," Numer. Math., 39 (1982), 119-137.

[48] N. I. M. Gould, "An algorithm for large-scale quadratic programming," IMA Journal of Numerical Analysis, 11(3) 1991, 299-324.

[49] M. Heinkenschloss, "On the solution of a two ball trust region subproblem," Universityät Trier, FB IV - Mathematik, Trier, Germany, 1993.

[50] Y. F. Hu and C. Storey, "On unconstrained conjugate gradient optimization methods and their interrelationships", Tech. Report, Department of Mathematical Sciences, Loughborough Institute of Technology, 1990.

[51] M. T. Jones and P. E. Plassman, "Graph Theory and Sparse Matrix Computation", The IMA Volumes in Mathematics and its Applications, Sprinter-Verlag, A. George, J. Gilbert, and W.H. Liu (editors), 56 (1993), 229-245.

[52] J. J. Jùdice and F. M. Pires, "Direct methods for convex quadratic programs subject to box constraints," Technical report, Universidade de Coimbra, 300 Coimbra, Portugal, 1989.

[53] C. Kanzow and H. Kleinmichel, "A Class of Newton-Type Methods for Equality and Inequality Constrained Optimization," Institut für Angewandte Mathematik der Universität Hamburg, 1992.

[54] C. T. Kelley, "A new incremental storage implementation of Broyden's method," Center for Research in Scientific Computation and Department of Mathematics, North Carolina State University, Raleigh, NC, 1994.

[55] M. Lalee, J. Nocedal, and T. Plantenga, "On the implementation of an algorithm for large-scale quality constrained optimization," Technical Report NAM 08, Department of EECS, Northwestern University, Evanston, IL, submitted for publication in SIAM Journal of Optimization, 1993.

[56] M. Lescrenier, "Convergence of trust region algorithms for optimization with bounds when strict complementarity does not hold," SIAM Journal on Numerical Analysis, 28(2), 476-495.

[57] D.C. Liu and J. Nocedal, "On the limited memory BFGS method for large scale optimization," Mathematical Programming, 45 (1989), 503-528.

[58] L. Lukšăn, "Computational experience with improved variable metric methods for unconstrained minimization," Kybernetika, 26 (1990), 415-431.

[59] J. J. Moré and G. Toraldo, "Algorithms for bound constrained quadratic programming problems," Numer. Math., 55 (1989), 377-400.

[60] J. J. Moré and G. Toraldo, "On the solution of large quadratic programming problems with bound constraints," SIAM Journal on Optimization, 1(1) 1991, 93-113.

[61] J. J. Moré and S. J. Wright, Optimization Software Guide, SIAM, Philadelphia, PA, 1993.

[62] W. Murray, Private communication, 1994.

[63] B. A. Murtagh and M. A. Saunders, "MINOS 5.1 User's Guide," Technical Report SOL83-20R, Department of Operations Research, Stanford University, Stanford, CA, 1987.

[64] S. G. Nash, R. Polyak, and A. Sofer, "A numerical comparison of barrier and modified barrier methods for large-scale constrained optimization," Technical Report 93-02, Department of Operations Research, George Mason University, Fairfax, VA, 1993.

[65] J. Nocedal, "Updating quasi-Newton matrices with limited storage," Mathematics of Computation, 35 (1980), 773-782.

[66] J. Nocedal and Y. Yuan, "Analysis of a self-scaling quasi-Newton method," Department of EECS, Northwestern University, Evanston, IL, 1991.

[67] E. O. Omojokun, "Trust Region Algorithms for Optimization with Nonlinear Equality and Inequality Constraints," Graduate School Thesis, Department of Computer Science, University of Colorado, 1989.

[68] M. J. D. Powell, Y. Yuan, "A trust region algorithm for equality constrained optimization," *Math. Prog. (Series A)*, **49** (1991), 189-211.

[69] M. B. Reed, "Reliable algorithms for a reduced storage quasi-Newton minimization," Department of Mathematics and Statistics, Brunel University, Uxbridge, UK, 1993.

[70] A. Sartenaer, "A class of trust region methods for nonlinear network optimization problems, including numerical experiments," Technical Report 93/21, Department of Mathematics, FUNDP, Namur, Belgium, to appear in *SIAM Journal on Optimization*, 1993.

[71] T. Schlick, "Modified Cholesky factorizations for sparse preconditioners," *SIAM Journal on Scientific and Statistical Computing*, **14**(2) 1993, 424-445.

[72] R. B. Schnabel and T. T. Chow, "Tensor methods for unconstrained optimization using second derivatives," *SIAM Journal on Optimization*, **1**(3) 1991, 293-315.

[73] R. B. Schnabel and P. D. Frank, "Tensor methods for nonlinear equations," *SIAM Journal on Numerical Analysis*, **21**(5) 1984, 815-843.

[74] D. Siegel, "Implementing and modifying Broyden class updates for large scale optimization," Report DAMTP 1992/NA12, Department of Applied Mathematics and Theoretical Physics, University of Cambridge, Cambridge, England, 1992.

[75] D. Siegel, "Modifying the BFGS update by a new column scaling technique," Department of Applied Mathematics and Theoretical Physics, University of Cambridge, Cambridge, England, 1993.

[76] S. Smith and L. Lasdon, "Solving large sparse nonlinear programs using GRG," *ORSA Journal on Computing*, **4** (1992), 2-15.

[77] J. Soares, J. J. Júdice, and F. Facchinei, "An active set Newton's algorithm for large-scale nonlinear programs with box constraints," Technical Report, Universidade de Coimbra, 300 Coimbra, Portugal, 1993.

[78] T. Steihaug, "The conjugate gradient method and trust regions in large scale optimization," *SIAM J. Num. Anal.*, **20** (1983), 626-637.

[79] T. Steihaug, "Local and superlinear convergence for truncated iterated projection methods," *Mathematical Programming*, **27** (1983), 199-223.

[80] Ph. L. Toint, "Towards an efficient sparsity exploiting Newton method for minimization," *Sparse Matrices and their Uses*, I. S. Duff (editor), Academic Press, New York, 1981, 57-87.

[81] A. Vardi, "A trust region algorithm for equality constrained minimization: convergence properties and implementation," *SIAM J. Numer. Anal.*, **22** (1985), 575-591.

[82] S. C. Eisenstat and H. F. Walker, "Globally convergent inexact Newton methods", Tech. Report 91/51, Utah State University, 1991.

[83] K. A. Williamson, "A robust trust region algorithm for nonlinear programming," Technical Report TR90-22, Department of Math. Sciences, Rice University, 1991.

[84] S. Wright, "An efficient representation of quasi-Newton matrices in the Broyden class", working paper, 1994.

[85] H. Yamashita, "A globally convergent primal-dual interior-point method for constrained optimization," Tech. Report, Mathematical Systems Institute Inc., Shinjuku, Tokyo, 1994.

[86] Y. Yuan, "On a subproblem of trust region algorithms for constrained optimization," *Math. Prog. (Series A)*, **47** (1990), 53-63.

[87] Y. Zhang, "Computing a Celis-Dennis-Tapia trust region step for equality constrained optimization," *Math. Prog. (Series A)*, **55** (1992), 109-124.

[88] C. Zhu, R. H. Byrd, P. Lu and J. Nocedal, "LBFGS-B: FORTRAN subroutines for nonlinear minimization subject to simple bounds," Tech. Report, EECS Department, Northwestern University, 1994.

On the Passage from Local to Global in Optimization

Panos M. Pardalos

Center for Applied Optimization and
Department of Industrial
and Systems Engineering
University of Florida
Gainesville, Forida

1 Introduction

Most of the existing methods in optimization focus on the very important problem of computing feasible points that satisfy necessary optimality conditions. Under certain convexity assumptions these points are also local optima. Local techniques are often the stepping-stones to the more difficult global optimization problems.

During the past decade several works have appeared on the subject of global optimization (see references at the end of this paper). Furthermore, the distinction of "local" versus "global" (or "optimal") with its various connotations has found a home on almost every branch of mathematical sciences.

Given a real valued function $f(x)$ defined on a set D in R^n, the *global optimization problem* to be considered is to find the function value f^* and a point $x^* \in D$ such that

$$f^* = f(x^*) = \text{ global min}_{x \in D} f(x) \quad (\text{or max}_{x \in D} f(x))$$

if such a point x^* exists. Under convexity assumptions on the feasible domain and the objective function, every local minimum is global.

If the objective function and/or the feasible domain are nonconvex, then we may have many local minima which are not global. Local techniques most often will not be able to compute the global optimum. Therefore, the task of computing the optimal solution to a general optimization problem is very difficult.

First attempts to compute global solutions can be found in combinatorial optimization algorithms such as enumeration and branch and bound methods. In fact, a number of combinatorial optimization problems such as integer programming and related graph problems (e.g. maximum clique problem, graph coloring), the quadratic assignment problem, and quadratic zero-one programming can be formulated and solved as global optimization problems [57].

Global optimization problems are widespread in the mathematical modeling of real world systems for a very broad range of applications. Such applications include structural optimization, engineering design, networks, transportation, chip design and database problems, nuclear and mechanical design, chemical engineering design and control, economies of scale, fixed charges, allocation and location problems, molecular biology and biochemistry.

Active research during the past decades has produced a variety of deterministic, stochastic methods and heuristics for determining global solutions to nonconvex nonlinear optimization problems. There are a dozen recent books that present in detail the developments of global optimization in all these directions. It is impossible to give here a full account and systematic treatment of the subject in its entirety. It is our modest intention to introduce the reader to a number of topics in order to show the spectrum of recent research activities and the richness of ideas in the development of algorithms and the applications of global optimization. While we were able to provide only a glimpse of this expansive field, we felt that the interested reader will be able to sense the breadth and the depth of global optimization. Recent results can be found in the "Journal of Global Optimization" (published since 1991), which deals with all theoretical, computational aspects and applications of global optimization.

2 Is Convexity Easy to Check?

In most classical optimization algorithms, the underlying theory is based on the assumption that the objective function is convex over some domain. Unless the function has constant Hessian (i.e. is quadratic) or has a very special structure, convexity is not easily recognizable. Even for multivariable polynomials there is no known computable procedure to decide convexity. Therefore, from the practical point of view, a general objective function can be assumed to be neither convex nor concave, having multiple local optima.

3 Complexity Issues

The main focus of computational complexity is to analyze the intrinsic difficulty of optimization problems and to decide which of these problems are likely to be tractable. The pursuit for developing efficient algorithms also leads to elegant general approaches for solving optimization problems, and reveals surprising connections among problems and their solutions.

Most classical optimization algorithms compute points that satisfy the Kuhn-Tucker (KT)–optimality conditions. When the feasible domain D is not bounded it is not easy to check existence of a KT–point. More precisely, consider the following quadratic problem

$$\min \quad f(x) = c^T x + \tfrac{1}{2} x^T Q x \qquad (1)$$
$$\text{s.t.} \qquad x \geq 0,$$

where Q is an $n \times n$ symmetric matrix, and $c \in R^n$. The Kuhn-Tucker optimality conditions for this problem become the following so–called linear complementarity problem (denoted by LCP(Q, c)): Find $x \in R^n$ (or prove that no such an x exists) such that

$$Qx + c \geq 0, \ x \geq 0 \tag{2}$$
$$x^T(Qx + c) = 0. \tag{3}$$

Hence, the complexity of finding (or proving existence) of Kuhn-Tucker points for the above simple quadratic problem is reduced to the complexity of solving the corresponding LCP. It has been shown in [37] that when Q is symmetric, the problem LCP (Q, c) is NP-hard. Therefore, in quadratic programming, the problem of "deciding whether a Kuhn-Tucker point exists" is NP-hard. When x is bounded, it is not known if a KT-point can be computed in polynomial time or not.

Computing locally optimal solutions is presumably easier than finding globally optimal solutions. However, from the complexity point of view it has been shown that the problem of checking local optimality for a feasible point and the problem of checking whether a local minimum is strict, are NP-hard, even for instances of quadratic problems with a simple structure in the constraints and the objective [59].

Most nonlinear optimization problems are NP-hard, therefore finding a global solution in polynomial time is unlikely. For this reason, there is a great interest in the development of efficient approximation algorithms, i.e. polynomial time algorithms that find an approximate solution within some tolerance from the global one. While approximation algorithms for some nonlinear problems exist, in most cases the complexity of approximation has not been established. However, it has been shown that there is no polynomial time approximation algorithm for the problem of finding the global optimum of a multivariable polynomial over a convex set, unless P=NP. Regarding complexity results in numerical optimization see the recent book [54].

4 Optimality Conditions

A major difficulty with nonconvex optimization problems is the lack of necessary and sufficient conditions for a feasible point to be a global minimum of the objective function. Motivated from the fact that in convex programming the Kuhn-Tucker conditions are necessary and sufficient for a point to be local (global) minimum, there are several efforts to convexify general problems.

Given $f : S \rightarrow R$, a lower semi-continuous function, where S is a nonempty convex set in R^n, the convex envelope of $f(x)$ taken over S is defined to be a function $C_f(x)$ such that

a) $C_f(x)$ is convex on S,

b) $C_f(x) \leq f(x)$ for all $x \in S$,

c) If $h(x)$ is any convex function defined on S such that $h(x) \leq f(x)$ for all $x \in S$, then $h(x) \leq C_f(x)$ for all $x \in S$.

Geometrically, $C_f(x)$ is the "best" convex underestimating function of $f(x)$ and is precisely the function whose epigraph coincides with the convex hull of the epigraph of f. The theoretical importance of the convex envelope is established by the following theorem.

Theorem 4.1 *Consider the problem*

$$global \min_{x \in S} f(x)$$

where S is a convex compact set in R^n. If $C_f(x)$ is the convex envelope of f(x) over S, then

$$f^* := \min\{f(x) : x \in S\} = \min\{C_f(x) : x \in S\}$$

and

$$\{y \in S : f(y) = f^*\} \subseteq \{y \in S : C_f(y) = f^*\}.$$

Therefore, f^* is the global minimum for the original problem iff $C_f^* = f^*(x^*)$, where C_f^* is the minimum of the convex envelope.

Several optimality conditions are based on the above result [35]. For example, if $f : R^n \to R$ is a differentiable function and $C_f(x)$ is its convex envelope, then x^* is a global minimum of $f(x)$ iff

$$\nabla f(x^*) = 0, \text{ and } C_f(x^*) = f(x^*).$$

Several generalizations to nondifferentiable problems can be found in [36].

Given an arbitrary function $\phi : R^n \to R \cup \{\infty\}$, a point x^* at which ϕ is finite, and an $\epsilon \geq 0$, the ϵ-subdifferential $\partial_\epsilon \phi(x^*)$ of ϕ at x^* is the following set

$$S = \{s : \phi(x) \geq \phi(x^*) + s^T(x - x^*), \text{ for all } x \in R^n\}.$$

Theorem 4.2 *Let $h : R^n \to R$ be a convex function and $f : R^n \to R \cup \{\infty\}$ be a lower semi-continuous convex function. The point x^* is a global minimum of the function $f = g - h$ iff*

$$\partial_\epsilon h(x^*) \subseteq \partial_\epsilon g(x^*), \text{ for all } \epsilon > 0.$$

For quadratic programming problems more practical optimality criteria can be derived by means of ϵ-subdifferential calculus or other techniques [7, 8, 34]. These optimality criteria can be used as stopping rules in branch and bound procedures.

It seems that the mathematical tools used in classical optimization analysis, are not adequate to devise global optimality conditions for general problems. However, this is not surprising since from the complexity theory point of view global optimization problems are intractable. Furthermore, at present, there are no known local criteria in deciding how good a local optimal solution is, in relation to the global optimum [49].

5 Nonconvex Quadratic Programming

After linear programming, quadratic programming is probably the most fundamental and most important optimization problem. Algorithms for the solution of nonconvex quadratic problems, are often basic subroutines to general nonlinear programming algorithms. Numerous problems in engineering, operations research, and computer science can be formulated as large-scale nonconvex quadratic problems of the form:

$$\text{min} \qquad \Psi(z,y) \qquad\qquad\qquad (4)$$

$$\text{s.t.} \qquad (z,y) \in \Omega = \{A_1 z + A_2 y = b,\ z \geq 0,\ y \geq 0\} \subseteq R^{n+k}, \qquad (5)$$

where $\Psi(z,y) = \phi(z) + d^T y$, and $\phi(z)$ is a nonconvex quadratic function, given by

$$\phi(z) = c^T z + \frac{1}{2} z^T Q z$$

with Q being a real indefinite symmetric $n \times n$ matrix. We assume that $A_1 \in IR^{m \times n}, A_2 \in IR^{m \times k}, b \in IR^m, d \in IR^k, c \in IR^n$, and that the set Ω is a nonempty polytope. In many practical applications $k \gg n$, and this reflects the large-scale characteristics of the problems.

Since Q is a real symmetric matrix, Q has n real eigenvalues, $\lambda_1, \ldots, \lambda_n$ with corresponding orthonormal eigenvectors $u_1, \ldots, u_n$. By the spectral decomposition theorem, $Q = UDU^T$ where U is the matrix with columns $u_1, \ldots, u_n$ and $D = \text{diag}(\lambda_1, \ldots, \lambda_n)$ (with $U^T U = I$). Using a simple affine transformation $z = Ux + s$ where s is an appropriate constant, we can transform the above problem into an equivalent one in separable (SP) form:

$$\text{min}\{\Psi(x,y) = \sum_{i=1}^{n} q_i(x_i) + d^T y : (x,y) \in \Omega,\ x \in R_x\}$$

where

$$
\begin{aligned}
q_i(x_i) &= c_i x_i + \tfrac{1}{2}\lambda_i x_i^2 \quad (i=1,\ldots,n), \\
\Omega &= \{(x,y) : A_1 x + A_2 y = b,\ y \geq 0\}, \\
R_x &= \{x : 0 \leq x_i \leq \beta_i \quad (i=1,\ldots,n)\},
\end{aligned}
$$

and $\lambda_i,\ (i=1,\ldots,n)$ are the eigenvalues of the original matrix Q. Therefore, the above affine transformation preserves the eigenstructure of the problem.

5.1 Piecewise Linear Approximations

Next, we describe an approach that uses piecewise linear approximations to solve the corresponding approximate problem. For each interval $[0,\beta_i]$, $i = 1,\ldots,n$ choose a fixed grid of points by partitioning it into k_i subintervals of equal length $h_i = \beta_i/k_i$. Let $p_i(x_i)$ be the piecewise linear function that interpolates $q_i(x_i)$ at the grid points $x_i = jh_i$, $j = 0,1,\ldots,k_i$. The following separable piecewise linear programming problem,

$$\min p(x) + d^T y = \sum_{i=1}^{n} p_i(x_i) + d^T y$$

$$\text{s.t.} \quad A_1 x + A_2 y = b, \ y \geq 0, \ 0 \leq x_i \leq \beta_i, \ i = 1, ..., n,$$

will provide an approximate solution to the original problem. The approximation error depends on the mesh size h_i and the curvatures of $q_i(x)$, $i = 1, \ldots, n$.

Piecewise linear interpolation preserves monotonicity and convexity. When the objective function is convex, the corresponding piecewise linear problem can be formulated as an equivalent linear program. However, in the absence of convexity, it can be formulated as an equivalent linear zero-one mixed integer problem. For simplicity, assume each $q_i(x_i)$ is a concave function. We will consider the following "method of bounded variables", first introduced by Pardalos and Rosen [57]. Let

$$x_i = h_i \sum_{j=1}^{k_i} \omega_{ij}, \ i = 1, ..., n,$$

$$\Delta q_{ij} = q_i(jh_i) - q_i((j-1)h_i), \ j = 1, ..., k_i, \ i = 1, ..., n.$$

The real variables ω_{ij} are bounded, $0 \leq \omega_{ij} \leq 1$, and are restricted to satisfy the condition: Every vector $\overline{\omega}_i = (\omega_{i1}, ..., \omega_{ik_i})$ can get the value $\overline{\omega}_i = (1, ..., 1, 0, ..., 0)$. In that way, we have a unique vector $\overline{\omega}_i$ representing $x_i \in [0, \beta_i]$. The equivalent linear mixed zero-one mixed integer program is given by:

$$\min \sum_{i=1}^{n} \sum_{j=1}^{k_i} \Delta q_{ij} \omega_{ij} + d^T y$$

$$\text{s.t.} \quad \sum_{i=1}^{n} h_i a_i \sum_{j=1}^{k_i} \omega_{ij} + A_2 y = b,$$

$$\omega_{i,j+1} \leq z_{ij} \leq \omega_{ij}, \ j = 1, ..., k_i - 1, \ i = 1, ..., n,$$

$$0 \leq \omega_{ij} \leq 1, y \geq 0, z_{ij} \in \{0, 1\},$$

where a_i is the ith column of A_1. It follows that the restriction on the integer and continuous variables,

$$\omega_{i,j+1} \leq z_{ij} \leq \omega_{ij},$$

implies that the vector $\overline{z}_i = (z_{i1}, ..., z_{ik_{i-1}})$ must satisfy

$$\overline{z}_i = (1, ..., 1, 0, ..., 0).$$

Because of these restrictions the number of all possible allowed combinations of the integer variables is given by

$$I_\omega = \prod_{i=1}^n k_i \ll 2^{N_\omega},$$

where $N_\omega = \sum_{i=1}^n k_i - n$, the total number of the zero-one integer variables introduced.

In the above analysis, we considered a fixed grid of equally spaced points to obtain the piecewise linear approximation. When each f_i has a constant second derivative (e.g. quadratic separable function), this approach minimizes the error in the approximation. When the second derivatives are variable, we may not consider equally spaced grid in $[0, \beta_i]$. Suppose

$$0 = a_{i1} < a_{i2} < ... < a_{i,k_{i+1}} = \beta_i.$$

Define $h_{il_i} = a_{il_{i+1}} - a_{il_i}$ to be the mesh size and let $h_i = \max_{l_i} h_{il_i}$. To ensure that $1/k_i$ and h_i are equivalent asymptotic scales, we may assume that $h_i \leq \min_{l_i} h_{il_i} c$ where c is a constant and $c \geq 1$. This implies that

$$\beta_i/k_i \leq h_i \leq c\beta_i/k_i.$$

As before, the piecewise linear program can be transformed to a mixed zero-one programming problem of special structure that can be solved with many existing mixed zero-one integer packages.

Development of global optimization codes for general problems is a very difficult task. Fortunately, many problems in applications have separable objective and constraints functions. For such problems we can obtain practically acceptable solutions by approximating the functions involved with piecewise linear functions and solving the corresponding problem. Many years ago, SCICONIC developed a package for these type of problems and this package is still in use today. The above alternative linearization scheme is much simpler to implement and obtain an a-priori error analysis. Computational results with this approach have been very encouraging [57, 56] for large classes of problems.

5.2 Remarks on Approximation

General global optimization techniques can be applied to solve quadratic programming problems. Such techniques include generalized Benders decomposition approaches, branch and bound algorithms, cutting plane methods, linear complementarity approaches, and partial enumeration [36].

Next we discuss some recent results on the problem of computing an approximate solution to quadratic programming problems in polynomial time. Vavasis (see paper in [54]) has shown that a solution satisfying weak bounds with respect to optimality can be computed in polynomial time.

Given a tolerance $\epsilon > 0$, if $f(x)$ denotes the objective function $1/2x^TQx + c^Tx$, we say that x° is an ϵ-approximate solution over the feasible domain D in R^n, if there exists another feasible point $x^{\#}$ such that $f(x^{\circ}) - f(x^*) \leq \epsilon[f(x^{\#}) - f(x^*)]$, where x^* is a global minimum of $f(x)$ over D.

Observe that any feasible point is a 1-approximation by this definition, and only a global solution is a 0-approximation. Thus, the definition makes sense only for ϵ in the interval $(0, 1)$. This definition has some useful properties. First, it is insensitive to translations or dilations of the objective function. In other words, if the objective function $f(x)$ is replaced by a new function $g(x) = af(x) + b$ where $a > 0$, a vector x° that was previously an ϵ-approximation will continue to have that property. A second useful property is that ϵ-approximation is preserved under affine transformations of the feasible region.

In earlier work by Vavasis [78], it was shown that if the feasible domain is compact and the matrix Q has k negative eigenvalues, then there is an algorithm that finds an ϵ-approximate solution to the quadratic problem in

$$O\left(\left\lceil \frac{n(n+1)}{\sqrt{\epsilon}} \right\rceil^k q\right)$$

steps. In this formula, q denotes the time to solve a convex quadratic programming problem of the same size n.

Note that as k gets close to n this running bound is not polynomial in the input size. Is it possible to construct a polynomial-time approximation algorithm for indefinite QP? Recent results suggest that such an algorithm, if it exists, could only satisfy weak approximation bounds. Specifically, Bellare and Rogaway [54] have shown the following result is true for some constant $\delta > 0$: Suppose there were an algorithm to approximate quadratic programming with $\epsilon = \left(2^{(\ln n)^{\delta}} - 1\right) / \left(2^{(\ln n)^{\delta}} + 1\right)$. Then any problem in NP can be solved in quasi-polynomial time, that is, time $O(n^{(\ln n)^k})$.

Since the concluding statement of the above result is thought to be unlikely, the supposition is probably false. In other words, we cannot hope to approximate indefinite quadratic programming in polynomial time unless we are willing to accept an approximation factor that tends to 1 asymptotically as the problem size increases.

Vavasis proposed a polynomial-time approximation algorithm for indefinite quadratic problems that satisfies a weak approximation bound of this sort. Assume the feasible domain D is compact, he proved that an approximate solution can be computed in polynomial time, where the approximation factor is $1 - \Theta(n^{-2})$. For more detailed discussion on these type of results see [54] and [36].

5.3 Combinatorial Optimization Problems

It has been observed that a variety of combinatorial optimization problems can be formulated and solved as nonconvex quadratic minimization problems. Studying discrete problems by formulating them as continuous quadratic optimization problems has motivated new efficient algorithmic approaches.

5.3.1 Nonlinear Integer Problems

Given a finite discrete set D and a function $f(x)$ defined on D, the general discrete (or combinatorial) optimization problem has the form:

$$\text{global min}_{x\in F\subseteq D}\ f(x),$$

where F is a set of additional constraints (alternatives). Problems of this form arise in many practical applications and are very difficult to solve. From the complexity point of view, most of these problems are NP-hard.

The unconstrained quadratic zero-one programming problem (QP) has the form:

$$\text{global min} f(x)\ =\ c^T x + x^T A x$$

subject to

$$x \in \{0,\ 1\}^n, \tag{6}$$

where A is an $n \times n$ rational matrix and c is a rational vector of length n. It can be shown that any nonlinear zero-one program can be reduced to a quadratic program [56] and that quadratic zero-one programming can be formulated as a continuous global concave minimization problem [57]. On the other hand, quadratic concave minimization problems with box constraints are equivalent to quadratic zero-one programming. Algorithms developed for zero-one programming can also be used to solve large clique problems [62]. Let $G = (V, E)$ be a graph with n vertices, where V denotes the set of vertices and E denotes the set of edges. Let $A_{\overline{G}}$ be the adjacency matrix of $\overline{G}$, and let I be the $n \times n$ identity matrix. Then, the maximum clique problem for the graph G is equivalent to solving the following quadratic zero-one program:

$$\min f(x) = x^T A x \tag{7}$$

$$\text{s.t. } x \in \{0,1\}^n,$$

where

$$A = A_{\overline{G}} - I.$$

A solution x^* to (7) defines a maximum clique C for G as follows. If $x_i^* = 1$ then $v_i \in C$, and if $x_i^* = 0$ then $v_i \notin C$ with $|C| = -z = -f(x^*)$. Note that in this formulation there is a one-to-one correspondence of discrete local minima of the quadratic zero-one problem and the maximal cliques of the graph G.

There are many alternative formulations of the maximum clique problem as an equivalent nonconvex quadratic optimization problem [37].

Consider the indefinite quadratic problem (Motzkin and Straus 1965):

$$\text{global max} f(x) = \frac{1}{2}x^T A_G x, \tag{8}$$

$$\text{s.t. } \sum_{i=1}^{n} x_i = 1,\ x_i \geq 0,\ i = 1,\ldots,n.$$

Let x^* and $\alpha = f(x^*)$ be the optimal solution and the corresponding objective value of problem (8). Then G has a maximum clique C of size $k = 1/(1 - 2\alpha)$. The global maximum of (8) can be attained by setting $x_i^* = \frac{1}{k}$ if $v_i \in C$, and $x_i^* = 0$ otherwise.

Using this formulation some interesting results can be proved. For example, if A_G has r negative eigenvalues, then at least $n - r$ constraints are active at any global maximum x^* of $f(x)$. Therefore, if A_G has r negative eigenvalues, then the size $|C|$ of the maximum clique is bounded by $|C| \leq r + 1$. Results of this type indicate the power of continuous global optimization formulations of discrete problems.

5.3.2 Quadratic Assignment Problems

Given a set $\mathcal{N} = \{1, 2, \ldots, n\}$ and $n \times n$ matrices $F = (f_{ij})$ and $D = (d_{kl})$, the quadratic assignment problem (QAP) can be stated as follows:

$$\min_{p \in \Pi_\mathcal{N}} \sum_{i=1}^{n} \sum_{j=1}^{n} f_{ij} d_{p(i)p(j)},$$

where $\Pi_\mathcal{N}$ is the set of all permutations of $\mathcal{N}$. One of the major applications of the QAP is in location theory where the matrix $F = (f_{ij})$ is the flow matrix, i.e. f_{ij} is the flow of materials from facility i to facility j, and $D = (d_{kl})$ is the distance matrix, i.e. d_{kl} represents the distance from location k to location l. The cost of simultaneously assigning facility i to location k and facility j to location l is $f_{ij}d_{kl}$. The objective is to find an assignment of all facilities to all locations (i.e. a permutation $p \in \Pi_\mathcal{N}$), such that the total cost of the assignment is minimized. Many difficult problems such as the traveling salesman problem, the maximum clique problem and the graph isomorphism problem are special cases of QAP.

The QAP can be formulated as an equivalent concave minimization problem. First, consider the following quadratic 0-1 formulation:

$$\min \quad \sum_{i=1}^{n} \sum_{j=1}^{n} \sum_{k=1}^{n} \sum_{l=1}^{n} f_{ij} d_{kl} x_{ik} x_{jl} \tag{9}$$

$$\text{s.t.} \quad \sum_{i=1}^{n} x_{ij} = 1, \ j = 1, \ldots, n, \ \sum_{j=1}^{n} x_{ij} = 1, \ i = 1, \ldots, n, \tag{10}$$

$$x_{ij} \in \{0, 1\}, \ i, j = 1, \ldots, n.$$

If we denote the feasible domain of the above problem by D, then the problem can be written as

$$\min \quad x^T S x$$
$$\text{s.t.} \quad x \in D, \tag{11}$$

where the $(n^2 \times n^2)$–matrix S has nonnegative entries. The entries of S are the products $f_{ij}d_{kl}$, and it is natural to define a row of S by i and j fixed, and a column of S by k and l fixed (or vice–versa).

The above quadratic 0-1 problem can be transformed into an equivalent quadratic concave minimization problem. Let $m = n^2$ and consider the row norm of the $(m \times m)$–matrix S which is defined by

$$||S||_\infty = \max\{\sum_{j=1}^{m} | s_{1j} |, \ldots, \sum_{j=1}^{m} | s_{mj} |\}.$$

Let $Q = S - \alpha I$, where I is the $(m \times m)$ unit matrix, and $\alpha > ||S||_\infty$. Moreover, let $x = (x_{11}, x_{12}, \ldots, x_{nn})^T = (x_1, x_2, \ldots, x_m)^T$, and consider the quadratic form $x^T Q x$. Assume, without loss of generality, that Q is symmetric. Q is symmetric when S is symmetric. If Q is not symmetric, then replace Q by $Q' = \frac{1}{2}(Q + Q^T)$, which is symmetric and satisfies $x^T Q' x = x^T Q x$ since $x^T Q^T x = (x^T Q^T x)^T = x^T Q x$. It is easy to check that $x^T Q x < 0$ for any $x \neq 0$; hence the matrix Q is negative definite.

Switching back to the original notation of the variables we now can formulate the following quadratic concave programming problem which is equivalent to the QAP.

$$\min \quad x^T Q x$$

$$\text{s.t.} \quad x \in \Omega,$$

where $Q = S - \alpha I, \alpha > ||S||_\infty$, and Ω is the set of all $x = (x_{11}, x_{12}, \ldots, x_{nn})^T \in R^{n^2}$ satisfying

$$\sum_{j=1}^{n} x_{ij} = 1, \ i = 1, \ldots, n,$$

$$\sum_{i=1}^{n} x_{ij} = 1, \ j = 1, \ldots, n,$$

$$x_{ij} \geq 0, \qquad i, j = 1, \ldots, n.$$

The recent book [64] contains many results which focus on computational approaches and applications of QAP and related problems. The new methods discussed include eigenvalue estimates and reduction techniques for lower bounds, parallelization, genetic algorithms, polyhedral approaches, greedy, and adaptive search algorithms. The applications include graph bandwidth problems, telecommunications network design, load balancing, VLSI design, data association problems, and multidimensional assignment problems.

6 Semidefinite Programming Problems

Let $\succeq$ denote the *Löwner* partial order, that is, for real symmetric matrices A and B, $A \succeq B$ (respectively $A \succ B$), whenever $A - B$ is positive semidefinite (respectively positive definite.)

The standard form of the semidefinite programming problem (SDP) is the following:

$$\min\{C \bullet X : A_i \bullet X = b_i \ i = 1, \cdots, m \ and \ X \succeq 0\} \tag{12}$$

where C, A_i's and X are $n \times n$ matrices, and X is symmetric; the "$\bullet$" operation is the inner product of matrices: $A \bullet B := \sum_{i,j} A_{ij} B_{ij} = \text{trace} A^T B$.

Semidefinite programs arise in a wide variety of applications including control theory, combinatorial optimization (e.g. maximum clique and graph partitioning problems) and structural computational complexity. One of the simplest cases of semidefinite programming is the evaluation of eigenvalues of a symmetric matrix. In fact, one can reformulate the classical theorems of Rayleigh-Ritz for the largest eigenvalue, and of Fan for the sum of the first few eigenvalues of a symmetric matrix, as semidefinite programs [1, 2].

The semidefinite programming problem has many interesting connections with spectral properties of graphs. Given a graph $G(V, E)$, let $\omega(G)$ be the clique number and $\chi(G)$ be the chromatic number of the graph. Lovász [43] discover a graph invariant, $\theta(G)$, which is polynomial time computable and simultaneously an upper bound for $\omega(G)$ and a lower bound for $\chi(G)$. This invariant $\theta(G)$ (which is a also called the Lovász number of G) can be defined by a pair of primal and dual semidefinite programs [1], [43].

7 Multiquadratic Programming

The Multiquadratic Programming Problem (MQP) is defined to be the problem of globally minimizing a quadratic function subject to quadratic equality and inequality constraints. The MQP offers a powerful unification of several mathematical optimization problems. For instance, it includes as special cases, the conventional quadratic programs and the binary integer programs, and allows compact formulations of problems such as the job-shop scheduling problem. Also, the more general problem of polynomial programming can be reduced to MQP.

Next we briefly discuss two approaches for solving the MQP based on [66]: a cutting plane approach and a relaxation approach, both of which can be introduced in a common framework to be presented below. First, some notation: Let $\mathcal{H}_n$ denote the space of $n \times n$ real symmetric matrices, and $\mathcal{U}_n$ denote the set of matrices of the form xx^T, called *Unary* matrices. The cone of positive semidefinite matrices is denoted by $\mathcal{P}_n$, and it is well known that $\mathcal{P}_n$ is the convex (or conical) hull of $\mathcal{U}_n$. Also, $\lambda_1(U) \leq \lambda_2(U) \leq \ldots \leq \lambda_n(U)$ denote the eigenvalues of a matrix $U \in \mathcal{H}_n$.

Now consider the MQP:

$$\min\{z^T Q_0 z + c_0^T z \,|\, z^T Q_i z + c_i^T z \leq b_i, \forall i = 1, .., m, z \in R^l\}.$$

By introducing a new matrix variable $U \in \mathcal{H}_l$, it is easy to show that the above MQP is equivalent to:

$$
\begin{aligned}
\max: \quad & U \bullet Q_0 + c_0^T z \\
\text{s.t.} \quad & U \bullet Q_i + c_i^T z \leq b_i \forall i = 1, .., m \\
& \begin{pmatrix} U & z \\ z^T & 1 \end{pmatrix} \in \mathcal{U}_{l+1}.
\end{aligned}
$$

With the above in mind, let us now define the Unary Programming Problem (UPP) as the following problem:

$$\begin{array}{rll}
\max & : & c^T x \\
\text{s.t.} & & Ax \le b \\
& & U(x) \in \mathcal{U}_n,
\end{array}$$

where $U : R^d \to \mathcal{H}_n$ is a given symmetric affine matrix map, A is an $m \times d$ matrix, and b, c are $m \times 1, d \times 1$ vectors respectively.

7.1   Cutting Planes for UPP

Suppose that we solve the LP Relaxation of the above UPP, i.e. the LP resulting from ignoring the constraint $U(x) \in \mathcal{U}_n$, and obtain an optimal vertex solution. By applying an appropriate linear transformation to the domain space, we may assume that the origin is this optimal solution, and the inequalities $x \ge 0$ form a subsystem of $Ax \le b$. If $U(0) \in \mathcal{U}_n$, then 0 is an optimal solution to UPP. Suppose this is not the case, which implies that at least one of the first $n - 1$ eigenvalues of $U(0)$ is not zero, say $\lambda_k(U(0)) < 0$ for some $1 \le k \le n - 1$. Now define the set

$$R = \{x | x \ge 0, \lambda_k(U(x)) \ge 0\},$$

and observe that R contains the feasible region of UPP, and that the origin is not in it. Recall the well known Weyl inequality for a pair of matrices $E, F \in \mathcal{H}_n$:

$$\lambda_j(E + F) \le \lambda_i(E) + \lambda_n(F) \; \forall \, j = 1, .., n.$$

By repeatedly applying this inequality, we deduce that the linear inequality,

$$0 \le \lambda_k(U(0) + \sum_i y_i \lambda_n(U(0)),$$

is valid for any y in R. Further, it is not satisfied by the origin, and hence it is a valid cut for R (**Weyl Cuts**). We may now add the cut to the linear inequality system, and repeat the above process. The convergence of the above method has been shown under some regularity conditions. An implementation of the method in a dual simplex framework has been carried and very encouraging results have been obtained. Randomly generated instances of UPPs of sizes up to $d = 25, m = 90, n = 15$ have been solved successfully, wherein about 100 cutting planes have been added and the CPU times (on a VAX system) of up to 26 seconds were observed [66].

7.2   The Relaxation Approach

It is well known that the MQP (and hence the UPP) is NP-Hard. Thus, it is a natural to ask if some relaxations of the MQP (or UPP) can be solved in polynomial time. This leads us to the following **Semidefinite Relaxation** of the UPP. Since $\mathcal{P}_n$ is the convex hull of $\mathcal{U}_n$, we relax the constraint $U(x) \in \mathcal{U}_n$ to $U(x) \in \mathcal{P}_n$, and the result is a Semidefinite program:

$$\begin{array}{rll}
\max & : & c^T x \\
\text{s.t.} & & Ax \le b \\
& & U(x) \succeq 0.
\end{array}$$

To explore this problem and its relation to MQP, we consider the feasibility problem of checking if the system $f(x) = b$ is consistent, where $f_i(x) = z^T Q_i z + c_i^T z \ \forall \ i = 1, .., m$. Then the above relaxation gives us the semidefinite system:

$$U \bullet Q_i + z^T c_i = b_i$$
$$U \succeq 0.$$

It can be easily shown that this system is consistent iff 0 lies in the convex hull of $Im(f) = f(R^n)$, the image of f. Further, if $\tilde{Q}_i$ denotes the matrix of the homogenized quadratic form $z^T Q_i z + \mu c_i^T z - \mu^2 b_i$, then, under some mild conditions, we have $0 \in Im(f)$ iff the following set if empty:

$$S = \{y \in R^m | \sum_{i=1}^m y_i \tilde{Q}_i \succ 0\}.$$

This problem is called the strict semidefinite feasibility problem(SSFP). There is an efficient separation oracle for spectrahedra (a set of the type $\{x|U(x) \succeq 0\}$ or $\{x|U(x) \succeq 0\}$ is called a spectrahedron), that employs a truncated Cholesky decomposition, and hence, in principle, one can apply the ellipsoid algorithm for the solution of SSFP. But, to guarantee polynomial time complexity of the resulting procedure, we need single-exponential lower bounds on the volumes of full-dimensional spectrahedra. Such bounds exist for the special case of linear programming, but, unfortunately, there are classes of spectrahedra, whose volumes may be doubly exponentially small. However, when m is fixed, one can obtain these lower bounds and show the polynomial time solvability of the SSFP for this case.

The complexity of the SSFP in the general case is still an open issue. It is also natural to ask when the above relaxation of systems of quadratic equations is exact. A map is said to be ICON if its image is convex. Clearly, the above relaxation is exact when the quadratic map f is ICON. It has been shown that the problem of checking if a given quadratic map is ICON is NP-Hard. However, if we require the image of every linear subspace of the R^n to be convex, we obtain a polynomial time procedure for detecting such maps [66].

We mention here the special case of MQP with only one quadratic constraint. This special problem, which is the key step in trust region optimization approaches, can be solved in polynomial time even when the quadratic functions are indefinite [80].

8 Concave Minimization

A large number of applications can be formulated as general concave minimization problems. In addition, concave minimization algorithms play an important role in designing algorithms for more general global optimization problems.

Consider the problem of minimizing a concave function $f(x)$ over a compact convex set defined by $D = \{x : g_i(x) \leq 0 \ i = 1, \ldots, m\}$, where the functions g_i are convex on a suitable set A satisfying $D \subseteq A \subseteq R^n$. When the objective function $f(x)$ is separable, the problem can be approximated by piecewise linear functions and linearization techniques discussed in Section 5, can be applied. Most of the techniques for solving the general concave minimization problem are based on cutting plane methods, outer (and inner) approximation algorithms and branch and bound techniques. The literature for algorithms and applications of concave minimization is so large that will cover an entire book. We refer the reader to the books [21], [38], [36], and [57].

Many problems in combinatorial optimization can be formulated and solved as concave minimization (or convex maximization) problems. Consider for example the 3-dimensional assignment problem (3-AP) of the following form:

$$\text{max} \quad \sum_{i=1}^{m} \sum_{j=1}^{n} \sum_{k=1}^{p} a_{ijk} x_{ijk}$$

$$\text{s.t.} \quad \sum_{j=1}^{n} \sum_{k=1}^{p} x_{ijk} = 1, \qquad i = 1, \ldots, m,$$

$$\sum_{i=1}^{m} \sum_{k=1}^{p} x_{ijk} \leq b_j, \qquad j = 1, \ldots, n,$$

$$\sum_{i=1}^{m} \sum_{j=1}^{n} x_{ijk} \leq c_k, \qquad k = 1, \ldots, p,$$

$$x_{ijk} \in \{0, 1\}, \text{ for all } i, j, k,$$

where $b_j > 0$ for $j = 1, \ldots, n$, $c_k > 0$ for $k = 1, \ldots, p$ and $a_{ijk} \geq 0$. There is a one-to-one correspondence between the optimal solution of the 3–AP and the global optimization problem:

$$\text{max} \quad \phi(y, z) = \frac{1}{2} \sum_{i=1}^{m} \sum_{j=1}^{n} \sum_{k=1}^{p} a_{ijk}(y_{ij} + z_{ik})(y_{ij} + z_{ik} - 1)$$

$$\text{s.t.} \quad \sum_{j=1}^{n} y_{ij} = 1, \sum_{k=1}^{p} z_{ik} = 1, i = 1, \ldots, m,$$

$$\sum_{i=1}^{m} y_{ij} \leq b_j, j = 1, \ldots, n, \sum_{i=1}^{m} z_{ik} \leq c_k, k = 1, \ldots, p,$$

$$y_{ij} \geq 0, z_{ik} \geq 0.$$

Since $a_{ijk} \geq 0$, this is a convex function, and the original (3-AP) is equivalent to a convex maximization problem.

As a second example, consider the complementarity problem(CP): Given the matrices $A_{m \times n}$, $B_{m \times n}$, $C_{m \times p}$ and a vector $q \in R^m$, find $x, y \in R^n$, $z \in R^p$ satisfying

$$Ax + By + Cz = q, x^T y = 0, x, y, z \geq 0.$$

It can be shown that, (x^*, y^*, z^*) is a solution of (CP) iff (x^*, y^*, z^*) is a solution of the following problem

$$\text{min } f(x, y, z) = \sum_{i=1}^{n} \min\{x_i, y_i\}$$

s.t. $Ax + By + Cz = q$, $x^T y = 0$, $x, y, z \geq 0$,

with $f(x^*, y^*, z^*) = 0$. Note that $f(x, y, z)$ is piecewise linear concave function.

9 Multiplicative and Fractional Programming

The term multiplicative programming refers to the class of minimization problems containing a product of several functions either in its objective or its constraints. Multiplicative programming problems appear in many applications including portfolio problems, geometric optimization, and economic analysis. The simplest example of multiplicative programming is the problem of minimizing the product of two linear functions subject to the linear constraints. Such a simple problem is in general nonconvex and still it is not known if it can be solved in polynomial time or not.

Recent work by Konno et al. concerns problems containing the product of several convex functions. Parametric and outer optimization techniques have been very efficient for solving many classes of multiplicative optimization problems.

The general problem has the form (MPC)

$$\min \quad f(x) = \prod_{j=1}^p f_j(x) \tag{13}$$
$$\text{s.t.} \quad g_i(x) \leq 0, \, i = 1, \ldots, m,$$

where f_j and g_i are all real valued functions defined on R^n. Under certain assumptions, the problem can be reduced to concave minimization (e.g. when all $f_j(x) > 0$ are concave and the feasible domain is convex). When all functions are convex and $f_j(x) > 0$, the problem can be solved by solving an auxiliary minimization problem of the form:

$$\min \quad F(x, \lambda) = \sum_{j=1}^p \lambda_j f_j(x) \tag{14}$$
$$\text{s.t.} \quad g_i(x) \leq 0, \, i = 1, \ldots, m \tag{15}$$
$$\text{and} \quad \prod_{j=1}^p \lambda_j \geq 1, \, \lambda \geq 0.$$

If (x^*, λ^*) is the solution of the above problem, then x^* is the global solution of (MPC). A parametric type algorithm based on this observation is proposed in [42].

In many applications of nonlinear optimization, a function is to be optimized which involved one or several ratios of functions. Such optimization problems are called fractional (or hyperbolic) programming problems. An up to date bibliography with more than one thousand papers regarding algorithms and applications can be found in [69].

The general problem has the form:

$$\max F(x) = \frac{f(x)}{g(x)}$$

$$\text{s.t.} \quad x \in S,$$

where $f(x)$ and $g(x)$ are real valued continuous functions and S is a convex subset of R^n.

With this problem we associate the "parametric" global optimization problem

$$\max G(x, \lambda) = f(x) - \lambda g(x)$$

$$\text{s.t. } x \in S.$$

The classical result that relates the two global optimization problems states that x^* is the global maximum of $F(x)$ iff x^* is the global maximum of $G(x, \lambda^*)$ with $\lambda^* = \frac{f(x^*)}{g(x^*)}$. A global optimization algorithm which solves a sequence of global optimization problems of the above form and provides a superlinear convergence of the lower and upper bounds to the global solution can be found in [55].

10   Nonconvex Potential Energy Functions

One of the most significant and challenging problems in molecular biophysics and biochemistry is that of computing the native 3 dimensional conformation (folded state) of a globular protein given its amino acid sequence [63], possibly in the presence of additional agents (e.g., drugs). The minimization of potential energy functions plays an important role in the determination of ground states or stable states of certain classes of molecular clusters and proteins. Since, in almost all the cases, the potential energy function is nonconvex and therefore has many local minimizers, the minimization of the potential energy function is a very hard problem.

Next, we describe some of the most commonly used potential energy functions in studying these problems. Given a cluster of N atoms in $3-$ dimensional space, the potential energy function of the cluster is defined as the summation (over all of the pairs) of the two-body interatomic pair potentials. Let the center of the N atoms be $a_1, \cdots, a_N$. The potential energy function is defined as follows.

$$V_N(a_1, \cdots, a_N) = \sum_{1 \leq i < j \leq N} v(\|a_i - a_j\|), \tag{16}$$

where $\|.\|$ is the Euclidean norm and $v(r)$ is the interatomic pair potential.

Although many types of the function $v(r)$ may be used in physical models, it is necessary to apply some restrictions if we want $v(r)$ to have satisfactory behavior. In general, we would like $v(r)$ to be continuous at least down to a hard core r_{min} and to possess derivatives up to the second order over the interval (r_{min}, ∞). More specifically, we are only interested in cases where $v(r)$ is a *well potential* satisfying the following conditions:

1. $v(r) \to 0^-$ as $r \to \infty$;
2. $v(r) \to \infty$ as $r \to r_{min}$ and $r_{min} \geq 0$;
3. $v'(r_0) = 0$ for a unique r_0 with $r_{min} < r_0 < \infty$;
4. $v''(r_0) > 0$ and $v(r_0) < 0$.

There are many potential functions that have been considered in the literature (see for example [19]). Pair potentials of interest in clusters include the following:

a. $v(r) = (n - m)^{-1}[nr^{-m} - mr^{-n}]$ (Mie)
b. $v(r) = r^{-12} - 2r^{-6}$ (Lennard-Jones)
c. $v(r) = [1 - e^{\alpha(1-r)}]^2 - 1$ (Morse)
d. $v(r) = Ae^{-ar^2} - Be^{-br^2}$ (Gaussian)
e. $v^{\alpha\beta}(r) = z^\alpha z^\beta / r + A exp(-r/\rho)$ (Born-Meyer)

From the above conditions (1-4), we can easily see that the potential energy function is continuous and always has a global minimizer with a minimum energy function value less than zero.

Many optimization methods have been proposed to find the global minimum of nonconvex potential energy functions. For a recent survey see [63]. Global minimization of nonconvex potential energy functions is one of the most active areas of research concerning challenging applications of global optimization. Furthermore, nonconvex energy functions appear in many other areas of science and engineering (see for example [48]).

11 Satisfiability Problems

The satisfiability (SAT) problem is a fundamental problem in mathematical logic, artificial intelligence, VLSI engineering, and computing theory [9, 25]. Traditional methods treat the SAT problem as a constrained decision problem. In recent years, researchers have applied global optimization methods to solve the SAT problem. In this approach, universal satisfiability models, i.e., *UniSAT*, are formulated that transform a discrete SAT problem on Boolean space $\{0, 1\}^m$ into a continuous SAT problem on real space R^m. Thus, this decision problem is transformed into a global optimization problem which can be solved by global optimization methods [30, 32].

In the typical *UniSAT* models [30, 31], using the universal DeMorgan laws, Boolean $\vee$ and $\wedge$ connectors in conjunctive normal form (CNF) formulas are transformed into $\times$ and $+$ of ordinary addition and multiplication operations, respectively. The *true* value of the CNF formula is thus converted to the minimum value, i.e., 0, of the objective function. With the *UniSAT5* and *UniSAT7* models [32], for example, a *CNF* formula

$$F(x) = (x_1 \vee \bar{x}_2) \wedge (\bar{x}_1 \vee x_2 \vee x_3)$$

is translated into

$$f(y) = |y_1 - 1||y_2 + 1| + |y_1 + 1||y_2 - 1||y_3 - 1|$$

and

$$f(y) = (y_1 - 1)^2(y_2 + 1)^2 + (y_1 + 1)^2(y_2 - 1)^2(y_3 - 1)^2,$$

respectively. The solution to the SAT problem corresponds to a set of global minimum points of the objective function. To find a *true* value of $F(x)$ is equivalent to find the minimum value, i.e., 0, of f(**y**).

The translation of SAT problem into a nonlinear program is different from the integer program approach. Although nonlinear problems are intrinsically more difficult to solve, an unconstrained optimization problem is conceptually simple and easy

to handle. Many powerful solution techniques have been developed to solve unconstrained optimization problems, which are based primarily upon calculus, rather than
upon algebra and pivoting, as in the simplex method.

Depending on the global optimization strategy used, the objective function can be
minimized in one dimension or in multiple dimensions. Presently many families of
global optimization algorithms have been developed for the *UniSAT* problem models
[32]. Computational results with thousands of variables and clauses are reported in
[30, 32] using global optimization algorithms. Efficient global optimization algorithms
in this area are of enormous practical significance in VLSI and computer engineering.

12   Minimax Optimization

Classical minimax theory initiated by Von Neumann, together with duality and saddle
point analysis, has played a critical role in optimization, game theory and best approximation. However, minimax appears in a very wide area of disciplines. Recently,
minimax theory has been applied in many diverse problems such as Steiner trees,
network flow, combinatorial group testing, and other combinatorial problems.

A new approach based on a nontrivial new minimax result was introduced in Du
and Hwang's proof of the Steiner tree conjecture [13]. The center part of this approach
is a new theorem about the following minimax problem:

$$\min_{x \in X} \max_{i \in I} f_i(x)$$

where X is a convex region in the n-dimensional Euclidean space R^n, I is a finite
index set, and $f_i(x)$'s are continuous functions over X.

A subset Z of X is called an *extreme subset* of X if

$$\left. \begin{array}{l} x, y \in X \\ \lambda x + (1 - \lambda)y \in Z \text{ for some } 0 < \lambda < 1 \end{array} \right\} \Rightarrow x, y \in Z.$$

With this definition, the Du-Hwang theorem can be stated as follows:

Theorem 12.1 *Let* $g(x) = \max_{i \in I} f_i(x)$. *If every* $f_i(x)$ *is a concave function, then the
minimum value of* $g(x)$ *over the polytope* X *is achieved at some point* x^* *satisfying
the following condition: There exists an extreme subset* Z *of* X *such that* $x^* \in Z$ *and
the* $I(x^*)$ $(= \{i \mid g(x^*) = f_i(x^*)\})$ *is maximal over* Z.

In addition, the following continuous version has been recently proved: Let $f(x, y)$
be a continuous function on $X \times Y$ where X is a polytope in R^m and Y is a compact
set in R^n. Let $g(x) = \max_{y \in Y} f(x, y)$. If $f(x, y)$ is concave with respect to x, then the
minimum value of $g(x)$ over X is achieved at some point $\hat{x}$ satisfying the following
condition: There exists an extreme subset Z of X such that $\hat{x} \in Z$ and the set $I(\hat{x})$
$(= \{y \mid g(\hat{x}) = f(\hat{x}, y)\})$ is maximal over Z.

These minimax results gave birth to new approaches for studying difficult combinatorial optimization problems [13, 15].

13   Network Flow Problems

Nonconvex-cost network flow problems can be stated formally as follows: Given a directed graph $G = (N, A)$ consisting of a set N of n nodes and a set A of m ordered pairs of distinct nodes called arcs, coupled with a n-vector $d = (d_i)$ (demand vector) and a cost function for each arc, $c_{ij}(x_{ij})$ (some of the functions are nonconvex), then solve

$$\text{global min} \sum_{(i,j)\in A} c_{ij}(x_{ij})$$

subject to

$$\sum_{(i,k)\in A} x_{ik} - \sum_{(k,i)\in A} x_{ki} = d_i, \; \forall i \in N, \tag{17}$$

and

$$0 \leq a_{ij} \leq x_{ij} \leq b_{ij}, \; (i,j) \in A. \tag{18}$$

All constraints and demands are assumed to be integral. A consistent system satisfies $\sum_{i=1}^{n} d_i = 0$, which states that total source demand equals total sink demand. The constraints in (17) are called the conservation-of-flow equations. The constraints in (18) are called capacity constraints on the arc flows. Nodes with $d_i < 0$ correspond to sinks. Nodes with $d_i > 0$ correspond to sources. The problem is uncapacitated if $a_{ij} = 0$ and $b_{ij} = \infty, \forall(i,j) \in A$. Problem features which vary from application to application include: size (number of nodes), density (arcs per node), structure (acyclic network, staged network, etc.), arc capacities, objective function type (concave, piecewise concave, fixed-charge, etc.), number of sources and sinks, and flow magnitudes.

Network problems arise in many different contexts. Numerous problems in production and inventory planning, transportation and communication network design, facilities location, hydroelectric power system scheduling, air traffic control, and other areas are modeled naturally using models with network constraints. Other problems without a natural network formulation have been solved efficiently after being reformulated as network optimization problems. In addition, network structures are imbedded in the feasible domain of many optimization problems; when the network structure is identified, more efficient algorithms can be used to solve these problems.

One of the simplest nonconvex network flow problems is the fixed-charge network flow problem (FC). Given a directed graph $G = (N_G, A_G)$, consider the (FC) network flow problem:

$$\min \sum_{(i,j)\in A_G} f_{ij}(x_{ij})$$

subject to

$$\sum_{(k,i)\in A_G} x_{ki} - \sum_{(i,k)\in A_G} x_{ik} = d_i, \; \forall i \in N_G, \tag{19}$$

and

$$0 \le x_{ij} \le b_{ij}, \ (i,j) \in A_G. \tag{20}$$

The objective function has the form $f(x) = \sum_{(i,j) \in A_G} f_{ij}(x_{ij})$, where

$$f_{ij}(x_{ij}) = \begin{cases} 0 & \text{if flow } x_{ij} = 0 \\ d_{ij} + c_{ij}x_{ij} & \text{if flow } x_{ij} > 0. \end{cases}$$

In many practical problems, the cost of some activity is the sum of a fixed cost and a cost proportional to the level of the activity. The fixed-charge network flow problem is obtained by imposing a fixed cost of $d_{ij} \ge 0$ if there is positive flow on arc (i,j) and a variable cost $c_{ij}(x_{ij})$.

Assuming $d_{ij} > 0$, $f_{ij}(x_{ij})$ can be replaced by

$$F_{ij}(x_{ij}, y_{ij}) = d_{ij}y_{ij} + c_{ij}x_{ij}$$

with

$$x_{ij} \ge 0, \ \text{and } y_{ij} = \begin{cases} 0 & \text{if } x_{ij} = 0 \\ 1 & \text{if } x_{ij} > 0. \end{cases} \tag{21}$$

The variables y_{ij} are $0-1$ variables, indicating whether the corresponding activity is being carried out or not. The above condition (21) can be incorporated into the capacity constraints to yield

$$x_{ij} \le b_{ij}y_{ij}, \ y_{ij} \in \{0,1\}.$$

Hence we obtain the following formulation of the fixed-charge network flow problem:

$$\min \sum_{(i,j) \in A_G} F_{ij}(x_{ij}, y_{ij}) = \sum_{(i,j) \in A_G} d_{ij}y_{ij} + c_{ij}x_{ij}$$

subject to

$$\sum_{(k,i) \in A_G} x_{ki} - \sum_{(i,k) \in A_G} x_{ik} = d_i, \ \forall i \in N_G, \tag{22}$$

and

$$0 \le x_{ij} \le b_{ij}y_{ij}, \ y_{ij} \in \{0,1\}, \ (i,j) \in A_G. \tag{23}$$

This is a linear mixed zero-one integer problem and can be solved by using any type of classical branch and bound algorithm that uses linear programming relaxations. The computational advantage of this approach is that linear network flow problems can be solved very efficiently by existing linear network algorithms.

As a second example we consider the general Transportation Problem (TP), formulated as follows:

$$\min f(x)$$

subject to

$$\sum_{j=1}^{n} x_{ij} = a_i > 0, \quad i = 1, \ldots, m,$$

$$\sum_{i=1}^{m} x_{ij} = b_j > 0, \quad j = 1, \ldots, n,$$

$$x_{ij} \geq 0, \quad \forall (i,j).$$

In this problem, all nodes are sources or sinks, and the network is a complete bipartite graph $G(V_1 \cup V_2, A_G)$. All nodes in V_1 are sources and all nodes in V_2 are sinks, and there is an arc from each source (supply node) to each sink source (demand node). Several well-known problems result from cases where f is separable, i.e., $f(x) = \sum_{i=1}^{m} \sum_{j=1}^{n} f_{ij}(x_{ij})$. If the f_{ij} are linear then the Hitchcock Transportation Problem results. If each f_{ij} is of the form

$$f_{ij}(x_{ij}) = \begin{cases} 0 & \text{if } x_{ij} = 0 \\ s_{ij} + b_{ij}x_{ij}, s_{ij} \geq 0 & \text{otherwise.} \end{cases}$$

then we obtain the Fixed-charge Transportation Problem. If the problem parameters change with time the Dynamic Transportation Problem results. For separable objective functions the problem is a special case of minimum concave cost network flow problem (MCNFP). However, due to the TP's special network structure specializations of MCNFP algorithms perform more efficiently.

Another example of nonconvex network flow problem is the Steiner Problem. The Steiner problem is an important network optimization problem due to its numerous applications in areas like manufacturing, communication, and transportation network design. The Steiner Problem in Networks (SPN) is as follows:

Given an undirected network $G = (V, E, c)$ with n nodes, m arcs, and the arc cost function $c : E \to R$, and a subset $Z \subseteq V$ with p nodes, find a subnetwork G_Z of G such that there is a path between every pair of Z-nodes, and the total cost of G_Z is a minimum. The SPN can be reformulated as an uncapacitated single-source separable MCNFP using the following transformation:

1. Nodes in Z correspond to sinks with a demand of one.

2. Each undirected arc (i, j) is replaced with directed arcs (i, j) and (j, i).

3. A source node S is added along with a directed arc from S to each node in V.

4. The fixed-charge cost function for each arc (i, j) is

$$c'_{ij} = \begin{cases} 0 & \text{if flow } x_{ij} = 0 \\ c_{ij} & \text{otherwise.} \end{cases}$$

The cost of the flow on arcs $(S, j), \forall j \in V$, are set to be greater than $\sum_{(i,j) \in E} c_{ij}$. This is to force all flow from the source through one arc into the original network.

By the properties of concave cost network flow problems the solution of an uncapacitated single-source concave network flow problem is a tree rooted at the source node. Given the solution tree, the desired Steiner subnetwork corresponds to arcs and nodes with nonzero flow (minus the added source node). The SPN is also a natural formulation for certain transportation problems. The Steiner network provides an optimum interconnection between facilities under the assumption of static fixed costs for transport. The fixed cost assumption could correspond to a distance between facilities, or possibly the type of transportation (e.g., mail, truck, air-freight, etc.).

In the last years, many algorithms have been proposed to globally solve nonlinear network flow problems. Most of the algorithms are concerned with the problems of solving concave cost network flow problems. This is a fundamental problem, since general nonlinear network flow problems can be reduced to networks with concave cost (see [14]).

Any general purpose algorithm for concave programming problems can be applied to the solution of concave cost network flow problems. Exact algorithms for concave network problems explicitly or implicitly enumerate the vertices of the feasible domain defined by the network constrains. General exact solution techniques include vertex ranking, branch-and-bound, dynamic programming, and decomposition techniques [28]. Since exact methods are computationally very expensive, many methods such as local search, approximate algorithms, and other heuristics using shortest path based solutions, have been proposed.

A dynamic programming approach to the general MCNFP, called the send-and-split method, was developed by Erickson et al [16]. This approach applies to uncapacitated networks but the classical Wagner's transformation can be used to convert capacitated arcs to an uncapacitated form. This approach has the feature that the running time is (for general networks) polynomial in the number of nodes and arcs, and exponential in the number of demand nodes $(d_i \neq 0)$ and capacitated arcs (which become demand nodes in the transformation).

In [72] a decomposition method for structured concave network problems has been proposed. This approach partitions the overall problem into a series of smaller subproblems which are solved in an ordered sequence. The basis of the method is a successive approximation technique called polyhedral annexation. The author in [72] applies this approach to two sets of problems, single-source uncapacitated circuitless networks, and networks with a staircase structure.

In [29] it has been shown that in network flow problems where all arcs have linear costs but one can be solved in polynomial time (see also [41]). This result contrasts the corresponding result without network constraints, in which the problem is known to be NP-hard [60]. Furthermore, if the numbers of nonlinear costs is fixed, the corresponding network flow problem can be still solved in polynomial time.

14 Multi-Objective Programming

Multiple objective mathematical programming problems (MOP) arise when two or more real-valued objective functions $f_1, \ldots, f_m$ are to be optimized over a nonempty set D in R^n. The concepts of "efficiency" and "weak efficiency" are frequently used in problem (MOP) in place of the usual concept of optimal solution in single-objective mathematical programming. In contrast to the single-objective case, there are usually many efficient and weakly efficient solutions for problem (MOP). Since the efficient and weakly efficient solutions describe the tradeoffs available in the objective functions f_i, $i = 1, \ldots, m$, of problem (MOP), the goal in (MOP) is to explore either the set $D(E)$ of efficient solutions or the set $D(WE)$ of weakly efficient solutions with the intent of choosing the most preferred solution [71].

The sets $D(E)$ and D(WE) are in general nonconvex sets. Furthermore, they are also disconnected, except for certain special cases. Therefore, the goal of exploring efficient and weakly efficient sets to find the most preferred solution generally involves searching complicated nonconvex, disconnected sets. Recently, motivated in part by this difficulty, researchers have focussed on problems in which a function g is available which the decision maker would like to optimize over either $D(E)$ or $D(WE)$. Problems of this type arise in many practical applications and in many of the formal and informal methods which attempt to explore the $D(E)$ and $D(WE)$ sets. Since $D(E)$ and $D(WE)$ are nonconvex, problems of optimizing functions over one of these sets are difficult global optimization problems. Furthermore, in contrast to most other types of global optimization problems, the usual difficulties of the existence of many local optima which are not global are enchanced by the fact that descriptions of the feasible regions $D(E)$ and $D(WE)$ of these problems as explicit systems of constraints are generally not available.

In spite of these difficulties, several algorithms and heuristics have been proposed for optimizing functions g over $D(E)$ or over $D(WE)$. Several of these apply only to the case where g is a linear function and the underlying multiple objective mathematical programming problem (MOP) is linear. A few of the more recent algorithms attempt to optimize certain types of nonlinear functions g over $D(E)$ or $D(WE)$. One of the most recent algorithms for the case where both g and problem (MOP) are linear works by solving a sequence of concave minimization problems [5]. For a brief review of these algorithms see [5], [6], [71], and references therein.

15 The Road Less Traveled

Some of the proposed algorithms for global optimization have been implemented and tested on certain classes of problems. However, there is no systematic implementation or testing of these algorithms. At present there are no widely available codes for solving global optimization problems. One hopes that in the future, various global optimization methods will find their way into common software packages.

Up to this date, most of the reported computational experiments regarding the performance of global optimization algorithms are using either randomly generated or standard test problems [20, 46, 51, 52]. Using such test problems, reported computational experiments can be documented in a manner that allows checking and reproducing the results.

The availability of collections of test problems and software will facilitate the efforts of comparing performance and correctness of the numerous proposed global optimization algorithms.

16 Further Remarks

We discussed a small fraction of research directions in global optimization. Several approaches have been proposed for Lipschitz optimization problems or dc problems, i.e. problems where the involved functions are differences of convex (dc) functions. In addition, there is active research in many other areas of global optimization such as bilevel (hierarchical) optimization, and solutions of systems of nonlinear equations [38]. Furthermore, the existence of commercial multiprocessing computers has created substantial interest in exploring the uses of parallel processing for solving global optimization problems [4], [53], [56].

There are also many approaches we did not discuss, such as interval analysis methods [68], homotopy methods [27], and stochastic techniques [45, 73] (including simulating annealing, pure random search techniques and clustering methods). The interested reader is referred to the Journal of Global Optimization and [36].

Acknowledgments

Thanks are due to Prof. G. Danninger for several discussions and the University of Vienna, Austria, for technical support during my visit in summer 1994.

References

[1] F. Alizadeh, "Interior point methods in semidefinite programming with applications to combinatorial optimization," to appear in *SIAM J. Optimization*, 1994.

[2] F. Alizadeh, "Optimization over positive semi-definite cone; interior-point methods and combinatorial applications," In *Advances in Optimization and Parallel Computing*, P. M. Pardalos (editor), North-Holland, 1992.

[3] F. Aluffi-Pentini, V. Parisis, and F.Zirilli, "A global optimization algorithm using stochastic differential equations," *ACM Trans. Math. Software*, **14** (1988), 345-365.

[4] G. Y. Ananth, V. Kumar and P. M. Pardalos, "Parallel processing of discrete optimization problems," in *Encyclopedia of Microcomputers*, **13** (1993), 129-147, Marcel Dekker Inc., New York.

[5] H. P. Benson, "A bisection-extreme point search algorithm for optimizing over the efficient set in the linear dependence case," *Journal of Global Optimization*, **3** (1993), 95-111.

[6] H. P. Benson, "Optimizing over the efficient set: Four special cases," *JOTA*, **80**(1) 1994, 3-18.

[7] I. M. Bomze and G. Danninger, "A Global optimization algorithm for concave quadratic programming problems," *SIAM J. Optimization*, **3**(4) 1993, 826-842.

[8] I. M. Bomze and G. Danninger, "A finite algorithm for solving general quadratic problems," *Journal of Global Optimization*, **4** (1994), 1-16.

[9] S. A. Cook, "The complexity of theorem-proving procedures," *Proceedings of the Third ACM Symposium on Theory of Computing*, 1971, 151-158.

[10] A. R. Conn, N. Gould and Ph. L. Toint, "Large-scale nonlinear constrained optimization," *Linear Algebra for Large Scale and Real-Time Applications*, M.S. Moonen et al. (editors), Kluwer Academic Publishers, 1993, 21-48.

[11] T. Csendes, "Nonlinear parameter estimation by global optimization: Efficiency and reliability," *Acta Cybernetica*, **8** (1988), 361-370.

[12] L. C. W. Dixon and G. P. Szego (editors), *Towards Global Optimization*, North Holland, 1978.

[13] D.-Z. Du and F. K. Hwang, "An approach for proving lower bounds: solution of Gilbert-Pollak's conjecture on Steiner ratio," Proceedings 31th FOCS, 1990, 76-85.

[14] D.-Z. Du and P. M. Pardalos (editors), *Network Optimization Problems:Algorithms, Applications and Complexity*, World Scientific, 1993.

[15] D.-Z. Du and P. M. Pardalos (editors), *Minimax and Applications*, Kluwer Academic Publishers, 1994.

[16] R. E. Erickson, C. L. Monma and A. F. Jr. Veinott, "Send-and-Split Method for Minimum-Concave-Cost Network Flows," *Mathematics of Operations Research*, **12**(4) 1987, 634-664.

[17] Y. G. Evtushenko, *Numerical Optimization Techniques*, New York, Optimization Software Inc., Springer-Verlag 1985.

[18] J. A. Filar, M. G. M. Oberije and P. M. Pardalos, "Hamiltonian cycle problem, controlled markov chains and quadratic programming," in *The Proceedings of The 12th National Conference of The Australian Society For Operations Research*, Adelaide July 7-9, 1993, 263-281.

[19] I. Z. Fisher, *Statistical Theory of Liquids*, University of Chicago Press, 1964.

[20] C. A. Floudas and P. M. Pardalos, "A Collection of Test Problems for Constrained Global Optimization Algorithms," *Lecture Notes in Computer Science*, **455**, Springer-Verlag, Berlin, 1990.

[21] C. A. Floudas and P. M. Pardalos, editors, *Recent Advances in Global Optimization*, Princeton University Press, 1992.

[22] C. A. Floudas and V. Visweswaran, "A primal-relaxed dual global optimization approach," *JOTA* **78** (2) 1993, 187-225.

[23] W. Forster, "Some computational methods for systems of nonlinear equations," *Journal of Global Optimization*, **2**(4) 1992, 317-356.

[24] W. Hager, R. Horst and P. M. Pardalos, "Mathematical programming: A computational prespective," in *Handbook of Statistics*, C. R. Rao (editor), **9** (1993), 201-278.

[25] M. R. Garey and D. S. Johnson, *Computers and Intractability: A Guide to the Theory of NP-Completeness*, W. H. Freeman and Company, San Francisco, 1979.

[26] P. E. Gill, W. Murray, M. A. Saunders, and M. H. Wright, "Inertia-controlling methods for general quadratic programming," *SIAM Review*, **33**(1) 1991, 1-36.

[27] J. Guddat, F. Guerra Vazquez, and H. Th. Jongen, *Parametric Optimization: Singularities, Pathfollowing and Jumps*, John Wiley & Sons, 1990.

[28] G. Guisewite and P. M. Pardalos, "Minimum concave-cost network flow problems: Applications, complexity, and algorithms," *Annals of Operations Research*, **25** (1990), 75-100.

[29] G. Guisewite and P. M. Pardalos, "A polynomial time solvable concave network flow problem," *Networks*, **23** (1993), 143-147.

[30] J. Gu, "Global optimization for satisfiability (SAT) problem," *IEEE Trans. on Knowledge and Data Engineering*, **6**(3) 1994.

[31] J. Gu, "Optimization Algorithms for the Satisfiability (SAT) Problem," *New Advances in Optimization and Approximation*, D. Z. Du and J. Sun (editors), Kluwer Academic Publishers, 1994, 72-154.

[32] J. Gu, *Constraint-Based Search*, Cambridge University Press, New York, to appear 1995.

[33] M. Hamami and S.E. Jacobsen, "Exhaustive nondegenerate conical processes for concave minimization on convex polytopes," *Mathematics of Operations Research*, **13**(3) 1988, 479-487.

[34] C. G. Han, P. M. Pardalos and Y. Ye, "On the solution of indefinite quadratic problems using an interior point algorithm," *Informatica*, **3**(4) 1992, 474-496.

[35] J. B. Hiriart-Urruty, "From convex optimization to nonconvex optimization. Part I: Necessary and sufficient conditions from global optimality," *Nonsmooth Optimization and Related Topics*, F. H. Clarke et al. (editors), Plenum Press, 1989, 219-239.

[36] R. Horst and P. M. Pardalos, editors), *Handbook of Global Optimization*, Kluwer Academic Publishers, 1994.

[37] R. Horst, P.M. Pardalos and N.V. Thoai, *Introduction to Global Optimization*, Kluwer Academic Publishers, 1994.

[38] R. Horst and H. Tuy, *Global Optimization, Deterministic Approaches*, Springer-Verlag, Berlin, 1993.

[39] T. C. Hu, V. Klee and D. Larman, "Optimization of globally convex functions," *SIAM J. Control and Optimization*, **27**(5) 1989, 1026-1047.

[40] N. Karmarkar, "An interior-point approach for NP-complete problems," *Contemporary Mathematics*, **114** (1990), 297-308, 1026-1047.

[41] B. Klinz and H. Tuy, "Minimum concave-cost network flow problems with a single nonlinear arc cost," *Network Optimization Problems*, DingZhu Du and P. M. Pardalos (editors), 1993, 125-146.

[42] H. Konno, T. Kuno and Y. Yajima, "Global minimization of a generalized convex multiplicative function," *Journal of Global Optimization*, **4** (1994), 47-62.

[43] L. Lovász, "On the shannon capacity of a graph," *IEEE Trans. Inf. Theory*, **25**(1979).

[44] G.P. McCormick, "Attempts to calculate global solutions of problems that may have local minima," *Numerical Methods for Nonlinear Optimization*, F. A. Lootsma (editor), Acad. Press, 1972, 209-221.

[45] J. Mockus, *Bayesian Approach to Global Optimization*, Kluwer Academic Publishers, Dordrecht-London-Boston, 1989.

[46] J. J. Moré and S. J. Wright, *Optimization Software Guide*, SIAM, Philadelphia, 1994.

[47] A. Neumaier, "An optimality criterion for global quadratic optimization," *Journal of Global Optimization*, **2** (1992), 201-208.

[48] P. D. Panagiotopoulos, *Inequality Problems in Mechanics and Applications, Convex and Nonconvex Energy Functions*, Birkhäuser, 1985.

[49] C. H. Papadimitriou and D. Wolfe, "The complexity of facets resolved," *Proceedings of the Foundations Of Computer Science*, 1985, 74-78.

[50] P. Y. Papalambros and D. J. Wilde, *Principles of Optimal Design, Modeling and Computation*, Cambridge University Press, 1988.

[51] P. M. Pardalos, "Generation of large-scale quadratic programs for use as global optimization test problems," *ACM Trans. Math. Software*, **13** (1987), 133-137 .

[52] P. M. Pardalos, "Construction of test problems in quadratic bivalent programming," *ACM Trans. Math. Software*, **17** (1991), 74-87 .

[53] P. M. Pardalos (editor), *Advances in Optimization and Parallel Computing*, North-Holland, 1992.

[54] P. M. Pardalos (editor), *Complexity in Numerical Optimization*, World Scientific, 1993.

[55] P. M. Pardalos and A. T. Phillips, "Global optimization of fractional programs," *Journal of Global Optimization*, **1** (1991), 173-182.

[56] P. M. Pardalos, A. T. Phillips and J. B. Rosen, *Topics in Parallel Computing in Mathematical Programming*, Science Press, 1993.

[57] P. M. Pardalos and J. B. Rosen, "Constrained Global Optimization: Algorithms and Applications," *Lecture Notes in Computer Science*, **268**, Springer-Verlag, Berlin, 1987.

[58] P. M. Pardalos and J. B. Rosen, "Global Optimization Approach to the Linear Complementarity Problem," *SIAM J. Scient. Stat. Computing*, **9**(2) 1988, 341-353.

[59] P. M. Pardalos and G. Schnitger, "Checking local optimality in constrained quadratic programming is NP-hard," *Operations Research Letters*, **7**(1) 1988, 33-35.

[60] P. M. Pardalos and S. Vavasis, "Quadratic programming with one negative eigenvalue is NP-hard," *Journal of Global Optimization*, **1** (1991), 15-23.

[61] P. M. Pardalos, Y. Ye, C. G. Han and J. Kalinski, "Solution of P-matrix linear complementarity problems using a potential reduction algorithm," *SIAM J. Matrix Anal. & Appl.*, **14**(4) 1993,1048-1060.

[62] P. M. Pardalos and J. Xue, "The maximum clique problem," *Journal of Global Optimization*, **4** (1994), 301-328.

[63] P. M. Pardalos, G. Xue and D. Shalloway, "Optimization Methods for Computing Global Minima of Nonconvex Potential Energy Functions," *Journal of Global Optimization*, **4** (1994), 117-133.

[64] P.M. Pardalos and H. Wolkowicz (editors), *Quadratic Assignment and Related Problems*, DIMACS Series, American Math. Society, 1994.

[65] C. Radin, "Global order from local sources," *Bulletin of the American Mathematical Society*, **25**(2) 1991, 335-364.

[66] M. V. Ramana, *An algorithmic analysis of multiplicative and semidefinite programming problems*, Ph.D. Thesis, The Johns Hopking University, Baltimore, 1993.

[67] H. Ratschek and R. L. Voller, "What can interval analysis do for global optimization?," *Journal of Global Optimization*, **1**(2) 1991,111-130.

[68] H. Ratschek and J. Rokne, *New Computer Methods for Global Optimization*, Ellis Horwood and John Wiley, 1988.

[69] S. Schaible, "Fractional Programming," *Handbook of Global Optimization*, R. Horst and P. M. Pardalos (editors), Kluwer Academic Publishers, 1994.

[70] N. Z. Shor, "Dual estimates in multiextremal problems," *Journal of Global Optimization*, 2(4) 1992, 411-418.

[71] R. E. Steuer, *Multiple Criteria Optimization: Theory, Computation and Applications*, Wiley, New York (1986).

[72] P. T. Thach, "A Decomposition Method Using a Pricing Mechanism for Min Concave Cost Flow Problems with a Hierarchical Structure," *Mathematical Programming*, 53 (1992), 339-359.

[73] A. Törn and A. Zilinskas, "Global Optimization," *Lecture Notes in Computer Science*, 350, Springer-Verlag, Berlin, 1989.

[74] H. Tuy, "Concave programming under linear constraints," *Soviet Math. Dokl.*, 5 (1964), 1437-1440.

[75] H. Tuy, "On polyhedral annexation method for concave minimization," *Functional Analysis, Optimization and Mathematical Economics*, L. Leifman (editor), Oxford University Press, 1990, 248-260.

[76] H. Tuy, S. Ghannadan, A. Migdalas and P. Värbrand, "Strongly polynomial algorithms for two special minimum concave cost network flow problems," Technical Report, Linkoping University, Sweden, 1993.

[77] S. A. Vavasis, *Nonlinear Optimization, Complexity Issues*, Oxford University Press, Oxford, 1991.

[78] S. A. Vavasis, "Approximation algorithms for indefinite quadratic programming," *Math. Programming*, 57 (1992), 279-311.

[79] Y. Ye, "Interior-point algorithms for global optimization," *Annals of Operations Research*, 25 (1990), 59-74.

[80] Y. Ye, "A new complexity result on minimization of a quadratic function with a sphere constraint," *Recent Advances in Global Optimization*, C. A. Floudas and P. M. Pardalos (editors), Princeton University Press, 1992, 19-21.

Nonsmooth Optimization

Terry Rockafellar
University of Washington
Seattle, Washington

1 Introduction

A function is *smooth* if it is differentiable and the derivatives are continuous. More specifically, this is first-order smoothness. Second-order smoothness means that second-derivatives exist and are continuous, and so forth, while infinite smoothness refers to continuous derivatives of all orders. From this perspective a *nonsmooth* function only has a negative description—it lacks some degree of properties traditionally relied upon in analysis. One could get the impression that "nonsmooth optimization" is a subject dedicated to overcoming handicaps which have to be faced in miscellaneous circumstances where mathematical structure might be poorer than what one would like. But this is far from right.

Instead, nonsmooth optimization typically deals with highly structured problems, but problems which arise differently, or are modeled or cast differently, from ones for which many of the mainline numerical methods, involving gradient vectors and Hessian matrices, have been designed. The nonsmoothness can be primary, in the sense of resulting from something deep in the nature of the application at hand, or secondary through the introduction of penalty expressions or various technical subproblems. Anyway, a strong argument can be made for the notion that nonsmoothness in optimization is very often a question of modeling, and due to the prevalence of inequality constraints, is present anyway in almost all problems of importance, at least in the background. The issue from that angle is simply how to make use of available structure in the best possible way. Nonsmooth optimization gives the message that many effective approaches are possible, and one need not be confined to a classical view of how functions are to be approximated and evaluated.

Because nonsmoothness has different manifestations and treatments, one should not imagine that numerical techniques in nonsmooth optimization can act as "black boxes." Techniques are developed for the particular structures that compensate for the absence of differentiability. It is important therefore to understand the source of any nonsmoothness, before deciding how it might be handled. Providing an overview of this issue is one of the main goals in these notes, along with painting a broad picture of the applications and computational ideas characteristic of nonsmooth optimization.

The central fact is that when functions are *defined* in terms of operations of maximization or minimization, in contrast to long-familiar operations of calculus like composition, addition and integration, they may well fail to inherit the smoothness enjoyed by the functions or mappings serving as "ingredients" in the definition. The theory of nonsmooth optimization is largely concerned therefore with extensions of calculus to cover such functions, for instance in terms of generalized directional derivatives and subgradients, and approximation methodology that can be substituted for nonexistent Taylor expansions of first or second order.

2 Functions with an Envelope Representation

One of the most common situations is that of minimizing a function f having a representation of the form

$$f(x) = \max_{s \in S} \phi_s(x) \text{ for } x \in R^n, \tag{1}$$

where S is some index set, finite or infinite—perhaps a subset of R^d as a parameter space—and $\phi_s(x)$ is smooth with respect to x. The likely nonsmoothness of f in such circumstances can be addressed in more than one way.

When S is finite, the minimization of f over a subset C of R^n can be approached in principle by reformulating the given problem in a higher dimensional space. From $x = (x_1, \ldots, x_n) \in R^n$ and an additional variable $x_0 \in R$, one can put together vectors $\tilde{x} = (x_0, x_1, \ldots, x_n) \in R^{n+1}$ and look instead to minimizing $f_0(\tilde{x}) = x_0$ over all $\tilde{x} \in \tilde{C} = R \times C$ that satisfy the constraints

$$f_s(\tilde{x}) \leq 0 \text{ for each } s \in S, \text{ where } f_s(\tilde{x}) = -x_0 + \phi_s(x_1, \ldots, x_n). \tag{2}$$

Here f_0 and the constraint functions f_s are smooth, so the problem has been placed in a standard setting. Additional constraints, only in the variables $x_1, \ldots, x_n$, may of course express the set C.

Although this is possible for any finite index set S, the question is whether it is the best way to proceed, and the answer hinges on the size of S as well as the viability of techniques for minimizing f directly. Clearly, when S is very large the proposed reformulation is no panacea. A huge number of constraints, especially nonlinear constraints, is not easy to cope with. The idea becomes attractive of working instead with subproblems in which a convenient local approximation to f, generated somehow from the envelope representation (1), is minimized over C.

When S is infinite, of course, the reformulation leads to an infinite constraint system and a problem of the kind known as *semi-infinite programming*. Indeed, semi-infinite programming could well be classified as the branch of nonsmooth optimization in which this tactic is applied to an objective function, or possibly an inequality constraint function, having an envelope representation.

The drawback to converting a problem with infinite or very large S to semi-infinite programming, or almost semi-infinite programming, is not only that dual dimensionality is increased, but that the focus is shifted away from properties of f that might otherwise be put to very good use. This is where ideas for generating approximations to f get interesting. For an introduction to direct numerical methods in this subject (about which we will have more to say later), the books of Kiwiel [1], Shor [2], and Hiriart-Urruty/Lemarechal [3] are suggested together with the paper of Zowe [4].

3 Eigenvalue Functions

Consider an $m \times m$ symmetric matrix $A(x)$ with entries depending smoothly on $x = (x_1, \ldots, x_n)$ as parameter vector, and let $\lambda_1(x) \geq \lambda_2(x) \geq \cdots \geq \lambda_m(x)$ be the associated eigenvalues (where multiple eigenvalues are repeated). Many applications of optimization involve minimizing a function

$$f(x) = g\big(\lambda_1(x), \cdots, \lambda_m(x)\big), \tag{3}$$

where g is smooth on R^m, or handling a constraint $f(x) \leq 0$ for such a function. A particularly important case is $f(x) = \lambda_1(x)$, where $g(u_1, \ldots, u_m) = u_1$.

Good insights into this situation are provided through the fact that the functions

$$\Lambda_k(x) = \lambda_1(x) + \cdots + \lambda_k(x) \quad \text{for} \quad k = 1, \ldots, m$$

have the envelope representation

$$\Lambda_k(x) = \max_{P \in \mathcal{P}_k} \operatorname{tr}\big(PA(x)P\big), \tag{4}$$

where $\mathcal{P}_k$ is the set of all symmetric $m \times m$ matrices P with rank k such that $P^2 = P$ (i.e., all matrices corresponding to projection mappings onto linear subspaces of R^m of dimension k), and "tr" denotes the trace of a matrix (the sum of its diagonal entries). This fits the pattern of (1) with $\mathcal{P}_k$ as the space S, the "indices" s being matrices P, and $\phi_P(x) = \operatorname{tr}(PA(x)P)$. Obviously one has

$$\lambda_1(x) = \Lambda_1(x), \qquad \lambda_k(x) = \Lambda_k(x) - \Lambda_{k-1}(x) \quad \text{for} \quad k = 2, \ldots, m,$$

so f can just as easily be expressed in the form $h\big(\Lambda_1(x), \ldots, \Lambda_m(x)\big)$ for $h(v_1, \ldots, v_m) = g(v_1, v_2 - v_1, \ldots, v_m - v_{m-1})$.

Especially to be noted is the case where the entries $a_{ij}(x)$ of $A(x)$ depend affinely on x, since then $\operatorname{tr}(PA(x)P)$ is affine in x, and it follows that $\Lambda_k(x)$ is convex in x. This implies $\lambda_1(x)$ is convex in x, while $\lambda_2(x), \ldots, \lambda_{m-1}(x)$ are difference-convex (the difference of two convex functions); $\lambda_m(x)$ is actually affine.

In envelope representations of type (3) the index set is a certain compact continuum within a finite-dimensional vector space. Simple discretization would be ill advised, since it would effectively remove the problem from the realm of eigenvalues, where the algebraic foundations are very rich.

Eigenvalue problems also arise for nonsymmetric matrices $A(x)$ and in this case are tougher, because envelope representations are not at hand. A deeper foray into nonsmooth analysis is required then in identifying the right properties to work with.

For a start on understanding recent work in this branch of nonsmooth optimization, papers of Overton [5] and Overton/Womersely [6] are helpful.

4 Lagrangian Relaxation and Decomposition

A major area leading to nonsmooth optimization is that of decomposition schemes for problems of convex type through Lagrange multipliers. These are closely related to Lagrangian relaxation schemes for getting lower bounds to the minimum in problems of nonconvex or combinatorial type.

Starting from a primal problem in which $f_0(x)$ is to be minimized over a subset $X \subset R^n$ subject to constraints $f_i(x) \leq 0$ for $i = 1, \ldots, m$, we suppose that X is compact and that the functions f_i are all smooth. (We stick to inequality constraints for simplicity, and suppose that a feasible solution exists.) The ordinary Lagrangian function associated with this problem is

$$L(x, y) = f_0(x) + y_1 f_1(x) + \cdots + y_m f_m(x) \text{ for } y = (y_1, \ldots, y_m) \in R_+^m,$$

and the ordinary Lagrangian *dual problem* takes the form

$$\text{maximize } g(y) \text{ over } y \in R_+^m, \text{ where } g(y) = \min_{x \in X} L(x, y).$$

In general, the optimal value in the primal problem (which is finite under the given assumptions) is bounded below by $g(y)$ for any $y \in R_+^m$; the supremum over all such lower bounds is the optimal value in the dual problem. In some circumstances, notably the case where X and the functions $f_0, f_1, \ldots, f_m$ are all convex, the primal optimal value is known to equal the dual optimal value—the greatest lower bound is then exact. When that holds, and $\bar{y}$ is an optimal solution to the dual, the solutions to the primal are precisely the vectors $\bar{x}$ among those that minimize $L(x, \bar{y})$ over $x \in X$ that happen to satisfy the other primal constraints as well.

Whether the primal problem exhibits convexity or not, there is incentive for possibly trying to solve the dual problem as a means of approaching the primal problem, or at least gaining information about it. This is especially true in situations where for some reason the primal problem is difficult to tackle directly because of the constraints $f_i(x) \leq 0$.

Subproblems in which $L(x, y)$ is minimized over $x \in X$ for some choice of y are called *relaxed* problems because, in comparison with the primal problem, they do not deal with the constraints $f_i(x) \leq 0$ but instead try to reflect them in a modified objective function. The optimal value in such a subproblem is, of course, the dual objective value $g(y)$. Solving a relaxed problem thus produces a lower bound to the primal objective value, which might be very useful. This is important for instance in combinatorial optimization problems where X is a discrete set.

To go from a lower bound $g(y)$ to a better lower bound $g(y')$, one obviously has to employ techniques for making an "ascent" on g. The important thing here is that g may be well be *nonsmooth*. On the other hand, the definition of g furnishes an envelope representation (of minimum instead of maximum type) in which the "indices" are the vectors $x \in X$ and the functions $\phi_x(y) = L(x, y)$ are affine in y. Thus, g is always *concave* in this situation, and the strategies that can be utilized toward maximizing g over R_+^m are those of convex programming as adapted to handling functions with an envelope representation.

Decomposition methodology puts an additional twist on this. The best-known case of decomposition uses Lagrange multipliers to take advantage of separability. Suppose in the primal problem that the vector $x \in R^n$ has a natural partition into a number of vector or scalar components: let us write $x = (x_1, \ldots, x_r)$, where the components x_k belong to spaces R^{n_k} (with $n_1 + \cdots + n_r = n$). Suppose further that

$$f_i(x) = f_{i1}(x_1) + \cdots + f_{ir}(x_r) \text{ for } i = 0, 1, \ldots, m,$$
$$X = X_1 \times \cdots \times X_r \text{ with } X_k \subset R^{n_k}.$$

The sets X_k could have constraint representations as well, but for now that kind of detail is unnecessary. The Lagrangian then enjoys the special structure

$$L(x,y) = L_1(x_1, y) + \cdots + L_r(x_r, y)$$
$$\text{with } L_k(x_k, y) = f_{0k}(x_k) + y_1 f_{1k}(x_k) + \cdots + y_m f_{mk}(x_k),$$

and in the dual problem one has

$$g(y) = g_1(y) + \cdots + g_r(y) \text{ with } g_k(y) = \min_{x_k \in X_k} L_k(x_k, y).$$

Solving the dual problem amounts therefore to maximizing $g_1(y) + \cdots + g_r(y)$ over $y \in R_+^m$ in a context where every function g_k has its own envelope representation with parameter index $x_k \in X_k$.

5 Penalty Expressions and Composite Optimization

Penalty terms have most often been viewed as a technical device for dealing with constraints in certain situations, such as within a numerical method. But in applications where caution must carefully be exercised when admitting hard constraints, such as stochastic programming, they have modeling advantages as well, cf. [7], [8].

In proceeding from a problem of minimizing $f_0(x)$ over all $x \in X \subset R^n$ such that $f_i(x) \leq 0$ for $i = 1, \ldots, m$, one can contemplate solving instead a problem of the form

$$\min \ f(x) = f_0(x) + \rho_1\big(f_1(x)\big) + \cdots + \rho_m\big(f_m(x)\big) \text{ over all } x \in X, \tag{5}$$

where each ρ_i is a convex function on R with $\rho_i(0) = 0$. It is helpful in this to allow ρ_i to take on the value ∞, with the understanding that (4) carries the implicit side condition that $f_i(x)$ should belong to the interval in R where $\rho_i < \infty$. The original problem can be identified with having $\rho_i(u_i) = 0$ when $u_i \leq 0$ but $\rho_i(u_i) = \infty$ when $u_i > 0$. The extreme discontinuity of ρ_i in this case underscores the fragility of modeling with hard constraints unless this is strictly necessary.

As alternatives to hard constraints there are rich possibilities. The first that come to mind are classical linear or quadratic penalty terms like $r_i \max\{0, f_i(x)\}$ or $r_i f_i(x)^2$ with $r_i > 0$ as penalty parameter. But ordinary Lagrangian terms $y_i f_i(x)$ fit the picture too, as do augmented Lagrangian terms, which combine multiplier expressions with ultimately quadratic expressions in a piecewise linear-quadratic function

ρ_i with y_i and r_i both as parameters. Still other possibilities for ρ_i are barrier expressions or piecewise linear expressions in f_i like $\rho_i\big(f_i(x)\big) = y_i^+ f_i(x)$ when $f_i(x) \geq 0$, $\rho_i\big(f_i(x)\big) = y_i^- f_i(x)$ when $f_i(x) \leq 0$, in which the parameter values y_i^+ and y_i^- (with $y_i^+ \geq y_i^-$) specify upper and lower bounds to the range of "shadow prices" to be allowed. Again, such a form of expression might be amalgamated with others.

In general, one can think of the usefulness of convex functions ρ_i that are finite on a certain interval, which is partitioned perhaps into subintervals on which ρ_i has different formulas. Although ρ_i is continuous over the entire interval, its first or second derivatives may not be. Then ρ_i exhibits nonsmoothness, and so too does the function f in (4) that needs to be minimized over X. (Constraints not softened by ρ expressions can be imagined here as incorporated into the specification of X.)

Beyond problems of type (4) there are formats involving composition in a broader manner:

$$\text{minimize} \quad f(x) = f_0(x) + \rho\big(f_1(x), \dots, f_m(x)\big) \quad \text{over} \quad x \in X, \tag{6}$$

where ρ is a convex but generally nonsmooth function on R^m. All such problem models belong to *composite* optimization, an important branch of nonsmooth optimization.

A nonsmooth function f of the kind in (4), in which the f_i's themselves are smooth and ρ_i's are relatively simple, has many nice properties which can easily be derived and put to use in minimization, for instance in mimicking something like steepest descent. But that is not the only way to go. Techniques of composite optimization focus instead on generating approximations to f in (4) or (5) by preserving ρ while making approximations to each f_i.

6   Subgradients and subderivatives

Numerical techniques in nonsmooth optimization can be divided roughly into two categories. In direct methods, local properties of the function to be minimized or maximized are developed through variational analysis, including convex analysis, and are utilized in a scheme that emulates well known primal approaches to optimization such as steepest descent, conjugate gradients, and so forth. In indirect methods, modes of approximation arising from Lagrangian functions of one sort or another dominate the scene.

Direct methods depend on the fact that most nonsmooth functions in applications are not just "bad" functions but have firm handles like envelope representations. We will sketch briefly what such a representation in the notation (1) provides. We assume in this that S is a compact space and $\phi_s(x)$ has first-order derivatives in x with these derivatives depending continuously on x and s together. These conditions are trivially satisfied when S is a finite set (in the "discrete topology") and $\phi_s(x)$ is continuously differentiable in s.

First of all, the assumptions guarantee the existence of one-sided directional derivatives that are especially well behaved. At each point $\bar{x}$ and for each vector

$\bar{w}$ the limit,

$$df(\bar{x})(w) = \lim_{t \searrow 0, \, w \to \bar{w}} \frac{f(\bar{x} + tw) - f(\bar{x})}{t}, \tag{7}$$

exists finitely and depends upper semicontinuously on $(\bar{x}, \bar{w})$, in fact continuously on $\bar{w}$. Moreover $df(\bar{x})$, as a function on R^n—called the *subderivative* function for f at $\bar{x}$)—is sublinear, hence convex:

$$df(\bar{x})(w_1 + w_2) \leq df(\bar{x})(w_1) + df(\bar{x})(w_2), \qquad df(\bar{x})(\lambda w) = \lambda df(\bar{x})(w) \text{ when } \lambda > 0.$$

The envelope representation furnishes moreover the formula

$$df(\bar{x})(\bar{w}) = \max_{s \in S_{\bar{x}}} \nabla \phi_s(\bar{x}) \cdot \bar{w} \text{ where } S_{\bar{x}} = \operatorname{argmax}_{s \in S} \phi_s(\bar{x}). \tag{8}$$

(In other words, $S_{\bar{x}}$ is the set of $s \in S$ at which the maximum in the envelope formula (1) for $f(\bar{x})$ is attained.) Secondly, the closed convex set,

$$\partial f(\bar{x}) = \{ v \in R^n \mid v \cdot w \leq df(\bar{x})(w) \text{ for all } w \in R^n \}, \tag{9}$$

which is called the *subgradient* set for f at $\bar{x}$, is nonempty and compact, and it and depends upper semicontinuously on $\bar{x}$, in the sense that the graph of the set-valued mapping $x \mapsto \partial f(x)$ is closed in $R^n \times R^n$. Furthermore, from the envelope representation one has (with "con" standing for convex hull)

$$\partial f(\bar{x}) = \operatorname{con}\{ \nabla \phi_s(\bar{x}) \mid x \in S_{\bar{x}} \}. \tag{10}$$

From these formulas it is evident that to calculate a subgradient of f at $\bar{x}$, all one has to do is determine a single element $\bar{s} \in S_{\bar{x}}$; then $v = \nabla \phi_{\bar{s}}(\bar{x})$. This requires carrying out the maximization of $\phi_x(\bar{x})$ with respect to $s \in S$, a process which yields the function value $f(\bar{x})$ simultaneously which may be easy or hard, depending on the circumstances. In the case of decomposition with Lagrange multipliers, for instance, where y is the variable and x is the "index" and max is replaced by min, it corresponds to solving a family of separate problems in which $L_k(x, \bar{y})$ is minimized with respect to $x_k \in X_k$ for $k = 1, \ldots, r$.

To calculate directional derivatives of f at $\bar{x}$ is harder. If (7) is to be utilized, *all* the elements $s \in S_{\bar{x}}$ may be needed in principle, not just one of them. It is no wonder, then, that direct methods of minimizing a nonsmooth function in terms of an envelope representation have concentrated on strategies that only require calculating a single subgradient at a time, regarding this as an "expensive" operation, although hardly more expensive than function evaluation. This is the pattern followed in [1]–[4]. Of course, in situations where formulas other than (7) are available for directional derivatives, such as many problem models in composite optimization, where function evaluation may be relatively easy as well, the picture is different and another range of techniques can be brought into play.

Background on the mathematics of subderivatives and subgradients can be found in [7] and the books of Clarke [10] and Rockafellar/Wets [11].

7 Approximations through Generalized Lagrangians

In contrast to direct methods in which a function f is minimized through its subderivatives or subgradients, it is possible often to follow a different path leading to the replacement of the given problem by a sequence of easier problems generated through Lagrangian expressions. The chief domain for this kind of approach is composite optimization, in particular the treatment of penalty expressions.

Consider again a problem expressed in the form (4), where the modeling functions ρ_i on R with values in $(-\infty, \infty]$ are convex, and ρ_i is continuous relative to the closure of the (nonempty) interval D_i where $\rho_i < \infty$. An interesting fact of convex analysis is that for such a function ρ_i there is a dual object, a uniquely determined function ψ_i on R having these same properties, and such that

$$\rho_i(u_i) = \sup_{y_i}\{y_iu_i - k_i(y_i)\}, \qquad k_i(y_i) = \sup_{u_i}\{y_iu_i - \rho_i(u_i)\}. \tag{11}$$

In terms of Y_i being the interval of R where $k_i < \infty$, the *generalized Lagrangian function* associated with problem (4) is

$$\mathcal{L}(x,y) = \quad f_0(x) + y_1f_1(x) + \cdots + y_mf_m(x) - k_1(y_1) - \cdots - k_m(y_m) \tag{12}$$
$$\text{for } (x,y) \in X \times Y, \quad \text{where } Y = (Y_1 \times \cdots \times Y_m).$$

This is not some abstraction; the specific form for k_i is well known for the common forms of ρ_i, and in the main cases k_i is smooth on Y_i, in fact typically just quadratic (with $k_i \equiv 0$ as a common special case, the specification of the interval Y_i then being primary). Extension to composite problems in the broader format (5) is easy, but we will not go into that here. An introduction to generalized Lagrangian functions is provided in Rockafellar [9].

The generalized Lagrangian in (11) has, through the first expression in (10), the property that

$$f(x) = \sup_{y \in Y} \mathcal{L}(x,y) \text{ for each } x.$$

This could be viewed as furnishing another kind of envelope representation for f to which optimization techniques already mentioned could be applied, and indeed it does if Y is compact. A valuable insight, however, is that *the generalized Lagrangian $\mathcal{L}$ well captures all the smoothness that might be used in working with f*. Although f may be a very complicated function, with its domain divided into numerous regions associated different formulas for $f(x)$, the function $\mathcal{L}$ is simple.

To understand what can be made of this, consider more closely the case where the functions f_i are twice continuously differentiable, the functions k_i are at most quadratic, and the set X is a box (perhaps all of R^n); this already covers a vast array of applications. Then $\mathcal{L}(x,y)$ is twice continuously differentiable in x and y, in particular concave quadratic or affine in y, and the set $X \times Y$ is a box in $R^n \times R^m$. First and second-order conditions for the optimality of $\bar{x}$ in the problem of minimizing $f(x)$ over $x \in X$ can be expressed in terms of first and second derivatives of $\mathcal{L}$ at $(\bar{x}, \bar{y})$, where $\bar{y}$ is a generalized Lagrange multiplier vector.

Analogs of sequential quadratic programming, for instance, can then be envisioned in which, in rawest form, the idea is to generate a sequence of primal-dual pairs (x^ν, y^ν) for $\nu = 0, 1, \ldots$ by taking $\mathcal{L}^\nu(x, y)$ to be the second-order expansion of $\mathcal{L}$ at (x^ν, y^ν), defining f^ν to be the approximation to f corresponding to this expansion, namely

$$f^\nu(x) = \sup_{y \in Y} \mathcal{L}^\nu(x, y) \quad \text{for each } x,$$

and then obtaining $(x^{\nu+1}, y^{\nu+1})$ as satisfying the optimality conditions for the subproblem minimizing $f^\nu(x)$ over X. (It would also be possible here to pursue notions of "trust region" in replacing $X \times Y$ iteratively by smaller boxes $X^\nu \times Y^\nu$.) In this kind of scheme $\mathcal{L}^\nu$ is linear-quadratic, and the problem of $f^\nu(x)$ over X is said to be one of *extended linear-quadratic programming*. Actually, one gets (for the same functions ρ_i) that

$$f^\nu(x) = f_0^\nu(x) + \rho_1\big(f_1^\nu(x)\big) + \cdots + \rho_m\big(f_m^\nu(x)\big),$$

where f_0^ν is the second-order expansion of $\mathcal{L}(x, y^\nu)$ in x at x^ν and, for $i = 1, \ldots, m$, f_i^ν is the first-order expansion of f_i at x^ν.

Although problems of extended linear-quadratic programming may still display rampant nonsmoothness in the primal objective function f^ν, they have their own characteristics which facilitate computation in other ways. When convexity is present, for example, they can be approached in terms of calculating a saddle point of $\mathcal{L}(x, y)$ with respect to $(x, y) \in X \times Y$. This is a subject in which many new computational ideas have recently been developed. See for instance [12] and its references.

8 Parametric Optimization

Yet another important source of nonsmoothness in optimization is found in decomposition schemes where a problem's variables are divided into "easy" and "hard." Suppose that the ultimate goal is to

$$\begin{aligned} \text{minimize} \quad & f_0(w, x) \text{ over all } (w, x) \in W \times X \\ \text{with} \quad & f_i(w, x) \leq 0 \text{ for } i = 1, \ldots, m, \end{aligned} \tag{13}$$

where $W \subset R^d$ and $X \subset R^n$. (Broader problem models on the order of (4) or (5) could be regarded in the same light.) Imagine that w stands for the "easy" variables, in the sense that, for any fixed $x \in X$, it is relatively easy to compute

$$\begin{aligned} f(x) = \quad & \text{minimum of} \quad & f_0(w, x) \text{ in } w \in W \\ & \text{subject to} \quad & f_i(w, x) \leq 0, \quad i = 1, \ldots, m. \end{aligned} \tag{14}$$

Then there is the residual problem of minimizing $f(x)$ over $x \in X$. If an optimal solution $\bar{x}$ can somehow be found for that, an optimal solution to the underlying problem (12) will be given by $(\bar{w}, \bar{x})$ for any $\bar{w}$ solving (13) for $x = \bar{x}$.

The main obstacle in this situation is, of course, the nonsmoothness of f. The hope is that information pertinent to minimizing f over X can be gleaned from calculations in the subproblems (13) for various choices of x. When W, X, and all the functions f_i are convex, f is at least convex, and special techniques can be used. It may be possible to proceed by dualizing (13) to obtain an envelope representation for f, which for instance is the approach of Benders decomposition. In general, though, an envelope representation may not be obtainable. This kind of nonsmoothness is then the most difficult to handle, because f does not have nice subderivatives and subgradients as described above in terms of such a representation. In certain cases such as those reviewed by Gauvin [13], Lagrange multipliers provide relatively accessible knowledge about directional derivatives. More generally the concepts of subderivative and subgradient have robust extensions to such a context (see [10] and [11]), but their utilization in effective methods of computation has hardly yet been explored.

9   Nonsmoothness of other orders

The discussion has revolved around nonsmoothness in a function f that one wishes to minimize, but other forms of nonsmoothness arise in areas of optimization where optimality conditions from one problem are introduced as constraints in another problem, or simply when attempts are made to solve first-order optimality conditions as if they resembled a system of nonlinear equations. This is the subject of *generalized equations*.

As a key example, a problem of the broad type (4), which covers traditional optimization problems as the case where ρ_i is 0 on $(-\infty, 0]$ or $[0,0]$ but ∞ elsewhere, has first-order optimality expressible in terms of the generalized Lagrangian (11) by

$$-\nabla_x \mathcal{L}(\bar{x}, \bar{y}) \in N_X(\bar{x}), \qquad \nabla_x \mathcal{L}(\bar{x}, \bar{y}) \in N_Y(\bar{y}), \tag{15}$$

where $N_X(\bar{x})$ is the normal cone to X at $\bar{x}$ and $N_Y(\bar{y})$ is the normal cone to Y at $\bar{y}$. When X and Y are boxes, for instance, these normal cone conditions reduce to sign conditions on the components of $\bar{x}$ and $\bar{y}$ and the partial derivatives of $\mathcal{L}$. The pairs $(\bar{x}, \bar{y})$ are the generalized Kuhn-Tucker points associated with the problem.

Consider in this vein the sets

$$
\begin{aligned}
G &= \big\{ (x, y, u, v) \in R^n \times R^n \times R^n \times R^n \,\big| \\
&\qquad\quad -\nabla_x \mathcal{L}(\bar{x}, \bar{y}) + v \in N_X(\bar{x}),\ \nabla_x \mathcal{L}(\bar{x}, \bar{y}) - u \in N_Y(\bar{y}) \big\}, \\
M &= \big\{ (x, y, u, v) \in R^n \times R^n \times R^n \times R^n \,\big|\, u = 0,\ v = 0 \big\}.
\end{aligned}
$$

Trying to determine $(\bar{x}, \bar{y})$ can be viewed as trying to find an element $(\bar{x}, \bar{y}, \bar{u}, \bar{v}) \in G \cap M$. The idea comes up then of devising algorithms patterned after ones that might work if G were a smooth manifold given by nice, nondegenerate equations. For instance, one can imagine creating a sequence of local first-order approximations

G^ν to G at points $(x^\nu, y^\nu, u^\nu, v^\nu)$, where in basic concept $(x^{\nu+1}, y^{\nu+1}, u^{\nu+1}, v^{\nu+1})$ is determined as a point of $G^\nu \cap M$ in a Newton-like scheme.

The challenge here is that G is not just a smooth manifold, and does not have first-order approximations in the sense of classical linearizations. It is a *nonsmooth* manifold, moreover of a kind requiring an advanced form of nonsmooth analysis. But actually the properties of G are convenient and attractive nevertheless. Natural and simple first-order approximations do exist. In particular, these can be obtained through linearizing $\nabla \mathcal{L}$, i.e., working in effect with quadratic approximations to $\mathcal{L}$ as already discussed.

An introduction to the methodology being developed for solving nonsmooth equations is furnished in Pang/Qi [14].

References

[1] K. C. Kiwiel, *Methods of Descent for Nondifferentiable Optimization*, Lecture Notes in Math. 1133, Springer-Verlag, Berlin, 1985.

[2] N. Z. Shor, *Minimization Methods for Non-Differentiable Functions*, Springer-Verlag, Berlin, 1985.

[3] J. B. Hiriart-Urruty and C. Lemaréchal, *Convex Analysis and Minimization Algorithms, I & II*, Springer-Verlag, Berlin, 1993.

[4] J. Zowe, "The BT-algorithm for minimizing a nonsmooth functional subject to linear constraints," *Nonsmooth Optimization and Related Topics*, F. H. Clarke, V. F. Dem'yanov and F. Giannessi (editors), Plenum Press, 1989, 459-480.

[5] M. L. Overton, "Large-scale optimization of eigenvalues," *SIAM J. Optimization*, 2 (1992), 88-120.

[6] M. Overton and R. S. Womersley, "Optimality conditions and duality theory for minimizing sums of the largest eigenvalues of symmetric matrices," *Math. Programming*, (1993).

[7] R. T. Rockafellar and R. J. B Wets, "A Lagrangian finite generation technique for solving linear-quadrataic problems in stochastic programing," *Math. Programming Studies*, 28 (1986), 63-93.

[8] R. T. Rockafellar and R. J. B Wets, "Linear-quadratic problems with stochastic penalties: the finite generation algorithm," in *Stochastic Optimization*, V. I. Arkin, A. Shiraev and R. J-B Wets (editors), Lecture Notes in Economics and Math. Systems, Springer-Verlag, Berlin, 255 1987, 545-560.

[9] R. T. Rockafellar, "Lagrange multipliers and optimality," *SIAM Review*, 35 (1993), 183-238.

[10] F. H. Clarke, *Optimization and Nonsmooth Analysis*, Wiley, New York, 1983; republished by the Centre de Recherches Mathématiques, Université de Montréal, C.P. 6128A, Montréal, Quebec, Canada.

[11] R. T. Rockafellar and R. J. B. Wets, *Variational Analysis*, forthcoming book.

[12] C. Zhu and R. T. Rockafellar, "Primal-dual projected gradient algorithms for extended linear-quadratic programming," *SIAM J. Optimization*, 3 (1993), 751-783.

[13] J. Gauvin, "Directional derivative for the value function in mathematical programming," *Nonsmooth Optimization and Related Topics*, F. H. Clarke, V. F. Dem'yanov and F. Giannessi (editors), Plenum Press, 1989, 167-183.

[14] J. S. Pang and L. Qi, "Nonsmooth equations: motivation and algorithms," *SIAM J. Optimization*, 3 (1993), 443-465.

Using Dynamical Chaos to Explain Voting and Statistical Paradoxes

Donald G. Saari

Department of Mathematics
Northwestern University
Evanston, Illinois

1 Introduction

Paradoxes are intriguing. By a paradox, I mean a mathematically counterintuitive conclusion. To illustrate, the likelihood of "Heads" with a flipped penny is approximately $\frac{1}{2}$. Presumably, the same answer holds for a penny spinning on edge, but, instead, it is only about 0.30. This is a surprise, not a paradox because an examination discloses that the slightly heavier "Head" tilts the axis of rotation. A "paradox" requires a more subtle mathematical structure.

Suppose 15 friends decide to choose a common beverage [31]. Their preferences
- six have "milk $\succ$ wine $\succ$ beer,"
- five have "beer $\succ$ wine $\succ$ milk," and
- four have "wine $\succ$ beer $\succ$ milk,"

define the plurality ranking "milk $\succ$ beer $\succ$ wine" with the vote $6 : 5 : 4$. If bottom-ranked wine is not available, we expect nothing to be changed; milk should remain the top-choice. However, 60% of this group prefers "beer $\succ$ milk" with the vote of $9 : 6$! They even prefer "wine $\succ$ milk" and "wine $\succ$ beer" with similarly large votes. These pairwise comparisons, then, suggest that the plurality ranking completely reverses what the voters really want! Instead of being the beverage of choice, these voters view milk as the inferior alternative. This is a paradox.

A probability example can be created by replacing the usual markings on three fair dice with the numbers (from a magic square)

$$
\begin{array}{cccc}
A & \boxed{8} & \boxed{1} & \boxed{6} \\
B & \boxed{3} & \boxed{5} & \boxed{7} \\
C & \boxed{4} & \boxed{9} & \boxed{2}
\end{array}
\qquad (1.1)
$$

where each number appears twice. Each person chooses and rolls a die where high score wins. The sample set for any two dice has nine pairs, so a die with the larger value in five or more pairs is the better die. A count proves that $A \succ B$, $B \succ C$, so it is reasonable to anticipate that $A \succ C$; instead, $C \succ A$.

Among the many statistical paradoxes, suppose a new cold remedy is tested in Evanston and in Recife where in both sites the data supports the new remedy. For figures, suppose out of the 144 subjects from Evanston who opt for the experimental approach, 54 regain health. From the 36 using the standard approach, only 12 regain health. As $\frac{54}{144} = \frac{3}{8} > \frac{12}{36} = \frac{1}{3}$, the experimental drug is better. In Recife, of the 36 who opted for the experimental approach, half regain health. Of the 144 in the control group, 66 regain health. As $\frac{1}{2} > \frac{66}{144}$, the experimental drug dominates again. In both locales, the experimental drug proved better than the standard approach, so, presumably, it should be chosen. In the aggregate with 180 subjects in each group, however, only 72 from the experimental group regained health compared to the 78 from the control group. This *Simpson's Paradox* (e.g., see [3, 10, 14]) demonstrates that the aggregated data can reverse the conclusion!

Paradoxes are amusing, but when they are recast in terms of an election for the departmental chair, the selection of the candidate for the sole tenure track position, a Presidential Candidate, the choice of economic or political decisions, or comparing medical tests it becomes evident that these mathematical puzzles have meaningful consequences that must be addressed. Indeed, with the vast numbers of daily decisions made in various contexts, one must expect that many of the mathematical paradoxes from the decision and statistical sciences have been manifested by groups unknowingly selecting inferior alternatives. This, in itself, underscores the importance of understanding these counterintuitive surprises.

These are well-studied issues, but progress has been severely limited. The reason is clear; a "paradox" is counterintuitive, so what does one look for? This is why, even with the huge literature for each of these fields (e.g., for voting, see the 72 page bibliography [15]), only relatively few paradoxes and properties of ranking systems have been found. In this article, I outline a new approach based on "chaotic dynamics" and symmetry groups that overcomes these limitations by extracting all possible ranking paradoxes and properties that can occur.

2   A Chaotic State of Affairs

The chaotic examination of aggregation procedures is based on the mathematical structures of "chaos" as outlined next. It is important to stress that those conclusions, such as period-doubling, homoclinic points, strange attractors, etc., are not, and cannot occur in an analysis of paradoxes. These results intimately depend upon particular structures of dynamical systems that are missing here. Instead, borrowed from dynamical systems is the change of perspective from a local to a global emphasis along with the mathematical insight motivating the development of "symbolic dynamics." Once modified, these conceptual ideas lead to the development of techniques to analyze the paradoxes of aggregation procedures. Moreover, this "borrowing" from dynamics also identifies a research program of parallel issues for discrete and "static" problems.

A way to review basic ideas from dynamics is to consider Newton's method [32] for finding a zero of a polynomial $f(x)$. (For a more general introduction, see [5].) If the initial guess, x_1, fails because $f(x_1) \neq 0$, then f is replaced by its linear approximation, $f(x_1) + (x - x_1)f'(x_1)$; the zero of this approximation, x_2, is the next iterate. As such, Newton's method has the standard geometric representation of Figure 1. Of course, the iteration procedure fails at a critical point of f because the horizontal linear approximation has no zeros.

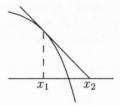

Figure 1: Newton's Method

Suppose we want to find the roots of the fifth-degree polynomial depicted in Figure 2 where the critical points of f define the endpoints of the three labeled intervals. We know that if x_1 is in an unbounded interval, the iteration converges to the zero of that interval. The remaining challenge, then, is to understand what happens should all iterates of an initial point remain in $a \cup b \cup c$. Some iterates converge to a zero of f in this region, but what else can happen?

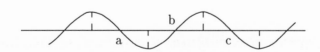

Figure 2: Regions for a fifth degree polynomial

One way to discover "nonconvergent" properties of Newton's method is to examine the behavior of the orbit $x_1, x_2, \ldots, x_n, \ldots$ for various choices of x_1. For instance, if x_{61} ends up near x_1, it is reasonable to expect from continuity considerations that x_1 can be modified to force $x_1 = x_{61}$; this means we should expect a period-sixty orbit to exist. This is a simplified, yet, not inaccurate description of how various properties were obtained. This approach, with its emphasis on local properties, can be technically difficult. Yet, by being property-specific, it does not tell us what else might happen. For instance, we must anticipate certain periodic orbits to exist; e.g., period-two orbits are easy to construct. Are there any other kinds? Can anything other than periodic orbits happen?

As an alternative way to address these questions, after $\{x_1, x_2, x_3, \ldots\}$ is determined, replace each iterate with a label identifying the interval in which it lives. For example, if $x_1 \in a, x_2 \in b, x_3 \in c, x_4 \in b, \ldots$, then the initial condition x_1 and its orbit defines the sequence $g_f(x_1) = (a, b, c, b, \ldots)$. As this listing of symbols is not random (it is ordered by the dynamics), call it a *word*. So, if $U^3 = \{a, b, c\}^N$ is the *universal set* consisting of all possible sequences where each entry is one of the three symbols, then Newton's method defines a mapping $g_f : a \cup b \cup c \to U^3$ where each initial iterate is identified with its word – an element of U^3. A word, then, specifies one kind of outcome. One way to measure the complexity of Newton's method is to find all possible words; that is, to find all of the entries in the *dictionary*

$$\mathcal{D}_f = \{g_f(x_1) \in U^3\} \subset U^3. \tag{2.1}$$

Common sense dictates that should $\mathcal{D}_f$ be a large subset of U^3, then Newton's method admits a rich supply of complex, chaotic dynamics. On the other hand, with a limited number of words, not much can happen, so if $\mathcal{D}_f$ is a small subset of U^3, a reasonably benign dynamic must be anticipated. Thus the size of $\mathcal{D}_f$ serves as a crude complexity measure of the process. A way to understand Newton's method, then, is to characterize its dictionary $\mathcal{D}_f$. Notice the change in emphasis; instead of detecting particular features of Newton's method, the more ambitious new goal is to completely catalogue all long term dynamical properties (i.e., all sequences in $\mathcal{D}_f$). This change in emphasis to favor global over local properties is key for our analysis and classification of aggregation paradoxes.

As developed in [32], Newton's method is as complex and chaotic as possible with this dictionary measure because

$$\mathcal{D}_f = U^3. \tag{2.2}$$

Namely, choose any sequence generated by the letters a, b, c – an entry can even be determined by rolling a die – and there is an initial iterate whereby the jth iterate is in the interval specified by the jth entry of the sequence. For instance, if the chosen sequence is $s = (b, a, c, a, c, b, b, \ldots,)$, Eq. 2.2 asserts there exists $x_1 \in b$ so that $x_2 \in a, x_3 \in c, x_4 \in a, x_5 \in c, \ldots$. With only slight extra effort, it follows that periodic orbits of any period following any pattern through the three intervals exist! Moreover, there also are orbits that bounce forever among the intervals in counterintuitive, non-periodic ways.

Particularly important for the analysis of paradoxes is that because this technique characterizes all orbits, it answers the puzzling problem of how to discover counterintuitive properties. So, by extending this dictionary approach to aggregation procedures, it may be realistic to search for "everything that can happen;" namely, we can hope to discover all possible ranking paradoxes and properties. With this objective in mind, we need to understand how to find the entries of $\mathcal{D}_f$.

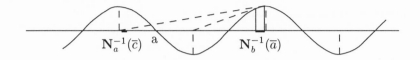

Figure 3: The iterated inverse image

The method of proof (to establish that g_f is surjective) uses an "iterated inverse image" approach that I will illustrate with the sequence $s = (b, a, c, a, c, b, b, \ldots,)$. The goal is to keep refining the set of initial iterates that accomplishes each portion of the designated future. To see how to do this, if N_k is Newton's map restricted to interval $k = a, b, c$, then the set of all iterates that start in b and end in a is $N_b^{-1}(\overline{a})$ where $\overline{a}$ is the closure of the interval a. The key fact is that N_k maps the interval k onto $(-\infty, \infty)$. To see why, notice that as x moves closer to the left-hand endpoint of interval k, the straight line approximation of f becomes horizontal forcing its zero (the next iterate $N_k(x)$) far to the right. Similarly, for values of $x \in k$ near the right-hand endpoint, the line again approaches an horizontal position, but now it crosses the x-axis far to the left. Because $N_k : k \to (-\infty, \infty)$ is surjective, $N_b(\overline{a})$ is a nonempty closed subset of b. Geometrically, $N_b(\overline{a})$ is easy to determine as depicted in Figure 3.

$N_b^{-1}(\overline{a})$ are the points in b that are mapped to $\overline{a}$. We do not care about reaching all points in a; only those for which the next iterate is in c. Using the same argument, we only want to reach $N_a^{-1}(\overline{c}) \subset a$. Thus, we want to find the subset of $N_b^{-1}(\overline{a})$ that restricts its next iterate to the much smaller target region $N_a^{-1}(\overline{c})$ (so that iterates start in b, go to a and then to c); clearly, this refinement is

$$b \supset N_b^{-1}(\overline{a}) \supset N_b^{-1}(N_a^{-1}(\overline{c})).$$

The idea now is obvious. To find all initial points satisfying the designated future, continue this iterated inverse image approach to obtain the nested sequence of nonempty, closed, bounded sets

$$b \supset N_b^{-1}(\overline{a}) \supset N_b^{-1}(N_a^{-1}(\overline{c})) \supset \ldots \supset N_b^{-1}(N_a^{-1}(N_c^{-1}(\ldots$$

$$\ldots (N_k^{-1}(\overline{k+1})\ldots)) \supset \ldots \tag{2.3}$$

A point in the intersection of all sets serves as an initial iterate satisfying the conditions of the sequence. As standard results from elementary analysis ensure that this intersection is nonempty, such an initial iterate exists.

Intuitively, the source of certain assertions, such as "sensitivity to initial conditions" and "Cantor set" constructions now are clear. Observe from Figure 3 that the expanding nature of N_b requires $N_b^{-1}(\bar{a})$ to be a small set and $N_b^{-1}(N_a^{-1}(\bar{c}))$ to be much smaller. Yet, this small set $N_b^{-1}(N_a^{-1}(\bar{c}))$ contains all b points that pass through a and then are mapped onto c. As such, $N_b^{-1}(N_a^{-1}(\bar{c}))$ contains a distinct subset for each of the uncountable ways $(b, a, c, \ldots)$ can be continued. Consequently, most points of $N_b^{-1}(N_a^{-1}(\bar{c}))$ must be near regions with radically different orbits, so a slightly varied initial point could change regions and embark on a dramatically different future. This is the source of the sensitivity. Similarly, for each extension of $(b, a, c, \ldots)$, each step of the iterated inverse image approach identifies all points that eventually are mapped onto the next specified interval. Among these points is an open set ensuring convergence to the zero of that interval. Thus, an open set is excised at each step in constructing the set of nonconvergent points – just as in the construction of the standard middle-thirds Cantor set.

3   Return to Paradoxes

In voting theory and statistics, success in discovering paradoxes and troublesome properties of procedures traditionally requires finding specific examples of voters' preferences or data. For instance, one of the better known voting paradoxes, designed by P. Fishburn [20], illustrates the interesting behavior where the voters' sincere plurality election ranking of $A \succ B \succ C \succ D$ is reversed to $C \succ B \succ A$ when D, the bottom ranked candidate, withdraws. Part of the difficulty is to even suspect such behavior could occur. The method of proof – finding a supporting example of voters' preferences – often leads to combinatoric difficulties.

Observe the close parallels between this traditional approach to analyze voting and an analysis of Newton's method guided by properties of particular initial iterates. As both approaches concentrate on local properties, it is difficult, if not impossible, to even suspect what else could occur. (For instance, with the above reversal example, what could happen if C also decides to withdraw; or, if D returns but now A withdraws?) Then, even with limited results, the supporting proofs can be technically difficult. To remove these obstacles, one might hope to mimic the dynamical systems approach (applied above to Newton's method) to transform these decision problems into a mathematical framework emphasizing "global properties." I will introduce the ideas with the beverage example.

In dynamics, the initial condition x_1 is the "starting point." Voting, on the other hand, begins with a listing of each voter's preferences, a *profile*. (For instance, the listing of voter's preferences in the beverage example defines a particular 15-voter profile.) Thus, *a profile serves as the "initial condition" for voting*; there are no restrictions on the number of voters.

Key to dynamics is the ordering of outcomes as imposed by "time;" in the Newton sequence this is indicated by the subscript of the jth term, x_j. A natural "time" variable does not exist for voting as the goal is to compare election outcomes over different subsets of candidates. But, we can invent one. For instance, with $n = 3$ candidates $\{c_1, c_2, c_3\}$, the four subsets with two or more candidates are $S_1 = \{c_1, c_2\}$, $S_2 = \{c_1, c_3\}$, $S_3 = \{c_2, c_3\}$, $S_4 = \{c_1, c_2, c_3\}$. In general with $n \geq 3$ candidates, list the $2^n - (n+1)$ subsets of two or more candidates (in some fashion) as $S_1, S_2, \ldots, S_{2^n - (n+1)}$. Replacing the ordering role of *"time"* from dynamics, then, is the integer indicating which subset of candidates is being considered.

In the Newton example, we are interested in the three subintervals labeled with the symbols a, b, c. Then, the precise value of the jth iterate, x_j, is replaced with the cruder information identifying the subinterval in which it lives. In voting, the subscript j represents the subset of candidates S_j. Corresponding to the precise value of an iterate is the precise election tally of the S_j candidates; replace this tally with the cruder information specifying the election rankings. Thus, replacing the set of symbols $\{a, b, c\}$ for Newton's method is $\mathcal{R}(S_j)$ – the set of all possible election rankings of S_j. For instance, $S_3 = \{c_2, c_3\}$ has the three symbols $\mathcal{R}(S_3) = \{c_2 \succ c_3, c_2 \sim c_3, c_3 \succ c_2\}$ indicating that either one candidate beats the other, or they tie. Similarly, $\mathcal{R}(S_4)$ has thirteen entries; six correspond to the six ways candidates can be ranked without ties, six are where there is one tie, and one corresponds to a complete tie.

In Newton's method, the Universal set consists of all possible sequences that can be constructed. Each entry is one of the three choices, so

$$U = \begin{pmatrix} a \\ b \\ c \end{pmatrix} \times \begin{pmatrix} a \\ b \\ c \end{pmatrix} \times \begin{pmatrix} a \\ b \\ c \end{pmatrix} \times \ldots = \begin{pmatrix} a \\ b \\ c \end{pmatrix}^N$$

In voting, the jth entry is the symbol from $\mathcal{R}(S_j)$ designating the ranking of the candidates from S_j. Therefore, the universal set is

$$\mathcal{U}^3 = \mathcal{R}(S_1) \times \mathcal{R}(S_2) \times \mathcal{R}(S_3) \times \mathcal{R}(S_4).$$

In the general case of n candidates, where there are $2^n - (n+1)$ subsets of two or more candidates, the universal set is

$$\mathcal{U}^n = \prod_{j=1}^{2^n - (n+1)} \mathcal{R}(S_j).$$

In Newton's method, an initial point determines a sequence $g_f(x_1)$. In voting, a given profile p determines the election ranking for each subset of candidates. Denote this listing, or *word,* by $F(p)$. For instance, if $c_1 = $ milk, $c_2 = $ beer, $c_3 = $ wine and p is the profile of the beverage example, then p defines the word

$$F(p) = (c_2 \succ c_1, c_3 \succ c_1, c_3 \succ c_2, c_1 \succ c_2 \succ c_3).$$

The complexity of Newton's method is analyzed by comparing the size of its dictionary with that of the universal set. Similarly, in voting, one way to understand "everything that can occur" is to characterize the *dictionary of election outcomes*

$$\mathcal{D}^n = \{F(p) \in \mathcal{U}^n \mid p \text{ is profile, no restriction on number of voters}\}. \tag{3.1}$$

By introducing the dictionary, which consists of all possible words (listings of sincere election outcomes), the emphasis is switched from local characteristics to a search for global properties. In particular, it now makes sense to search for everything that possibly could happen; i.e., to find all words in the dictionary. If this could be done, the entries of the dictionary would specify *all possible relationships and paradoxes of plurality election rankings.*

4 Comparing Dictionaries

As with Newton's method, the first goal is to determine whether $\mathcal{D}^n$ is a large or small subset of $\mathcal{U}^n$. Clearly, $\mathcal{D}^n$ contains all well-behaved words where the election ranking of each subset behaves as expected because they agree with one another; e.g., $(c_1 \succ c_2, c_1 \succ c_3, c_2 \succ c_3, c_1 \succ c_2 \succ c_3) \in \mathcal{D}^3$. All remaining words lack this orderly property, so they identify "paradoxes." Consequently, a dictionary with only a few words represents an orderly situation. Conversely, a large dictionary, with its rich selection of words specifying different ways election outcomes can vary over the subsets of candidates, suggests a chaotic state of affairs admitting many new election paradoxes. So, is $\mathcal{D}^n$ is a large or small subset of $\mathcal{U}^n$?

Theorem 4.1 ([24]) *For the plurality voting system, where each voter votes for his or her top-ranked candidate, and for any $n \geq 3$, we have that*

$$\mathcal{D}^n = \mathcal{U}^n. \tag{4.1}$$

This disturbing conclusion allows *any* sequence of rankings, one ranking chosen for each subset of candidates, to be admissible election outcomes! As such, there is a profile supporting the election outcomes $c_1 \succ c_2, c_3 \succ c_2 \succ c_1, c_1 \succ c_2 \succ c_3 \succ c_4$, $c_5 \succ c_4 \succ c_3 \succ c_2 \succ c_1, \ldots$ where one kind of ranking applies with an even number of candidates but the reverse holds with an odd number. In fact, this example extends to require all subsets with an even number of candidates to be ranked consistent with $c_1 \succ c_2 \succ \ldots \succ c_n$, and all other subsets to be ranked in the opposite manner. As another illustration, this theorem allows us to extend the beverage example to the extreme setting where the group's ranking is $c_1 \succ c_2 \succ c_3 \succ \ldots \succ c_n$, yet their ranking of *all proper subsets of candidates* is the exact reversal! To help the reader relate to this example, suppose this occurs when your department ranks n candidates for a single tenure track position. The first vote suggests that c_1 is the favorite, but, is she? Should *any* candidate withdraw, the department's sincere ranking is completely reversed! Who should get the offer; c_1 or c_n? Even worse, rarely do we hold another election when some candidates withdraw, so we might never discover that we have chosen badly due to this behavior.

4.1 Extension of the Literature

Because the traditional literature tends to be property- and example-specific, it is reasonably clear how this theorem extends what was previously known. (I recommend the nice survey by Niemi and Riker [20] and Ordeshook's book [22].) Namely, with the traditional emphasis on local properties, much of the literature consists of special examples of the Fishburn type mentioned above. Recall, his reversal example proves there exists a profile defining the contradictory plurality rankings $A \succ B \succ C \succ D$ and $C \succ B \succ A$. But, there is no way to guess from this example what else can accompany this behavior. For instance, it does not address what might happen should C, or B, or A, rather than D, withdraw from the competition. The above theorem, however, provides the answer; it asserts that the subsequent ranking can be whatever you want it to be! Just one possibility is where *the ranking is reversed whenever any candidate (not just the bottom ranked one) withdraws*. In fact, the theorem allows you to extend this example in any imaginable way over all other subsets of candidates including the pairs. Similarly, all other examples of this type in the literature now can be significantly extended in previously unimaginable ways.

As another illustration, a troubling behavior known for two centuries and exploited in interesting, unexpected ways by contemporary authors including Arrow [1], McKelevey [18], and Kramer [16], is where the sincere pairwise elections define cycles such as $c_1 \succ c_2$, $c_2 \succ c_3$, $\ldots$, $c_{n-1} \succ c_n$, $c_n \succ c_1$. (As shown in [31], cycles and their supporting profiles can be understood in terms of the orbit of a cyclic group of order n.) To the best of my knowledge, cycles involving larger subsets of candidates have not been discovered, and for a good reason – it is difficult to find supporting profiles. It follows trivially from the theorem, however, that such phenomenon exists. The theorem requires, for instance, that there exist voters' preferences defining the plurality cycle $c_1 \succ c_2 \succ c_3$, $c_2 \succ c_3 \succ c_4$, $c_3 \succ c_4 \succ c_1$, $c_4 \succ c_1 \succ c_2$. In fact, this example can be made more complicated by requiring these voters' election ranking for the pairs to cycle in the "other" direction $c_2 \succ c_1$, $c_1 \succ c_n$, $\ldots$. Again, this holds because whatever rankings are specified for the different subsets, the theorem ensures that the plurality method allows them to coexist!

Another theme receiving considerable attention (e.g., see [20, 21] and the references they list) is to understand what can happen with procedures, such as runoff elections or tournaments, where after eliminating certain candidates another plurality vote is taken. For instance, could a preferred candidate be dropped at the very first step? How about a candidate who always is favored when compared with any other candidate; could she lose with a runoff election? Thanks to the theorem, we now can answer almost all questions of this type about almost all procedures. To illustrate with elimination methods, just rank the initial set of candidates so that c_1 is dropped. Then, for all other subsets containing c_1, have her top-ranked. Such examples, where c_1 is arguably the voters' favorite even though she is dropped at the first stage, exist because the theorem admits *any listing of rankings* for different subsets of candidates. So, while the literature answers certain specific issues, this theorem motivated by dynamics provides a far more general, yet almost a trivial way to resolve a much wider spectrum of questions.

A related theme is to determine whether the choice of a procedure can influence the outcomes. (Again, see [15, 20, 21, 22].) For instance, to choose who to hire for the one departmental opening we could use the plurality ranking, or we could plurality rank the candidates in subsets of five and then plurality rank the resulting set of winners. But, just because these procedures involve different subsets of candidates, it is immediate from the theorem that their outcomes can be as different as desired. Proving this conclusion only involves assigning appropriate rankings to specified subsets of candidates. Once done, we can further exploit the many other subsets of candidates without an assigned ranking. For instance, by choosing appropriate rankings for these sets, we prove immediately that the outcome is sensitive to how the candidates are assigned to the five-candidate subsets, or whether three, or four, or six, ... candidate subsets are used, or whether procedures involve pairwise votes, or whether a runoff election is used. What a chaotic state of affairs! (General results of this new type, of course, are not possible with the traditional local analysis.)

As a final troublesome corollary, note that we can choose the rankings for each of the $2^n - (n + 1)$ subsets in a completely random fashion. Even though these election rankings need not have anything to do with one another, the theorem ensures there is a profile where the sincere election outcome for each subset of candidates is the randomly selected one! This conclusion, which again demonstrates the power of this global perspective, does not instill much confidence in our standard tool of democracy.

4.2 Other Election Methods

Why use the plurality vote? Instead of just voting for our top-ranked candidates, maybe we should recognize voters' lesser ranked candidates. This approach was pioneered by the French mathematician J.C. Borda in 1770. He argued that with k candidates, a voter's top-ranked candidate should receive $k - 1$ points, the second should receive $k - 2$ points, ..., the jth ranked candidate should receive $k - j$ points,..., $j = 1, 2, \ldots, k$. More generally, the *Borda Count* (BC) is where the difference between points assigned to successive candidates is the same positive scalar. (So, for $n = 4$ candidates, both $(3, 2, 1, 0)$ and $(25, 20, 15, 10)$ define Borda Counts.)

Why this particular choice? Why not some other arrangement? After all, any vector $w = (w_1, w_2, \ldots, w_k = 0)$ suffices as long as $w_j \geq w_{j+1}$ for $j = 1, \ldots, k - 1$, and $w_1 > 0$. With such a *voting vector*, w_j points are assigned to a voter's jth ranked candidate and a candidate's election ranking is based on how many points she receives. The responses of Borda, Laplace, and other mathematicians from the late eighteenth century supporting the BC are more philosophical than mathematical so they are unsatisfactory. Nevertheless, the BC was used for years in the French Academy until changed by a new member Napoleon Bonaparte.

To investigate these questions, for each subset of candidates S_j, assign a voting vector w_j. This defines a *system voting vector*

$$W^n = (w_1, w_2, \ldots, w_{2^n - (n+1)}).$$

Thus, with an assigned system voting vector and a profile, the ranking of the jth subset of candidates S_j is found by tallying the ballots with w_j. (Think of w_j as defining the "dynamic" on the S_j portion of the product space.) Let $F(p, W^n)$ be the resulting listing of these election rankings; $F(p, W^n)$ is the W^n word defined by p.

By admitting all possible profiles, we obtain the W^n dictionary

$$\mathcal{D}^n(W^n) = \{F(p, W^n) \mid p \text{ is a profile}\}.$$

Again, to understand the consequences of different system voting vectors, use the dictionary measure. Because system voting vectors can be identified with points in an appropriate dimensional Euclidean space, we use these mathematical structures to describe sets of system vectors. Let B^n be the system vector where the BC is used with each subset of candidates.

Theorem 4.2 ([24, 25]) *For $n \geq 3$, with the exception of a lower dimensional algebraic set α^n, all system voting vectors have the property that*

$$\mathcal{D}^n(W^n) = \mathcal{U}^n.$$

If $W^n \neq B^n$, then

$$\mathcal{D}^n(B^n) \subset \mathcal{D}^n(W^n). \tag{4.2}$$

In general, then, the election outcomes can be as chaotic as desired! Only those system voting vectors belonging to a particular algebraic set (which recently has been characterized [26, 27] and used to define a partial ordering for system vectors) can avoid certain paradoxes. Also, because Eq. 4.2 identifies the BC as the unique method to minimize both the number and kinds of paradoxes that can occur (for any $n \geq 3$), we finally have the sought after mathematical justification supporting the optimal status of the BC. With the beverage paradox, for instance, the BC ranking is the more reasonable "wine $\succ$ beer $\succ$ milk" with the tally $19 : 14 : 12$. Indeed, the BC is the *only* voting method where its outcomes must be related to the pairwise rankings. For any other method, choose the rankings of the $\binom{n}{2}$ pairs and the set of all n candidates in any desired manner and there is a supporting profile; only the BC imposes order upon these electoral conclusions [25, 31].

But, does this difference in dictionaries matter? If the BC avoids only a couple of paradoxes, then who cares? However, as dramatically illustrated with the inequality $10^{50}|\mathcal{D}(B^6)| < |\mathcal{D}(\text{Plurality}^6)|$, already with just 6 candidates the Borda Dictionary allows shockingly fewer paradoxes than, say, the plurality method. Many other arguments along this line, along with the characterization of the entries of $\mathcal{D}(B^n)$ [25], can be advanced to prove that the BC is, by far, the superior choice. (See, for example, [26, 31].) This includes the single transferable vote used by the AMS, Approval Voting by the MAA, and plurality voting by SIAM.

4.3 Extending the Literature

A natural mathematical theme is to discover invariants – here the invariants are relationships among election rankings. Of the comparatively few that were previously known is one asserting that the top-ranked BC candidate never could be beaten in all possible pairwise elections. (Most surely, Borda knew this result; it definitely was understood by Nanson [19] in the nineteenth century and then rediscovered by several others (including me) in recent years starting with Smith's nice paper [33].) As already indicated, this perverse phenomenon does occur with all other positional procedures!

An "election relationship," then, is the complement of the dictionary; it can be thought of as specifying what "paradoxes" cannot occur. Therefore, these ranking relationships are completely determined by the set $\mathcal{U}^n \backslash \mathcal{D}^n(W^n)$. So, now that we know ([26, 27]) how to find the entries of all dictionaries $\mathcal{D}^n(W^n)$, *we also know all possible ranking relationships.* (Again, just as in dynamics, this is a consequence of changing from a local to a global perspective.) And, just by the number count given above for $n = 6$ candidates, it is clear that this dictionary approach uncovers an incredible number of new relationships previously not even suspected. One new Borda relationship, for instance, is that if there is an integer k, $2 \le k < n$, so that c_1 is top-ranked in all k-candidate elections, then she is not bottom-ranked in the full election. Other new kinds of Borda relationships impose restrictions on the possible k-candidate rankings for each $k > 2$, etc.

Further insight into how this theorem extends the literature exploits the assertion that, in general, "anything can happen." As such, all of the earlier comments about runoffs, comparing procedures, etc. immediately extend from plurality vote to almost all other positional voting methods. Again, most results (i.e., all that I know about) using the traditional local emphasis can compare only limited classes of procedures addressing highly restricted questions. With this global approach (as true for dynamical systems), many of these conclusions can be extended in almost all ways with minimal effort.

To further illustrate the advantages of this global approach, I point to the attention focussed on the BC. (See almost any article listed in [15] with the word "Borda" in the title.) In part, this is because the BC is well known and easy to use. Often, by discovering specific profiles, BC faults are identified. (This means they have discovered portions of particular words from $\mathcal{D}^n(B^n)$.) The obvious problem with this approach is that we do not know whether the conclusion is specific to the BC, or whether the identical fault holds for other procedures. Equation 4.2, however, gives the answer. *Any BC fault illustrating undesirable changes in how the candidates are sincerely ranked over different subsets must be shared by all other procedures.* Namely, a BC flaw is universal; it must be suffered by all procedures. On the other hand, the strict containment requires all other procedures to admit faults that are not possible with the BC.

4.4 The Choice of a Procedure Matters

Before examining topics other than voting, one might wonder whether the choice of a voting method matters. For a given profile, will not all procedures give essentially the same outcome? As the beverage example already proves (the BC reverses the plurality outcome), the answer is no. (In fact, this example admits *seven* different rankings as the weights change [31].) Fishburn [6] proved that two different tallying methods can have opposite outcomes. Now, with this more general perspective, we can find all possible rankings that occur with changes in the procedure [27, 31]. As a dramatic example, instead of just two reversed outcomes, it is possible to have *over 84 million different rankings of 10 candidates with one profile!* [27]. Remember, the voters' preferences remain fixed as marked on the ballots, so these millions upon millions of outcomes are due to changes in the tallying method. In fact, each of the ten candidates can be top-ranked when some procedures are used and then bottom-ranked with others! So, which of these contradictory but "sincere" rankings is the correct one? No wonder we need a mathematical theory to understand voting procedures!

4.5 Statistics

To illustrate this dictionary theme with statistics, start with 2^j pairs of urns marked I_k^1, II_k^1. Place red and blue balls in each urn; the choice determines whether $P(R|I_k^1)$, the probability of choosing red when selecting at random from urn I_k^1, is larger, smaller, or equal to $P(R|II_k^1)$. Thus, associated with each pair are the three symbols $\{>, =, <\}$. In the empty urn I_k^2, combine the contents of I_{2k-1}^1 with I_{2k}^1; similarly, let II_k^2 hold the contents of II_{2k-1}^1 and II_{2k}^1, $k = 1, \ldots, 2^{j-1}$. Again, for each pair of urns, one of the three symbols applies.

Continue this aggregation process where, at the sth stage, the contents of I_{2k-1}^s and I_{2k}^s are combined to create I_k^{s+1}, while the contents of II_{2k-1}^s and II_{2k}^s are combined to create II_k^{s+1}, $s = 1, \ldots, j$, $k = 1, \ldots, 2^{j-s}$. For each of the $2^{j+1} - 1 = \sum_{i=0}^{j} 2^i$ pairs of urns, one of the symbols $\{>, =, <\}$ applies. Therefore, the universal set for the aggregation of urns, $\mathcal{U}_{agg}^j$, consists of $3^{2^{j+1}-1}$ listings of symbols. An initial condition, p, is the initial allocation of red and blue balls in the first layer of urns, and the listing of inequalities for the different urns defined by p is a word. The dictionary for the aggregation of urns, $\mathcal{D}_{agg}^j$ consists of all words. Again, the complexity of the aggregation process is measured by the following.

Theorem 4.3 *For all $j \geq 2$ with the aggregation of urns,*

$$\mathcal{D}_{agg}^j = \mathcal{U}_{agg}^j.$$

Again, anything can happen. For instance, start with $2^{50} \approx 1.125 \times 10^{15}$ urns, and keep combining the contents in the above manner until only one pair is left. The initial contents can be chosen so that $P(R|I_k^s) > P(R|II_k^s)$ holds for each pair during the first 49 levels, but there is a reversal at the last stage. Figure 4 (designed by A. Konchan as a homework exercise) illustrates a $j = 2$ example where the inequalities flip at each level.

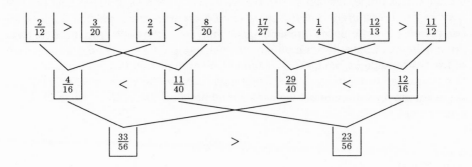

Figure 4: A three level Simpson paradox

Another illustration of the power of this dictionary approach comes from the widely used *Kruskal-Wallis* test of nonparametric statistics. This natural method replaces data with rankings. For instance, in comparing firms producing light bulbs according to the hundreds of hours a bulb lasted, the outcome

Firm 1	Firm 2	Firm 3
6.01	5.90	5.85
6.10	6.05	6.15

defines the KW matrix of ranks

	Firm 1	Firm 2	Firm 3
	3	2	1
	5	4	6
Total	$\overline{8}$	$\overline{6}$	$\overline{7}$

and the KW ranking of Firm 1 $\succ$ Firm 3 $\succ$ Firm 2.

If bottom-ranked Firm 2 goes out of business, we want to compare Firms 1 and 3. The data remains fixed, but the relative ranking does not; now the KW ranking is Firm 1 $\sim$ Firm 3. While this is a reasonably innocuous change, we might wonder whether more radical outcomes can occur. This question was answered by Deanna Haunsperger in her thesis [12] (part of which is in [13]).

With $n \geq 3$ firms, there are $2^n - (n + 1)$ subsets of two or more firms. Thus, the subsets and the symbols are the same as those for the voting problem. In place of voters' preferences, the initial data represents the initial condition, and the list of rankings for each subset of alternatives defines the KW word. The KW dictionary, $\mathcal{D}^n_{KW}$, consists of all possible words admitted by the KW test. In characterizing $\mathcal{D}^n_{KW}$, Haunsperger proved that the Kruskal-Wallis test admits far more paradoxes and difficulties than previously imagined. For instance, for any $n \geq 3$, examples of data can be found so that even though the KW rankings of pairs is Firm i $\succ$ Firm j iff $i < j$, the KW ranking comes perilously close to reversing this conclusion as it is Firm $n-1$ $\succ$ Firm $n-2 \succ \ldots \succ$ Firm 1 $\succ$ Firm n. (The argument in [31] discussing pairwise voting rankings extends to show that the real difficulty is not the KW ranking, but the pairwise rankings.)

On the other hand, even though the KW test admits previously unknown difficulties, Haunsperger also proved that of all possible nonparametric ranking methods, the KW test is by far the best choice. She did this by showing that the dictionary for most nonparametric methods agrees with the universal set. Then she proved that the KW dictionary is a proper subset of the dictionary for any other method. In addition, she proved that in certain settings, the KW dictionary agrees with the Borda dictionary; the "symmetry" reasons for this are suggested below.

4.6 Probability

The cyclic dice example can be generalized to all levels as follows. (See [7, 11].) For $j \geq 2$, start with 3^j triplets of dice, $< A_k^1, B_k^1, C_k^1 >$. There is a constructive approach to mark these dice so that $A_k^1 \succ B_k^1, B_k^1 \succ C_k^1$, and $C_k^1 \succ A_k^1$ for all k. Now, treat each triplet as a set of dice; e.g., denote by A_k^2 the three dice in the triplet $3k-2$ (so $A_k^2 = < A_{3k-2}^1, B_{3k-2}^1, C_{3k-2}^1 >$), B_k^2 is the triplet $3k-1$ and C_k^2 is the triplet $3k$, $k = 1, \ldots, 3^{j-1}$. Instead of "high score wins," with the new triplets, the larger sum of the set of dice wins. Again, a cycle occurs. Indeed, continue the aggregation process; at each aggregation level the dice define the cycle

$$A_k^s \succ B_k^s, B_k^s \succ C_k^s, \text{ and } C_k^1 \succ A_k^1; s = 1, \ldots, j; k = 1, \ldots, 3^{j-s+1}.$$

Each pair of dice admits the three possible symbols $\succ, \sim, \prec$. Thus, a triplet admits 3^3 possible symbols. The universal space for the dice, $\mathcal{U}_{dice}^j$, is the space of listings of these symbols for each of the $3^{j+1} - 1$ triplets. The initial condition corresponds to how the dice are marked. Each choice of markings defines a sequence of symbols – a word in the dictionary $\mathcal{D}_{dice}^j$. A measure of the complexity of these dice games, which significantly extends our earlier construction, follows.

Theorem 4.4 *For the dice problem*

$$\mathcal{D}_{dice}^j = \mathcal{U}_{dice}^j.$$

Again, anything can happen! Indeed, to show how wild the process can be, Funkenbusch created an example [11] showing how the final stack of dice can be split in many different ways allowing for all sorts of cyclic outcomes.

4.7 Complexity of Aggregation

The point is made. Clearly, when this dictionary construction is applied to other aggregation and classification procedures, whether they come from voting, probability, statistics, programming, or the decision sciences, examples emerge demonstrating a similar complexity while exposing many new paradoxes. In fact, related assertions have been found even for processes involving function spaces and vector fields (rather than finite discrete objects); assertions which raise doubts about such commonly accepted themes as the "supply and demand" story and other techniques from economics [29]. What these examples illustrate is that aggregation processes, the basis of statistics, probability, and much of the social sciences, are far more complex than previously expected.

To provide an overview of the rest of the story, I briefly outline how other structures from dynamical chaos can be identified with standard concerns from aggregation processes. For instance, it is natural to worry about the effects of small data errors for statistics and small numbers of voters trying to manipulate the outcome. Typically (at least for voting), these issues are studied by using computer simulations. (See, for instance, [4].) In our formulation, these concerns are identified with the "sensitivity with respect to initial conditions." This motivates the development of analytic techniques to analyze these issues. For instance, it is natural to try to determine which procedures are least susceptible to these "small change," negative effects. From this analytic approach we learn that the voting system least susceptible to small manipulation problems is the BC [31]; for nonparametric systems, the answer is the KW test.

Another natural concern is to determine the likelihood of paradoxes. (While most of this analysis involves computer simulations, [8, 9] have analytic voting results for $n = 3$.) The parallel theme from dynamics is where the "dictionary" and "topological" approaches of chaos have been refined by using measure theory and topological entropy to indicate the likelihood of various behaviors. (See, for instance, [2, 23].) Again, using the approach motivated by dynamics, work by Van Newenhizen [35] and [31] significantly extend these earlier efforts and again prove that the BC is the least likely to admit different paradoxes. One must anticipate a similar result for the KW statistical test.

As a final illustration, there is an enormous literature (e.g., see the expository books [21, 34]) identifying perverse behavior such as where a candidate can be hurt by receiving more electoral support. In our formulation, this turns out to be identified with the Cantor set structure from dynamics. Cantor sets do not occur for these aggregation procedures (because finite, not infinite intersections are involved), but the geometry of data generated by the iterated inverse images can be complicated. In fact, most (again, all that I know of) of the identified difficulties plus many new ones can be explained when convexity, connectiveness, and other geometric properties are lost by this iterative process. In this way it becomes easy to discover and explain examples such as where department splits into two subcommittees to select one of three candidates for Chair. With the standard runoff (where the top two candidates from a plurality election have a runoff) each subcommittee could choose c_1 only to have c_1 lose when the full department uses the same procedure. (For an example and a geometric explanation showing that this is just a convexity issue, see [31].) A similar result holds for other aggregation methods from statistics and elsewhere that involve the rankings from different subsets of the alternatives.

One message from this approach is very clear. For subtle reasons, these issues are far more complex than most of us have expected!

References

[1] K. J. Arrow, *Social choice and individual values*, 2nd edition, Wiley, New York,1963.

[2] L. Alseda, J. Llibre, and M. Misiurewicz, *Combinatorial Dynamics and Entropy in Dimension One*, World Scientific, 1993.

[3] C. Blyth, "On Simpson's paradox and the Sure-Thing Principle," *JASA*, **67** (1972), 364-366.

[4] JR. Chamberlin, "An investigation into the relative manipulability of four voting systems," *Behav. Sci.* **30** (1985), 195-203.

[5] R. Devaney, *An introduction to chaotic dynamical systems*, 2nd ed., Addison-Wesley, 1989.

[6] P. C. Fishburn, "Inverted orders for monotone scoring rules," *Discrete Appl. Math.*, **3** (1981), 27-36.

[7] B. Funkenbusch, and D. G. Saari, "Preferences among preferences or nested cyclic stochastic inequalities," *Congressus Numerantium*, **39** (1983), 419-432.

[8] W. V. Gehrlein, "Expected probability of Condorcet's paradox," *Econ. Lett.*, **7** (1981), 33-37.

[9] W. V. Gehrlein, P. C. Fishburn, "Probabilities of election outcomes for large electorates," *J. Econ Theory*, **19** (1978), 38-49.

[10] I. J. Good and Y. Mittal, "The amalgamation and geometry of two-by-two contingency tables," *The Annals of Statistics*, **15** (1987), 694-711.

[11] R. Honsberger, *More mathematical morsels*, MAA, 1991.

[12] D. Haunsperger, "Projection and aggregation paradoxes in nonparametric statistical tests," Northwestern University Ph.D. Dissertation, 1991.

[13] D. Haunsperger, "Dictionaries of paradoxes for statistical tests on *k* samples," *JASA*, **87** (1992), 149-155.

[14] D. Haunsperger and D. G. Saari, "The lack of consistency for statistical decision procedures," *The American Statistician*, **45** (1991), 252-255.

[15] J. S. Kelly, "Social choice bibliography," *Soc. Choice Welfare*, **8** (1991), 97-169.

[16] G. H. Kramer, "A dynamical model of political equilibrium," *J. Econ. Theory*, **12** (1977), 310-344.

[17] T.-Y. Li and J. A. Yorke, "Period three implies chaos," *Amer. Math Monthly*, **82** (1975), 985-992.

[18] R. D. McKelevey, "Intrasitivities in mulitdimensional voting models and some implications for agenda control," *J. Econ. Theory*, **12** (1976), 472-482.

[19] R. G. Niemi and W. H. Riker, "The choice of voting systems," *Scientific American*, **234** (1976), 21-27.

[20] H. Nurmi, *Comparing Voting Systems*, D. Reidel, Dordrecht, 1987.

[21] P. Ordeshook, *Game Theory and Political Theory*, Cambridge University Press, 1986.

[22] C. Robinson, *Dynamical Systems*, Northwestern University book manuscript, 1994.

[23] D. G. Saari, "A dictionary for voting paradoxes," *Jour. Econ Theory*, **48** (1989), 443-475.

[24] D. G. Saari, "The Borda Dictionary," *Soc. Choice Welfare*, **7** (1990), 279-317.

[25] D. G. Saari, "Symmetry extensions of "neutrality" I. Advantage to the Condorcet loser," *Soc. Choice Welfare*, **9** (1992), 307-336

[26] D. G. Saari, "Symmetry extensions of "neutrality" II. Partial ordering of dictionaries," *Soc. Choice Welfare*, 1993.

[27] D. G. Saari, "Millions of election rankings from a single profile," *Soc. Choice Welfare*, **9** (1992), 277-306.

[28] D. G. Saari, "The aggregate excess demand function and other aggregation procedures," *Economic Theory*, **2** (1992), 359-388.

[29] D. G. Saari, "The ultimate of chaos resulting from weighted voting systems," *Advances in Applied Math*, **5** (1984), 286-308.

[30] D. G. Saari, *Geometry of Voting*, Springer-Verlag, New York, 1994.

[31] D. G. Saari and J. Urenko, "Newton's method, circle maps, and chaos," *Amer. Math Monthly*, **91** (1984), 3-17.

[32] J. H. Smith, "Aggregation of preferences with variable electorate," *Econometrica*, **41** (1973), 1027-1041.

[33] P. Straffin, *Topics in the Theory of Voting*, Birhauser, Boston, 1980.

[34] J. Van Newenhizen, "The Borda Count is most likely to respect the Condorcet principle," *Economic Theory* **2** (1992), 69-83.

Positive-Definite Programming

Lieven Vandenberghe
Electrical Engineering Department
K.U. Leuven
Leuven, Belgium

Stephen Boyd
Electrical Engineering Department
Stanford University
Stanford, California

1 Introduction

1.1 Positive-definite programming

We consider the problem of minimizing a linear function of a variable $x \in \mathbf{R}^m$ subject to a *linear matrix inequality:*

$$\begin{aligned} \text{minimize} \quad & c^T x, \\ \text{subject to} \quad & F(x) \geq 0, \end{aligned} \tag{1}$$

where

$$F(x) \stackrel{\Delta}{=} F_0 + \sum_{i=1}^{m} x_i F_i. \tag{2}$$

The problem data are the vector $c \in \mathbf{R}^m$ and $m+1$ symmetric matrices $F_0, \ldots, F_m \in \mathbf{R}^{n \times n}$. The inequality sign in $F(x) \geq 0$ means that $F(x)$ is positive-semidefinite, i.e., $z^T F(x) z \geq 0$ for all $z \in \mathbf{R}^n$. This problem is called a *positive-definite program* (PDP), following Nesterov and Nemirovsky [43].

Problem (2) is a convex optimization problem since its objective and constraint are convex: if $F(x) \geq 0$ and $F(y) \geq 0$, then, for all $\lambda, 0 \leq \lambda \leq 1$,

$$F(\lambda x + (1 - \lambda y)) = \lambda F(x) + (1 - \lambda) F(y) \geq 0.$$

Although the PDP (1) may appear quite specialized, we will see that it includes many important optimization problems as special cases. For instance, consider the linear program

$$\begin{aligned} \text{minimize} \quad & c^T x, \\ \text{subject to} \quad & Ax \geq b, \end{aligned} \tag{3}$$

276

where the inequality denotes *componentwise* inequality. Since a vector $v \geq 0$ (componentwise) if and only if the matrix $diag(v)$ (i.e., the diagonal matrix with the components of v on its diagonal) is positive-semidefinite, we can express the LP (3) as a PDP with $F(x) = diag(Ax + b)$, i.e.,

$$F_0 = -diag(b), \quad F_i = diag(a_i), \quad i = 1, \ldots, m,$$

where $A = [a_1 \ldots a_m] \in \mathbf{R}^{n \times m}$.

Positive-definite programming can therefore be regarded as an extension of linear programming where the componentwise inequalities between vectors are replaced by matrix inequalities, or, equivalently, the first orthant is replaced by the cone of positive-semidefinite matrices. We can also view the PDP (1) as a semi-infinite linear program, since the matrix inequality $F(x) \geq 0$ is equivalent to an infinite set of linear constraints on x, i.e., $z^T F(x) z \geq 0$ for each $z \in \mathbf{R}^n$. It is therefore not surprising that the theory of positive-definite programming closely parallels linear programming theory, or that many algorithms for solving linear programs should have generalizations that handle PDPs. There are many important differences, however. For instance, the duality results are weaker for PDPs than for LPs. As another important difference, there is no simple or obvious analog of the simplex method for PDPs.

Before proceeding further we give a simple example of a *nonlinear* (convex) optimization problem that can be cast as a PDP, but not as a linear program. Consider the problem

$$\begin{array}{ll} \text{minimize} & \frac{(c^T x)^2}{d^T x} \\ \text{subject to} & Ax + b \geq 0, \end{array} \tag{4}$$

where we assume that $d^T x > 0$ whenever $Ax + b \geq 0$. We start with the standard trick of introducing an auxiliary variable t that serves as an upper bound on the objective:

$$\begin{array}{ll} \text{minimize} & t \\ \text{subject to} & Ax + b \geq 0 \\ & \frac{(c^T x)^2}{d^T x} \leq t. \end{array} \tag{5}$$

In this formulation, the objective is a linear function of the variables x and t; the nonlinear (convex) objective in (4) shows up as a nonlinear (convex) constraint in (5). These constraints, in turn, can be expressed as a linear matrix inequality in the variables x and t:

$$\begin{array}{ll} \text{minimize} & t \\ \text{subject to} & \begin{bmatrix} diag(Ax + b) & 0 & 0 \\ 0 & t & c^T x \\ 0 & c^T x & d^T x \end{bmatrix} \geq 0. \end{array} \tag{6}$$

(Here, again, $diag(v)$ represents the diagonal matrix with the elements of v on its diagonal.) Thus we have reformulated the nonlinear (convex) problem (4) as the PDP (6). More examples and applications will be given in the next section.

There are good reasons for studying positive-definite programming problems. First, positive-definiteness constraints arise directly in a number of important applications. Secondly, many convex optimization problems, e.g., linear programming and (convex) quadratically constrained quadratic programming, can be cast as PDPs. Positive-definite programming therefore offers a unified way to study the properties of and derive algorithms for a wide variety of convex optimization problems. Most importantly, however, *PDPs can be solved very efficiently, both in theory and in practice.*

Theoretical tractability follows from convexity, along with the observation that we can construct, in polynomial time, a cutting plane for the constraint set through any given infeasible point (see, e.g., [8, §2.3]). One can therefore apply the ellipsoid method of Yudin and Nemirovsky, and Shor (see [64, 56]) to solve problem (1) in polynomial time. In practice, however, the ellipsoid method is slow.

In this paper we concentrate on recently developed interior-point methods for positive-definite programming. Of course general-purpose nonlinear optimization methods (trust region methods, sequential quadratic optimization, . . .) could be used, possibly after modification, to solve PDPs. Interior-point methods, however, enjoy several properties that make them especially interesting.

- It is now generally accepted that interior-point methods for linear programming are competitive with the simplex method and even faster for problems with more than $10,000$ variables or constraints (see, e.g., [32]). One can expect to see the same trend more generally, especially since the very efficient simplex method has no counterpart in positive-definite programming. In our experience with positive-definite programming for control applications, we have found interior-point methods to be very efficient.

- Interior-point methods have a polynomial worst-case complexity.

- Interior-point methods are ideally suited for structured problems. We will see that every iteration of an interior-point method involves the solution of a least-squares problem with the same structure as $F(x)$ in (1). These matrices are often highly structured but not necessarily sparse. The structure can be exploited by combining interior-point methods with iterative least-squares methods such as conjugate-gradients [21] or LSQR [49]. This is not possible in the simplex method, for instance, nor in many other classical methods.

1.2 Examples and Applications
In this section we list a few examples and applications. The list is not exhaustive, and the purpose is more to give an idea of the generality of the problem. More examples are described in [43] and [8].

Quadratically Constrained Quadratic Programming
A convex quadratic constraint $(Ax + b)^T (Ax + b) - c^T x - d \leq 0$ can be written as

$$\begin{bmatrix} I & Ax + b \\ (Ax + b)^T & c^T x + d \end{bmatrix} \geq 0.$$

The left-hand side depends affinely on the vector x: it can be expressed as $F_0 + x_1 F_1 + \cdots + x_m F_m \geq 0$, with

$$F_0 = \begin{bmatrix} I & b \\ b^T & d \end{bmatrix}, \quad F_i = \begin{bmatrix} 0 & a_i \\ a_i^T & c_i \end{bmatrix}, \quad i = 1, \ldots, m,$$

where $A = [a_1 \ldots a_m]$. Therefore, a general *quadratically constrained quadratic program*

$$\begin{aligned} \text{minimize} \quad & f_0(x) \\ \text{subject to} \quad & f_i(x) \leq 0, \quad i = 1, \ldots, L, \end{aligned}$$

where each f_i is a convex quadratic function $f_i(x) = (A_i x + b)^T (A_i x + b) - c_i^T x - d_i$, can be written as

$$\begin{aligned} \text{minimize} \quad & t \\ \text{subject to} \quad & \begin{bmatrix} I & A_0 x + b_0 \\ (A_0 x + b_0)^T & c_0^T x_0 + d_0 + t \end{bmatrix} \geq 0, \\ & \begin{bmatrix} I & A_i x + b_i \\ (A_i x + b_i)^T & c_i^T x + d_i \end{bmatrix} \geq 0, \quad i = 1, \ldots, L. \end{aligned}$$

This is a PDP in x and t, since one can think of the $L+1$ matrix inequalities as diagonal blocks of one block diagonal matrix inequality $F(x,t) \geq 0$.

Matrix Norm and Maximum Eigenvalue Minimization

Suppose $A(x)$ is a (possibly rectangular) matrix that depends affinely on x: $A(x) = A_0 + x_1 A_1 + \cdots + x_m A_m$. The problem of minimizing the (spectral, or maximum singular value) norm $\|A(x)\|$ over x is a PDP:

$$\begin{aligned} \text{minimize} \quad & t \\ \text{subject to} \quad & \begin{bmatrix} tI & A(x) \\ A(x)^T & tI \end{bmatrix} \geq 0. \end{aligned} \qquad (7)$$

If $A(x)$ is a symmetric matrix, a related problem is to minimize the maximum eigenvalue of A:

$$\begin{aligned} \text{minimize} \quad & t \\ \text{subject to} \quad & tI - A(x) \geq 0. \end{aligned}$$

Note that both $\|A(x)\|$ and the maximum eigenvalue $\lambda_{max}(A(x))$ are nondifferentiable functions of x.

Logarithmic Chebychev Approximation

Suppose we wish to solve $Ax \approx b$ approximately, where $A = [a_1 \cdots a_n]^T \in \mathbf{R}^{n \times m}$ and $b \in \mathbf{R}^n$. In Chebychev approximation we minimize the infinity norm of the residual, i.e., we solve

$$\text{minimize} \quad \max_i |a_i^T x - b_i|.$$

This can be cast as a linear program, with x and an auxiliary variable t as variables:

$$\text{minimize} \quad t$$
$$\text{subject to} \quad -t \leq a_i^T x - b_i \leq t, \quad i = 1, \ldots, n.$$

In some applications b_i has the dimension of a power or intensity, and is typically expressed on a logarithmic scale. In such cases the more natural optimization problem is

$$\text{minimize} \ \max_i |\, log(a_i^T x) - log(b_i)| \tag{8}$$

(assuming $b_i > 0$, and interpreting $log(a_i^T x)$ as $-\infty$ when $a_i^T x \leq 0$).

This *logarithmic Chebychev approximation* problem can be cast as a PDP. To see this, note that

$$|\, log(a_i^T x) - log(b_i)| = log \max(a_i^T x / b_i, b_i / a_i^T x)$$

(assuming $a_i^T x > 0$). Problem (8) is therefore equivalent to

$$\text{minimize} \quad t$$
$$\text{subject to} \quad 1/t \leq a_i^T x / b_i \leq t, \quad i = 1, \ldots, n,$$

or,

$$\text{minimize} \quad t$$
$$\text{subject to} \quad \begin{bmatrix} t - a_i^T x / b_i & 0 & 0 \\ 0 & a_i^T x / b_i & 1 \\ 0 & 1 & t \end{bmatrix} \geq 0, \quad i = 1, \ldots, n,$$

which is a PDP. This example illustrates two important points. It shows that positive-definite programming includes many optimization problems that do not look like (1) at first sight. And secondly, it shows that the problem is much more general than linear programming, despite the close analogy.

Control and System Theory

Positive-definite programming problems arise frequently in control and system theory. Boyd, El Ghaoui, Feron and Balakrishnan catalog many examples in [8]. We will describe one simple example here.

Consider the *differential inclusion*

$$\frac{dx}{dt} \in Co\{A_1, \ldots, A_L\} x(t), \tag{9}$$

where $x(t) \in \mathbf{R}^n$ and the matrices $A_1, \ldots, A_L$ are given, and $Co\{A_1, \ldots, A_L\}$ denotes the *convex hull* of $A_1, \ldots, A_L$. We seek an ellipsoidal invariant set, i.e., an ellipsoid $\mathcal{E}$ such that for any x that satisfies (9), $x(T) \in \mathcal{E}$ implies $x(t) \in \mathcal{E}$ for all $t \geq T$. The existence of such an ellipsoid implies, for example, that all solutions of the differential inclusion (9) are bounded.

The ellipsoid $\mathcal{E} = \{x \mid x^T P x \leq 1\}$, where $P = P^T > 0$, is invariant if and only if the function $V(t) = x(t)^T P x(t)$ is nonincreasing for any solution x of (9). (In this case we say that V is a quadratic Lyapunov function that proves stability of the differential inclusion (9).) Thus, $\mathcal{E}$ is invariant if and only if

$$\frac{d}{dt}V(x(t)) = x(t)^T \left(A(t)^T P + PA(t)\right) x(t) \leq 0,$$

for any $x(t) \in \mathbf{R}^n$ and $A(t) \in Co\{A_1, \ldots, A_L\}$. This is equivalent to $A^T P + PA \leq 0$ for all $A \in Co\{A_1, \ldots, A_L\}$, which in turn is equivalent to the condition

$$A_k^T P + PA_k \leq 0, \quad k = 1, \ldots, L.$$

This is a linear matrix inequality constraint in the matrix P, considered as the variable.

To find an invariant ellipsoid for the differential inclusion (9) (or verify that none exists), we need to solve the feasibility problem

$$P > 0, \quad A_k^T P + PA_k \leq 0, \quad k = 1, \ldots, L \tag{10}$$

for the (matrix) variable P. Several standard methods can be used to convert this feasibility problem into a PDP that has an obvious initial feasible point. For instance, we can solve the PDP with variables $P = P^T \in \mathbf{R}^{n \times n}$ and $t \in \mathbf{R}$,

$$\begin{aligned}
\text{minimize} \quad & t \\
\text{subject to} \quad & A_k^T P + PA_k \leq 0, \quad k = 1, \ldots, L, \\
& P \geq -tI, \\
& P \leq I.
\end{aligned}$$

(The last constraint is added, without loss of generality, to normalize the otherwise homogeneous problem.) This PDP can be initialized with $P = 0$, $t = 1$ and then solved; the optimum value of t is negative if and only if (10) is feasible.

Structural Optimization

Ben-Tal and Bendsøe in [11] consider the following problem from structural optimization. A structure of L linear elastic bars connect a set of N nodes. The geometry (topology and lengths of the bars) and the material (Young's modulus) are fixed; the task is to size the bars, i.e., determine appropriate cross-sectional areas for the bars. In the simplest version of the problem we consider one fixed set of externally applied nodal forces f_i, $i = 1, \ldots, N$. (More complicated versions consider mulitple loading scenarios.) The vector of (small) node displacements resulting from the load forces f will be denoted d. The objective is the elastic stored energy $\frac{1}{2}f^T d$, which is a measure of the inverse of the stiffness of the structure. We also need to take into account constraints on the total volume (or equivalently, weight), and upper and lower bounds on the cross-sectional area of each bar.

The design variables are the cross-sectional areas x_i. The relation between f and d is linear: $A(x)d = f$, where

$$A(x) \triangleq \sum_{i=1}^{N} x_i A_i$$

is called the stiffness matrix. The matrices A_i are all symmetric positive-semidefinite and depend only on fixed parameters (Young's modulus, length of the bars, and geometry). The optimization problem then becomes (see [11])

$$
\begin{aligned}
\text{minimize} \quad & f^T d \\
\text{subject to} \quad & A(x)d = f, \\
& \sum_{i=1}^{L} l_i x_i \leq v, \\
& \underline{x}_i \leq x_i \leq \overline{x}_i \quad i = 1, \ldots, L,
\end{aligned}
$$

where d and x are the variables, v is maximum volume, l_i are the lengths of the bars, and $\underline{x}_i$, $\overline{x}_i$ the upper and lower bounds on the cross-sectional areas. For simplicity, we assume that $\underline{x}_i > 0$, and that $A(x)$ is positive-definite for all positive values of x_i. We can then eliminate d and write

$$
\begin{aligned}
\text{minimize} \quad & f^T A(x)^{-1} f \\
\text{subject to} \quad & \sum_{i=1}^{L} l_i x_i \leq v, \\
& \underline{x}_i \leq x_i \leq \overline{x}_i, \quad i = 1, \ldots, L,
\end{aligned}
$$

or,

$$
\begin{aligned}
\text{minimize} \quad & t \\
\text{subject to} \quad & \begin{bmatrix} t & f^T \\ f & A(x) \end{bmatrix} \geq 0, \\
& \sum_{i=1}^{L} l_i x_i \leq v, \\
& \underline{x}_i \leq x_i \leq \overline{x}_i, \quad i = 1, \ldots, L,
\end{aligned}
$$

which is a PDP in x and t.

Pattern Separation by Ellipsoids

The simplest classifiers in pattern recognition use hyperplanes to separate two sets. A hyperplane $a^T x + b = 0$ separates two sets of points $\{x_i\}$ and $\{y_j\}$ if

$$
\begin{aligned}
a^T x_i + b < 0 \qquad & \text{for all } i, \\
a^T y_j + b > 0 \qquad & \text{for all } j.
\end{aligned}
$$

This is a set of linear inequalities in $a \in \mathbf{R}^n$ and $b \in \mathbf{R}$, and a solution can be found by linear programming. If the two sets cannot be separated by a hyperplane, we can try to separate them by a quadratic surface. In other words we seek a quadratic function $f(x) = x^T P x + b^T x + c$ that satisfies

$$
(x_i)^T P x_i + b^T x_i + c < 0 \qquad \text{for all } i, \tag{11}
$$
$$
(y_j)^T P y_j + b^T y_j + c > 0 \qquad \text{for all } j. \tag{12}
$$

These inequalities are a set of linear inequalities in the variables $P = P^T \in \mathbf{R}^n$, $b \in \mathbf{R}^n$, and $c \in \mathbf{R}$, and again can be solved using linear programming.

We can put further restrictions on the quadratic surface separating the two sets. For instance, for cluster analysis we might try to find an ellipsoid that contains all the points x_i and none of the y_j (see [53]). This constraint imposes the condition $P > 0$ in addition to the linear inequalities (11) and (12) on the variables P, b, and c. Thus finding an ellipsoid that contains all the x_i variables but none of the y_i variables (or determining that no such ellipsoid exists) can be done by solving a linear matrix inequality feasibility problem.

We can optimize the shape and the size of the ellipsoid by adding an objective function and other constraints. For instance, the ratio of the largest to the smallest semi-axis length is the square root of the condition number of P. In order to make the ellipsoid as spherical as possible, one can introduce an additional variable γ, add the constraint

$$I \leq P \leq \gamma I, \tag{13}$$

and minimize γ over (11), (12) and (13). This is a PDP in the variables γ, P, b and c. This PDP will be feasible if and only if there is an ellipsoid that contains all the x_i and none of the y_i; its optimum value is one if and only there is a sphere that separates the two sets of points.

Geometrical Problems Involving Quadratic Forms

Many geometrical problems involving quadratic functions can be expressed as PDPs. We will give one simple example. Suppose we are given m ellipsoids $\mathcal{E}_1, \ldots, \mathcal{E}_m$ described as the sublevel sets of the quadratic functions

$$f_i(x) = x^T A_i x + 2b_i^T x + c_i, \quad i = 1, \ldots, m,$$

i.e., $\mathcal{E}_i = \{x | f_i(x) \leq 0\}$. The goal is to find the smallest sphere that contains all m of these ellipsoids (or equivalently, contains the convex hull of their union).

The condition that one ellipsoid contain another can be expressed in terms of a matrix inequality. Suppose that the ellipsoids $\mathcal{E} = \{x | f(x) \leq 0\}$ and $\tilde{\mathcal{E}} = \{x | \tilde{f}(x) \leq 0\}$, with

$$f(x) = x^T A x + 2b^T x + c, \quad \tilde{f}(x) = x^T \tilde{A} x + 2\tilde{b}^T x + \tilde{c},$$

have nonempty interior. Then it can be shown that $\mathcal{E}$ contains $\tilde{\mathcal{E}}$ if and only if there is a $\tau \geq 0$ such that

$$\begin{bmatrix} A & b \\ b^T & c \end{bmatrix} \leq \tau \begin{bmatrix} \tilde{A} & \tilde{b} \\ \tilde{b}^T & \tilde{c} \end{bmatrix}.$$

(The 'if' part is trivial; the 'only if' part is less trivial. See [8, 59]).

Returning to our problem, consider the sphere $\mathcal{S}$ represented by $f(x) = x^T x - 2x_c^T x + \gamma \leq 0$. $\mathcal{S}$ contains the ellipsoids $\mathcal{E}_1, \ldots, \mathcal{E}_m$ if and only if there are nonnegative $\tau_1, \ldots, \tau_m$ such that

$$\begin{bmatrix} I & -x_c \\ -x_c^T & \gamma \end{bmatrix} \leq \tau_i \begin{bmatrix} A_i & b_i \\ b_i^T & c_i \end{bmatrix}.$$

Note that these conditions can be considered one large linear matrix inequality in the variables x_c, γ, and $\tau_1, \ldots, \tau_m$.

Our goal is to minimize the radius of the sphere $\mathcal{S}$, which is $r = \sqrt{x_c^T x_c - \gamma}$. To do this we express the condition $r^2 \leq t$ as the matrix inequality

$$\left[\begin{array}{cc} I & x_c \\ x_c^T & t + \gamma \end{array} \right] \geq 0$$

and minimize the variable t.

Putting it all together we see that we can find the smallest sphere containing the ellipsoids $\mathcal{E}_1, \ldots, \mathcal{E}_m$ by solving the PDP

$$\begin{aligned}
&\text{minimize} \quad t \\
&\text{subject to} \quad \left[\begin{array}{cc} I & -x_c \\ -x_c^T & \gamma \end{array} \right] \leq \tau_i \left[\begin{array}{cc} A_i & b_i \\ b_i^T & c_i \end{array} \right], \quad i = 1, \ldots, m, \\
&\qquad\qquad \tau_i \geq 0, \quad i = 1, \ldots, m, \\
&\qquad\qquad \left[\begin{array}{cc} I & x_c \\ x_c^T & t + \gamma \end{array} \right] \geq 0.
\end{aligned}$$

The variables here are $x_c, \tau_1, \ldots, \tau_m, \gamma$, and t.

This example demonstrates once again the breadth of problems that can be re-formulated as PDPs. It also demonstrates that the task of this reformulation can be nontrivial.

Other Fields

- PDPs occur in statistics, in minimum trace factor analysis (see Watson [61]), as the *educational testing problem* (see [17, 18]), and in optimum experiment design (see Pukelsheim [50]).

- PDPs have been used to compute upper or lower bounds for combinatorial optimization problems. Examples are Lovász's famous upper bound on the Shannon capacity of a graph [33], Shor's bounds for integer programming, and Alizadeh's work [4, 2, 3].

- The problem of minimizing the maximum eigenvalue of a matrix has been studied extensively by Overton; see [45] for a list of applications.

1.3 Historical Overview

A very early paper on the theoretical properties of PDPs is Bellman and Fan [9]. Other references discussing optimality conditions are Craven and Mond [13], Shapiro [55], Fletcher [18], and Allwright [5].

Many researchers have worked on the problem of minimizing the maximum eigenvalue of a symmetric matrix. See, for instance, Cullum, Donath and Wolfe [12],

Goh and Teo [24], Panier [48], Allwright [6], Overton [44, 45], Overton and Womersley [47, 46], Ringertz [51], Fan and Nekooie [20], Fan [16], and Hiriart-Urruty and Ye [26].

The history of interior-point methods is relatively young. Interior-point methods for linear programming were introduced by Karmarkar in 1984 [28], although many of the underlying principles are older (see, e.g., Fiacco and McCormick [19], Lieu and Huard [31], and Dikin [15]). Karmarkar's algorithm, and the interior-point methods developed afterwards, combine a very low, polynomial, worst-case complexity with excellent behavior in practice.

Karmarkar's paper has had an enormous impact, and several variants of his method have been developed (see, e.g., the survey by Gonzaga [23]). Interior-point methods have also been extended and generalized to convex quadratic programming, and to certain linear complementarity problems (see Kojima, Megiddo, Noma and Yoshise [29]).

An important breakthrough was achieved by Nesterov and Nemirovsky in 1988 [38, 40, 39, 41, 41]. They showed that the interior-point methods for linear programming can be generalized to all convex optimization problems. The key element is the knowledge of a barrier function with certain properties (*self-concordance*). Unfortunately, although Nesterov and Nemirovsky prove that a self-concordant barrier function exists for every convex set, it is not always known how to compute it in practice.

PDPs are an important class of convex optimization problems for which self-concordant barrier functions are known, and, therefore, interior-point methods are applicable. At the same time, they offer a simple conceptual framework and make possible a self-contained treatment of interior-point methods for many convex optimization problems.

Independently of Nesterov and Nemirovsky, Alizadeh [4] has generalized interior-point methods from linear programming to positive-definite programming. Other recent articles are Jarre [27], Vandenberghe and Boyd [60], Rendl, Vanderbei and Wolkowicz [54], Yoshise [65], and Alizadeh, Haeberly and Overton [1]. An excellent reference on interior-point methods for general convex problems is Den Hertog [14].

1.4 Outline

This paper gives a survey of interior-point methods for positive-definite programming. We start with a section on duality theory. In Section 3 we introduce the barrier function for PDPs, and the concepts of central points and central path. The notion of central path is heavily used in Section 4, which discusses primal-dual methods.

This survey is not meant to be exhaustive and emphasizes primal-dual methods. The most important omissions are the projective methods of Karmarkar, and of Nesterov and Nemirovsky [43]. Our motivation for the restriction to primal-dual methods is twofold. Primal-dual methods are commonly held to be more efficient in practice, and, secondly, their behavior is often easier to analyze. Moreover, all interior-point methods are based on similar principles, and we hope that the material discussed here is sufficient as a tutorial introduction to the entire field.

2 Duality

2.1 The Dual PDP

The *dual* problem associated with the PDP (1) is

$$\begin{aligned} \text{maximize} \quad & -TrF_0Z \\ \text{subject to} \quad & TrF_iZ = c_i, \ i = 1, \ldots, m, \\ & Z \geq 0. \end{aligned} \tag{14}$$

Here the variable is the symmetric matrix Z, which is subject to m equality constraints and the nonnegativity condition. We write TrX for the trace of a symmetric matrix, i.e., $TrX = X_{11} + \cdots + X_{nn}$. Note that the objective function in (14) is a linear function of Z.

The dual problem (14) is also a PDP, i.e., it can be put in the same form as the *primal* problem (1). Let us assume for simplicity that the matrices $F_1, \ldots, F_m$ are linearly independent. Then we can express

$$\left\{ \ Z \mid Z = Z^T \in \mathbf{R}^{n \times n}, \ TrF_iZ = c_i \ \right\}$$

in the form

$$\left\{ \ G(y) = G_0 + y_1G_1 + \cdots + y_pG_p \mid y \in \mathbf{R}^p, \right\}$$

where $p = \frac{n(n+1)}{2} - m$ and the G_i are appropriate matrices. We define $d \in \mathbf{R}^p$ by $d_i = TrF_0G_i$, so that $d^Ty = TrF_0(G(y) - G_0)$. Then the dual problem becomes (except for a constant term in the objective and a change of sign to transform maximization into minimization)

$$\begin{aligned} \text{minimize} \quad & d^Ty \\ \text{subject to} \quad & G(y) \geq 0, \end{aligned}$$

which is a PDP. It is possible to use notation that, unlike ours, emphasizes the complete symmetry between the primal and dual problems (see, e.g., Nesterov and Nemirovsky). Our notation was designed to make the primal problem as "explicit" as possible, with x denoting a "free" variable.

As an example of the dual PDP, let us apply the definition to the linear program (3), i.e., take $F_0 = -diag(b)$ and $F_i = diag(a_i)$. In this case, the diagonal structure makes it possible to simplify the dual problem. The objective function and the equality constraints only involve the diagonal elements of Z, and, obviously, replacing the off-diagonal of a positive-definite matrix by zeros does not alter its positive-definiteness either. Instead of optimizing over all symmetric $n \times n$ matrices Z, we can therefore limit ourselves to diagonal matrices $Z = diag(z)$. Problem (14) then reduces to

$$\begin{aligned} \text{maximize} \quad & b^Tz \\ \text{subject to} \quad & z \geq 0, \\ & a_i^Tz = c_i, \ i = 1, \ldots, m, \end{aligned} \tag{15}$$

which is the familiar dual of the LP (3).

This example demonstrates an important point. In general, it is often the case that the dual problem can be simplified when the matrices F_i are structured. For example, if the matrix $F(x)$ is block diagonal, the dual variables Z can be assumed to have the same block diagonal structure.

Linear programming duality is very strong owing to the polyhedral character of the feasible set: The optimum values of (3) and (15) are always equal, except in the pathological case where both problems are infeasible. (We adopt the standard convention that the optimum value of (3) is $+\infty$ if the problem is infeasible, and the optimum value of (15) is $-\infty$ if the dual problem is infeasible.) Duality results for general PDPs are weaker, as we will see below.

Let us return to our discussion of the dual PDP. The key property of the dual PDP is that it yields bounds on the optimal value of the primal PDP, and vice versa. Suppose that Z is dual feasible, and x is primal feasible. Then we have:

$$-TrF_0Z \leq c^Tx. \tag{16}$$

The inequality follows from the simple calculation

$$c^Tx + TrZF_0 = \sum_{i=1}^{m} TrZF_ix_i + TrZF_0 = TrZF(x) \geq 0.$$

(We used the fact that $TrAB \geq 0$ when $A = A^T \geq 0$ and $B = B^T \geq 0$.)

Since (16) holds for *any* feasible x, we conclude that $p^* \geq -TrZF_0$, where p^* is the optimal value of the PDP (1),

$$p^* \stackrel{\Delta}{=} inf\left\{ c^Tx \mid F(x) \geq 0 \right\}. \tag{17}$$

In other words: *Dual feasible matrices yield lower bounds for the primal problem.* We can interpret x as a *suboptimal point* which gives the upper bound $p^* \leq c^Tx$ and Z as a *certificate* that proves the lower bound $p^* \geq -TrF_0Z$.

If x is primal feasible and Z is dual feasible, we refer to the quantity $\eta \stackrel{\Delta}{=} c^Tx + TrF_0Z$ as the *duality gap* associated with x and Z. The duality gap is the difference between the upper and lower bound; it is the width of the interval in which we have localized p^*. If we define d^* to be the optimal value in the dual problem,

$$d^* \stackrel{\Delta}{=} sup\left\{ -TrF_0Z \mid Z = Z^T \geq 0, TrF_iZ = c_i, i = 1,\ldots,m \right\}, \tag{18}$$

then we can restate the result (16) as $p^* \geq d^*$, i.e., the optimal value of the dual problem is less than or equal to the optimal value of the primal problem. Note that when the primal (or dual) problem is infeasible, the right-hand (left-hand) side becomes ∞ ($-\infty$) so the inequality trivially holds. In fact equality usually obtains, as stated in the following theorem (see Nesterov and Nemirovsky [43], or Rockafellar [52]).

Theorem 1 *We have $p^* = d^*$ if any of the following conditions holds.*

1. *The primal problem (1) is strictly feasible.*

2. *The dual problem (14) is strictly feasible.*

3. *The primal solution set*

$$X_{opt} \triangleq \left\{ x \mid F(x) \geq 0 \text{ and } c^T x = p^* \right\}$$

is nonempty and bounded.

4. *The dual solution set*

$$Z_{opt} \triangleq \left\{ Z \mid Z \geq 0, \ Tr F_i Z = c_i, \text{ and } -Tr F_0 Z = d^* \right\}$$

is nonempty and bounded.

Example

Consider the PDP

$$\text{minimize} \quad x_1$$
$$\text{subject to} \quad \begin{bmatrix} 0 & x_1 & 0 \\ x_1 & x_2 & 0 \\ 0 & 0 & x_1 + 1 \end{bmatrix} \geq 0.$$

The feasible set is $\{(x_1, x_2) \mid x_1 = 0, x_2 \geq 0\}$, and therefore $p^* = 0$. The dual problem can be simplified as

$$\text{maximize} \quad -z_2$$
$$\text{subject to} \quad \begin{bmatrix} z_1 & (1 - z_2)/2 & 0 \\ (1 - z_2)/2 & 0 & 0 \\ 0 & 0 & z_2 \end{bmatrix} \geq 0,$$

and the feasible set is $\{(z_1, z_2) \mid z_1 \geq 0, z_2 = 1\}$. The dual problem therefore has optimal value $d^* = -1$. This PDP violates all four conditions mentioned in the theorem. Both problems are feasible, but not strictly feasible, and the optimal sets X_{opt} and Z_{opt} are both unbounded. Note also the contrast with linear programming, where it is impossible to have a finite nonzero duality gap at the optimum.

Example

We take the matrix norm minimization problem mentioned in Section 1.2:

$$\text{minimize} \quad \|A(x)\| \qquad (19)$$
$$x \in \mathbf{R}^m$$

where $A(x) = A_0 + x_1 A_1 + \cdots + x_m A_m$, and we remind the reader that $\|A(x)\|$ is the maximum singular value of $A(x)$.

The problem (19) is a basic problem in the theory of Banach spaces; its optimum value is the norm of (the image of) A_0 in the quotient space of all $p \times q$ matrices modulo the span of $A_1, \ldots, A_m$. In this theory we encounter the following dual of (19):

$$\text{maximize} \quad Tr A_0^T Q \qquad (20)$$
$$\text{subject to} \quad Tr A_i^T Q = 0, \quad i = 1, \ldots, m,$$
$$\|Q\|_* \leq 1,$$

where $\|Q\|_* = \sum \sigma_i(Q)$ is the *nuclear norm* of Q, which is the norm dual to the maximum singular value. It is also known that the optimal values of (19) and (20) are always equal.

Let us verify that this (Banach space) notion of duality coincides with PDP duality. The dual PDP of problem (7) is

$$\begin{array}{ll}
\text{maximize} & -2TrA_0^T Z_{12} \\
\text{subject to} & TrA_i^T Z_{12} = 0, \quad i = 1, \ldots, m, \\
& TrZ_{11} + TrZ_{22} = 1, \\
& \begin{bmatrix} Z_{11} & Z_{12} \\ Z_{12}^T & Z_{22} \end{bmatrix} \geq 0.
\end{array} \tag{21}$$

This can be simplified. The positive-definite constraint can be rewritten as

$$\begin{bmatrix} Z_{11} & 0 \\ 0 & Z_{22} \end{bmatrix} \geq \begin{bmatrix} 0 & -Z_{12} \\ -Z_{12}^T & 0 \end{bmatrix}. \tag{22}$$

The eigenvalues of the matrix on the right are the singular values of Z_{12}, each singular value appearing twice. It is well known that if A and B are two symmetric matrices, then $A \geq B$ implies $\lambda_k(A) \geq \lambda_k(B)$, assuming the eigenvalues of A and B are taken in the same order. As a consequence, inequality (22) implies

$$2\sum \sigma_i(Z_{12}) \leq TrZ_{11} + TrZ_{22} \leq 1.$$

Since the matrices Z_{11} and Z_{22} do not appear in any other constraint, nor in the objective, we see that problem (21) reduces to

$$\begin{array}{ll}
\text{maximize} & -2TrA_0^T Z_{12} \\
\text{subject to} & TrA_i^T Z_{12} = 0, \quad i = 1, \ldots, m, \\
& 2\sum \sigma_i(Z_{12}) \leq 1,
\end{array}$$

which is the same as (20) with $Q = 2Z_{12}$.

Problem (19) is always strictly feasible; it suffices to choose $x = 0$ and $t > \|A_0\|$. Applying Theorem 1, we conclude that the optimal duality gap is always zero.

2.2   The Primal-Dual Formulation

Theorem 1 has important consequences for PDP algorithms. It gives conditions under which the primal-dual optimization problem

$$\begin{array}{ll}
\text{minimize} & c^T x + TrF_0 Z \\
\text{subject to} & F(x) \geq 0, \\
& Z \geq 0, \\
& TrF_i Z = c_i, \quad i = 1, \ldots, m,
\end{array} \tag{23}$$

has optimum value zero. Here we minimize the duality gap $c^T x + TrF_0 Z$ over all primal and dual feasible points. The duality gap is a linear function of x and Z, and therefore problem (23) is a PDP in x and Z.

Primal-dual methods for PDPs solve (23), assuming the primal and dual problems are strictly feasible. They generate a sequence of feasible points $x^{(k)}$ and $Z^{(k)}$, and in each step use the dual information in $Z^{(k)}$ to find good updates for $x^{(k)}$ and vice-versa.

This means that at every stage of the algorithm, we have available suboptimal primal and dual solutions x, Z. The primal solution x proves an upper bound $c^T x > p^*$ on the optimal value; the dual solution proves a lower bound $p^* > -Tr F_0 Z$. The iteration continues until the duality gap is less than a given tolerance ϵ.

3 The Barrier Function

In this section, we introduce a barrier function for linear matrix inequality constraints and discuss its properties. This leads us to the fundamental concept of *centrality*, and the definition of *central points* and *central path*. From now on we will assume that the matrices F_i are independent.

3.1 Definition

The function

$$\phi(x) \triangleq \begin{cases} log\, det\, F(x)^{-1} & \text{if } F(x) > 0 \\ +\infty & \text{otherwise} \end{cases} \tag{24}$$

is a *barrier function* for $X \triangleq \{x \mid F(x) \geq 0\}$, i.e., $\phi(x)$ is finite if and only if $F(x) > 0$, and becomes infinite as x approaches the boundary of X. There are many other barrier functions for X (for example, trace can be substituted for determinant in (24)), but this one enjoys many special properties (see [43]). In particular, when $F(x) > 0$, it is analytic and strictly convex.

In the case of a set of linear inequalities $Ax \geq b$, where $A = [a_1 \dots a_n]^T$, we have $F(x) = diag(Ax - b)$, and the definition reduces to the familiar logarithmic barrier function

$$\phi(x) = \begin{cases} -\sum_{i=1}^n log(a_i^T x - b_i) & \text{if } Ax \geq b, \\ +\infty & \text{otherwise.} \end{cases}$$

We first give formulas for the gradient $g(x)$ and Hessian $H(x)$ of ϕ. Recall that $Tr AB$ is the standard inner product of two symmetric matrices A and B; the corresponding norm is the Frobenius norm, $\|A\|_F = (Tr A^2)^{1/2}$.

The gradient and the Hessian of ϕ can be readily derived from the following second order approximation of the function $-log\, det\, X$. If $X > 0$ is $n \times n$ and symmetric, then

$$log\, det(X + Y)^{-1} = log\, det\, X^{-1} - Tr X^{-1} Y + \tfrac{1}{2} Tr X^{-1} Y X^{-1} Y + o\left(\|Y\|^2\right). \tag{25}$$

From equation (25), one can immediately derive a second order approximation for $\phi(x)$:

$$\phi(x + v) \approx \phi(x) - Tr F(x)^{-1} \left(\sum_{i=1}^m v_i F_i \right) +$$

$$\tfrac{1}{2}TrF(x)^{-1}\left(\sum_{i=1}^{m}v_iF_i\right)F(x)^{-1}\left(\sum_{j=1}^{m}v_jF_j\right)$$

$$=\ \phi(x)-\sum_{i=1}^{m}v_iTrF(x)^{-1}F_i+\tfrac{1}{2}\sum_{i=1}^{m}\sum_{j=1}^{m}v_iv_jTrF(x)^{-1}F_iF(x)^{-1}F_j.$$

We conclude that the gradient $g(x)$ and the Hessian $H(x)$ of $\phi(x)$ are given by

$$g_i(x)=-TrF(x)^{-1}F_i=-TrF(x)^{-1/2}F_iF(x)^{-1/2}, \tag{26}$$

and

$$\begin{aligned}H_{ij}(x)&=\ TrF(x)^{-1}F_iF(x)^{-1}F_j\\&=\ Tr\left(F(x)^{-1/2}F_iF(x)^{-1/2}\right)\left(F(x)^{-1/2}F_jF(x)^{-1/2}\right),\end{aligned} \tag{27}$$

for $i,j=1,\ldots,m$.

From expression (27) we can verify that ϕ is strictly convex for strictly feasible x. For $x,y\in\mathbf{R}^m$ with $F(x)>0$,

$$\begin{aligned}y^TH(x)y&=\ \sum_{i,j=1}^{m}y_iy_jTr\left(F(x)^{-1/2}F_iF(x)^{-1/2}\right)\left(F(x)^{-1/2}F_jF(x)^{-1/2}\right)\\&=\ Tr\left(F(x)^{-1/2}\left(\sum_{i=1}^{m}y_iF_i\right)F(x)^{-1/2}\right)^2\\&=\ \left\|F(x)^{-1/2}\left(\sum_{i=1}^{m}y_iF_i\right)F(x)^{-1/2}\right\|_F^2\ \ge\ 0.\end{aligned} \tag{28}$$

We see that $y^TH(x)y=0$ if and only if $\sum_{i=1}^{m}y_iF_i=0$. By independence of $F_1,\ldots,F_m$, we conclude that $H(x)>0$, i.e., ϕ is strictly convex.

Finally, we note that the barrier function ϕ is bounded below if and only if the feasible set X is bounded.

3.2 Analytic Center

3.2.1 Definition

We suppose now that the linear matrix inequality $F(x)\ge 0$ is strictly feasible and that its feasible set is bounded. Since ϕ is strictly convex, it has a unique minimizer, which we denote

$$x^*\stackrel{\Delta}{=}\ \underset{x}{\text{argmin}}\ \ \phi(x). \tag{29}$$

We will refer to x^* as the *analytic center* of the linear matrix inequality $F(x)>0$. It is important to note that the analytic center is a property of a linear matrix inequality and not of its solution set X. The same set X can be represented by different matrix inequalities, which have different analytic centers.

From (26) we see that x^* is characterized by

$$TrF(x^*)^{-1}F_i = 0, \quad i = 1, \ldots, m. \tag{30}$$

Thus, $F(x^*)^{-1}$ is orthogonal to the span of $F_1, \ldots, F_m$.

In the case of a set of linear inequalities, the definition coincides with Sonnevend's definition [57, 58], i.e.,

$$x^* = \mathop{\mathrm{argmax}}_{Ax \geq b} \prod_{i=1}^{n}(a_i^T x - b_i).$$

3.2.2 Computing the Analytic Center

Newton's method, with appropriate step length selection, can be used to efficiently compute the analytic center. Starting with a strictly feasible point $x^{(0)}$, the algorithm follows the iteration:

$$x^{(k+1)} := x^{(k)} - \alpha^{(k)}H(x^{(k)})^{-1}g(x^{(k)}), \tag{31}$$

where $\alpha^{(k)}$ is the damping factor at the kth iteration, and $g(x)$ and $H(x)$ are the gradient and Hessian of the barrier function in x.

Nesterov and Nemirovsky [43] give a simple step length rule appropriate for the general class of self-concordant barrier functions mentioned earlier. The damping factor depends on a quantity called the *Newton decrement* of ϕ at x:

$$\delta(x) \stackrel{\Delta}{=} \left\| H(x)^{-1/2}g(x) \right\|.$$

The name comes from the observation that $\delta(x)^2/2$ is the difference between $\phi(x)$ and the minimum value of the quadratic approximation of ϕ at x. Alternatively, $\delta(x)$ is the length of the Newton step $-H(x)^{-1}g(x)$ measured in the norm induced by the Hessian $H(x)$.

The damping factor is:

$$\alpha^{(k)} := \begin{cases} 1 & \text{if } \delta(x^{(k)}) \leq 1/4, \\ 1/(1 + \delta(x^{(k)})) & \text{if } \delta(x^{(k)}) > 1/4. \end{cases} \tag{32}$$

Nesterov and Nemirovsky show that this step length always results in $F(x^{(k+1)}) > 0$. Moreover, for $\delta(x^{(k)}) < 1/4$, we have $\delta(x^{(k+1)}) \leq 2\delta(x^{(k)})^2$, i.e., the algorithm converges quadratically. They also give a complete convergence analysis. The main results can be summarized as follows.

- Until the region of quadratic convergence ($\delta(x) < 1/4$) is reached, the objective $\log \det F(x)^{-1}$ decreases at least by the absolute constant 0.3068 at each New-ton step. (By absolute constant we mean it does not depend on n, the problem data, or the required accuracy of computing x^*.)

- Once the region of quadratic convergence is reached, at most a *constant* number c of Newton steps is required to compute x^* to a given accuracy. (The constant c does not depend on n or the problem data, but only on the required accuracy ϵ. Since the convergence is quadratic in this region, c grows as $\log \log 1/\epsilon$ if ϵ decreases.)

In other words, the number of Newton steps required to compute x^* given x can be bounded in terms of $-\log \det F(x) + \log \det F(x^*)$:

$$\#\text{Newton steps} \leq c + 3.26(-\log \det F(x) + \log \det F(x^*)), \qquad (33)$$

where c depends only on the required accuracy of computing x^* and grows extremely slowly.

Therefore, the quantity $\psi(x) \triangleq -\log \det F(x) + \log \det F(x^*)$ has a very natural interpretation as the 'deviation from centrality' of a point x. In general, however, $\psi(x)$ can only be evaluated by computing the center x^*.

3.3 The Primal-Dual Central Path

We now return to the primal-dual formulation of Section 2.2.

For $\alpha > 0$ consider the set of strictly feasible pairs x, Z with duality gap α, i.e., $c^T x + Tr F_0 Z = \alpha$. The *analytic center* of this set is the minimizer of the barrier term $-\log \det F(x) - \log \det Z$. We denote the analytic center as $x^*(\alpha), Z^*(\alpha)$:

$$
\begin{aligned}
(x^*(\alpha), Z^*(\alpha)) = \quad &\text{argmin} \quad -\log \det F(x) - \log \det Z \\
&\text{subject to} \quad F(x) \geq 0, \\
&\qquad\qquad\quad Z \geq 0, \\
&\qquad\qquad\quad Tr F_i Z = c_i, i = 1, \ldots, m, \\
&\qquad\qquad\quad c^T x + Tr F_0 Z = \alpha.
\end{aligned}
\qquad (34)
$$

Thus, among all feasible pairs x, Z with the duality gap α, the pair x^*, Z^* maximizes $det(F(x)Z)$. The pair (x^*, Z^*) converges to a primal and dual optimal pair as $\alpha \to 0$, and the curve given by (x^*, Z^*) for $\alpha > 0$ is called the *central path* for the problem (23).

The central pair (x^*, Z^*) has many important properties. For our purposes here we need:

Theorem 2 $F(x^*(\alpha))Z^*(\alpha) = (\alpha/n)I$. *Conversely, if x and Z are a feasible pair and $F(x)Z = (\alpha/n)I$ then $x = x^*(\alpha)$ and $Z = Z^*(\alpha)$.*

In other words, centrality is characterized by $F(x)$ and Z being inverses of each other, up to a constant.

Now consider a feasible pair (x, Z), and define $\alpha = Tr F(x)Z$. Then $(x^*(\alpha), Z^*(\alpha))$ is the central pair with the same duality gap as x, Z. Therefore

$$\log \det F(x)Z \geq \log \det F(x^*(\alpha))Z^*(\alpha) = n \log n - n \log Tr F(x)Z$$

with equality holding only when x, Z are central.

As in Section 3.2.2 we can say that the difference

$$
\begin{aligned}
\psi(x, Z) \;&\overset{\triangle}{=}\; -\log\det F(x)Z + \log\det F(x^*(\alpha))Z^*(\alpha) \\
&=\; -\log\det F(x)Z + n\log Tr F(x)Z - n\log n
\end{aligned}
\tag{35}
$$

is a measure of the deviation of (x, Z) from centrality: $\psi(x, Z)$ is, up to a constant, an upper bound on the computational effort required to "center" (x, Z) (meaning, compute the central pair with the same duality gap). From (35), we also note that $\psi(x, Z)$ can be evaluated as a function of x and Z, without knowing x^* and Z^*.

The function ψ is not convex or quasiconvex (except of course when restricted to $Tr F(x)Z$ constant). We also note that ψ depends only on the eigenvalues $\lambda_1, \ldots, \lambda_n$ of $F(x)Z$:

$$
\psi(x, Z) = n\log \frac{\left(\sum_{i=1}^{n} \lambda_i\right)/n}{\left(\prod_{i=1}^{n} \lambda_i\right)^{1/n}}.
$$

Thus, $\psi(x, Z)$ is n times the logarithm of the ratio of the arithmetic to the geometric mean of the eigenvalues of $F(x)Z$. (From which we see again that ψ is nonnegative, and zero only when $F(x)Z$ is a multiple of the identity.) We can also think of ψ as a smooth measure of condition number of the matrix $F(x)Z$ since

$$
\log \kappa - 2\log 2 \leq \psi(x, Z) \leq (n-1)\log \kappa,
\tag{36}
$$

where $\kappa = \lambda_{max}/\lambda_{min}$ is the condition number of $F(x)Z$.

4   Primal-Dual Potential Reduction Methods

The methods in this section take full advantage of duality theory by updating primal and dual feasible points in each iteration. The basic idea is the following. We choose a potential function $\varphi(x, Z)$ such that

- φ is smooth on the interior of the feasible set

- φ is unbounded below if x and Z approach the optimal solution

- φ increases unboundedly as x or Z approach all other points on the boundary of the feasible set

The algorithms minimize the potential function by some modification of Newton's method. If the potential function is properly chosen, this leads to algorithms with a polynomial worst-case complexity.

Theoretically there is a perfect symmetry between primal and dual problems. In practice, it is important to keep in mind that the dimension of the dual problem is usually much larger than the dimension of the primal problem. Computations involving the dual problem will therefore have to be arranged carefully.

4.1 Primal-Dual Potential Function

Define the primal-dual potential function

$$
\begin{aligned}
\varphi(x, Z) &\triangleq \nu\sqrt{n}\,log\left(c^T x + TrF_0 Z\right) + \psi(x, Z) \\
&= q\,log\left(c^T x + TrF_0 Z\right) - log\,det\,F(x) - log\,det\,Z - n\,log\,n, \quad (37)
\end{aligned}
$$

where $q = n + \nu\sqrt{n}$. The first term, $\nu\sqrt{n}\,log\left(c^T x + TrF_0 Z\right)$, measures a decrease in objective function. The second term, $\psi(x, Z)$, is the deviation from centrality defined in the previous section. The parameter ν controls the relative weight of both terms. To minimize the worst-case complexity it has to be of $O(1)$, but in practice a larger value can be more efficient. Note that along the central path, $\varphi(x^*(\alpha), Z^*(\alpha)) = \nu\sqrt{n}\,log\,\alpha$ which decreases to $-\infty$ as α converges to zero.

A fixed decrease in the first term corresponds to a fixed fractional reduction of the duality gap. A fixed decrease in the second term corresponds to a fixed amount of "centering" in the following sense: up to a constant, it is the reduction in the (bound on) computational effort required to "center" the current pair.

If ψ decreases by one, the new pair is 3.26 Newton steps closer to centrality than the original pair (or more precisely, the upper bound on the number of Newton steps required to center the new pair is 3.26 smaller than the upper bound for the original pair). If the other term, $\nu\sqrt{n}\,log\,TrF(x)Z$, decreases by one, then the duality gap is reduced by the factor $exp(-1/\nu\sqrt{n}) \approx 1 - 1/(\nu\sqrt{n})$. In other words,

$\nu\sqrt{n}$ fewer Newton steps to center $\sim 31\%$ duality gap reduction,

where $\sim$ means that the left and right-hand sides result in an equal decrease in φ.

By minimizing the smooth function φ, we solve the primal and dual problems. Indeed since $\psi(x, Z) \geq 0$ for feasible x and Z, we have

$$
TrF(x)Z \leq exp\,\frac{\varphi(x, Z)}{\nu\sqrt{n}}, \quad (38)
$$

which shows that small potential implies small duality gap.

The basic idea of the primal-dual algorithm is to generate iterations of primal and dual feasible matrices satisfying

$$
\varphi(x^{(k+1)}, Z^{(k+1)}) \leq \varphi(x^{(k)}, Z^{(k)}) - \delta, \quad (39)
$$

for some absolute positive constant δ. By (38) we therefore have:

$$
TrF(x^{(k)})Z^{(k)} \leq exp\,\frac{\varphi(x^{(0)}, Z^{(0)}) - k\delta}{\nu\sqrt{n}} = c_0 c_1^k TrF(x^{(0)})Z^{(0)}, \quad (40)
$$

where

$$
c_0 = exp\,\frac{\psi(x^{(0)}, Z^{(0)})}{\nu\sqrt{n}}, \qquad c_1 = exp\,\frac{-\delta}{\nu\sqrt{n}}.
$$

We can interpret the result (40) as follows: the duality gap converges to zero at least exponentially at a rate given by the constant c_1. The constant c_0 depends only on the centrality of the initial pair, and is one if the initial pair is central.

In other words, we have polynomial convergence:

Theorem 3 *Assume that (39) holds with some $\delta > 0$ that does not depend on n or ϵ, where $0 < \epsilon < 1$. Then for*

$$k \geq \frac{\nu\sqrt{n}\,log(1/\epsilon) + \psi(x^{(0)}, Z^{(0)})}{\delta},$$

we have $TrF(x^{(k)})Z^{(k)} < \epsilon TrF(x^{(0)})Z^{(0)}$.

Roughly speaking, we have convergence in $O(\sqrt{n})$ steps, provided the initial pair is sufficiently centered.

We conclude this section with a conceptual outline of the potential reduction algorithm.

Potential Reduction Algorithm

given strictly feasible x and Z.

repeat

    1. Find primal and dual feasible directions δx and δZ.
    2. Find $p, q \in \mathbf{R}$ that minimize $\varphi(x + p\delta x, Z + q\delta Z)$.
    3. Update: $x := x + p\delta x$ and $Z := Z + q\delta Z$.

until duality gap $\leq \epsilon$.

The key task, then, is to show how to update $(x^{(k)}, Z^{(k)})$ into $(x^{(k+1)}, Z^{(k+1)})$ such that (39) holds.

4.2 Plane Search

In the next sections we will describe several possibilities for computing directions δx and δZ. Here we assume that these search directions are given, and we will show how the potential function $\varphi(x + p\delta x, Z + q\delta Z)$ can be minimized over the plane defined by these two directions. We will use the notation $F \triangleq F(x)$ and $\delta F \triangleq \sum_{i=1}^{m} F_i \delta x_i$.

The two-dimensional minimization can be done very efficiently if we first compute the eigenvalues $\mu_1, \ldots, \mu_n$ of $F^{-1/2}\delta F F^{-1/2}$ and the eigenvalues $\nu_1, \ldots, \nu_n$ of $Z^{-1/2}\delta Z Z^{-1/2}$. The potential function can then be written in terms of p and q as

$$\varphi(p, q) = (n + \nu\sqrt{n})\,log(c_1 + c_2 p + c_3 q) + f(p, q) - n\,log\,n,$$

where $c_1 = TrFZ$, $c_2 = TrZ\delta F$, and $c_3 = TrF\delta Z$, and f is the restriction of the barrier term to the plane, i.e.,

$$
\begin{aligned}
f(p, q) &= -log\,det(F + p\delta F) - log\,det(Z + q\delta Z) \\
&= -\sum_{i=1}^{n} log(1 + p\mu_i) - \sum_{i=1}^{n} log(1 + q\nu_i) - log\,det(FZ).
\end{aligned}
$$

Note that once we have the eigenvalues μ_i and ν_i, we can compute $\varphi(p,q)$ and its derivatives in $O(n)$ operations.

An efficient way to minimize φ is Newton's method with Nesterov and Nemirovsky's step length. Although $\varphi(p,q)$ is not convex (it is the sum of a concave and a convex term), it can be replaced by a convex upper bound at every iteration of the plane search by linearizing the concave term. Newton's method can then be safely applied to this convex approximation (see next section for more details).

There is no need to calculate the minimum of $\varphi(p,q)$ very accurately. One can take a fixed number of steps, for example, as Ben Tal and Nemirovsky in [10].

The main cost of this scheme is in the initial computation of the eigenvalues μ_i and ν_i. Once these are known, each step in the plane search can be carried out at a cost of $O(n)$ operations.

4.3 Potential Reduction Method 1

The first, and most obvious, method for computing search directions δx and δZ, is by Newton's method. We need a slight modification to Newton's method, however, because the potential function $\varphi(x,Z)$ is not convex itself, but the sum of a concave and a convex function. The modification can be interpreted in several ways.

- We use exactly Newton's method, but apply it to a different potential function $\tilde{\varphi}$. The modified function $\tilde{\varphi}$ is obtained from φ by linearizing the concave term $q\,log(c^Tx + TrF_0Z)$ around the current iterate $x^{(k)}, Z^{(k)}$. Therefore, $\tilde{\varphi}$ is a convex function. It has the property that

$$\tilde{\varphi}(x,Z) \geq \varphi(x,Z),$$

for all x and Z, and $\tilde{\varphi}(x^{(k)}, Z^{(k)}) = \varphi(x^{(k)}, Z^{(k)})$. In other words, if a particular choice of $x^{(k+1)}, Z^{(k+1)}$ reduces $\tilde{\varphi}$ by a certain amount, then the reduction in φ itself will be even bigger.

- It is a *quasi-Newton method* applied to φ. A quasi-Newton method uses the exact gradient of φ, but replaces the Hessian by an approximation. In this case the approximation is to drop the second derivatives of the concave term, and to consider only the barrier term when forming the Hessian.

Following the first interpretation, let us linearize the concave term:

$$\tilde{\varphi}(x,Z) \;\triangleq\; \varphi(x^{(k)}, Z^{(k)}) + \rho\left(c^T(x - x^{(k)}) + TrF_0(Z - Z^{(k)}) - log\,det\,F(x) - log\,det\,Z - n\,log\,n,\right.$$

where $\rho \triangleq q\big/\left(c^tx^{(k)} + TrF_0Z^{(k)}\right)$. Note that the function $\tilde{\varphi}$ is separable in x and Z:

$$\tilde{\varphi}(x,Z) = \tilde{\varphi}^p(x) + \tilde{\varphi}^d(Z) + \text{constant},$$

where

$$\tilde{\varphi}^p(x) \;\triangleq\; \rho c^Tx - log\,det\,F(x), \tag{41}$$

$$\tilde{\varphi}^d(Z) \;\triangleq\; \rho TrF_0Z - log\,det\,Z. \tag{42}$$

In the derivation of the Newton directions δx^p and δZ^d, we can therefore consider the primal and dual parts separately.

4.3.1 Primal Least-Squares Problem

The Newton direction for $\tilde{\varphi}^p$ is the direction that minimizes the best quadratic approximation to the function. Recall from Section 3 that the best quadratic approximation to the barrier function $-\log \det V$ in a given matrix X, is given by

$$-\log \det(X + V) = -TrX^{-1}V + \tfrac{1}{2}TrVX^{-1}VX^{-1}.$$

We therefore have (with $F \triangleq F(x)$):

$$\delta x^p = \underset{v \in \mathbf{R}^m}{\operatorname{argmin}} \ \rho c^T v - TrF^{-1}\left(\sum_{i=1}^m F_i v_i\right) + \tag{43}$$

$$\tfrac{1}{2}Tr\left(\sum_{i=1}^m F_i v_i\right) F^{-1}\left(\sum_{j=1}^m F_j v_j\right) F^{-1}$$

$$= \underset{v \in \mathbf{R}^m}{\operatorname{argmin}} \ \rho c^T v - \sum_{i=1}^m v_i\left(TrF^{-1}F_i\right) +$$

$$\tfrac{1}{2}\sum_{i=1}^m\sum_{j=1}^m v_i v_j\left(TrF_i F^{-1}F_j F^{-1}\right). \tag{44}$$

Let $\tilde{F}_i = F^{-1/2}F_i F^{-1/2}$, for $i = 1,\dots,m$, and C be an $n \times n$ matrix satisfying $TrF_i C = c_i$, $i = 1,\dots,m$ (take $C = Z$, for instance). Then the expression for δx^p can be further simplified to

$$\delta x^p = \underset{v \in \mathbf{R}^m}{\operatorname{argmin}} \ \sum_{i=1}^m v_i TrF_i\left(\rho C - F^{-1}\right) + \tag{45}$$

$$\tfrac{1}{2}\sum_{i=1}^m\sum_{j=1}^m v_i v_j Tr\left(F^{-1/2}F_i F^{-1/2}F^{-1/2}F_j F^{-1/2}\right)$$

$$= \underset{v \in \mathbf{R}^m}{\operatorname{argmin}} \ \sum_{i=1}^m v_i Tr\tilde{F}_i\left(\rho F^{1/2}CF^{1/2} - I\right) + \tfrac{1}{2}\sum_{i=1}^m\sum_{j=1}^m v_i v_j Tr\tilde{F}_i\tilde{F}_j$$

$$= \underset{v \in \mathbf{R}^m}{\operatorname{argmin}} \ \left\|\rho F^{1/2}CF^{1/2} - I + \sum_{i=1}^m \tilde{F}_i v_i\right\|_F. \tag{46}$$

This is a least squares problem with m variables and $n(n+1)/2$ equations.

Let us consider the LP (3) as an illustration. In this case, all matrices in (46) are diagonal, and $F = diag(Ax - b)$. The least-squares problem then reduces to minimizing the norm of the diagonal elements

$$\text{minimize} \left\|\rho X\tilde{c} - e + X^{-1}Av\right\|,$$

where $X = diag(Ax - b)$, e is a vector with all components one, and $\tilde{c}$ is any vector that solves $A^T\tilde{c} = c$.

4.3.2 Dual Least-Squares Problem

In a similar way, among all feasible directions, δZ^d will be the direction that minimizes the second order approximation of $\tilde{\varphi}^d(Z)$:

$$\delta Z^d = \underset{\substack{V : TrF_iV = 0 \\ i = 1,\ldots,m}}{\text{argmin}} \quad \rho TrF_0V - TrZ^{-1}V + \tfrac{1}{2}TrVZ^{-1}VZ^{-1}. \tag{47}$$

This involves a least-squares problem with exactly the same dimension as the primal problem. The easiest way to see this, is to introduce Lagrange multipliers v_i, $i = 1,\ldots,m$ for the constraints on V, and to write the optimality conditions for (47):

$$\rho F_0 - Z^{-1} + Z^{-1}\delta Z^d Z^{-1} + \sum_{i=1}^{m} v_i F_i = 0,$$

$$TrF_i\delta Z^d = 0, \quad i = 1,\ldots,m,$$

or, equivalently, if we write $\tilde{F}_i = Z^{1/2}F_iZ^{1/2}$, $i = 1,\ldots,m$:

$$\rho Z^{1/2}F_0Z^{1/2} - I + Z^{-1/2}\delta Z^d Z^{-1/2} + \sum_{i=1}^{m} v_i\tilde{F}_i = 0, \tag{48}$$

$$Tr\tilde{F}_iZ^{-1/2}\delta Z^d Z^{-1/2} = 0, \quad i = 1,\ldots,m.$$

Hence the multipliers v_i, $i = 1,\ldots,m$, are the solution of the least squares problem

$$\underset{v \in R^m}{\text{minimize}} \quad \left\| \rho Z^{1/2}F_0Z^{1/2} - I + \sum_{i=1}^{m} \tilde{F}_iv_i \right\|_F, \tag{49}$$

which, again, has $n(n + 1)/2$ equations and m variables. From the least-squares solution v, one then obtains δZ^d by substitution in equation (48).

In the case of an LP where $F(x) = diag(Ax - b)$ and $Z = diag(z)$, problem (49) becomes

$$\text{minimize} \, \|-\rho Zb - e + ZAv\|.$$

4.3.3 Summary

The following outline summarizes the two preceding sections. The method is due to Vandenberghe and Boyd [60], and is a generalization of Gonzaga and Todd's method for linear programming [25].

Potential Reduction Algorithm 1

given strictly feasible x and Z.

repeat

1. Solve the least-squares problem (46) for δx^p.

2. Solve the least-squares problem (49), and compute δZ^d from (48).
3. Find $p, q \in \mathbf{R}$ that minimize $\varphi(x + p\delta x^p, Z + q\delta Z^d)$.
4. Update: $x := x + p\delta x^p$ and $Z := Z + q\delta Z^d$.

until duality gap $\leq \epsilon$.

We refer to [60] for a complexity analysis. The basic ideas are summarized in the following two theorems. Define

$$\lambda^p \triangleq \left\| \sum_{i=1}^{m} F^{-1/2} F_i F^{-1/2} \delta x_i^p \right\|_F \quad , \quad \lambda^d \triangleq \left\| Z^{-1/2} \delta Z^d Z^{-1/2} \right\|_F .$$

The first theorem states that it is always possible to reduce the potential as long as λ^p and λ^d are not both small.

Theorem 4 *Assume that x and Z are strictly feasible, and that δx^p and δZ^d are the corresponding Newton directions. Let $p = 1/(1 + \lambda^p)$ and $q = 1/(1 + \lambda^q)$. Then*

$$\varphi(x + p\delta x^p, Z) \leq \varphi(x, Z) - \lambda^p + log(1 + \lambda^p),$$
$$\varphi(x, Z + q\delta Z^d) \leq \varphi(x, Z) - \lambda^d + log(1 + \lambda^d).$$

The second theorem states that λ^p and λ^d never become small at the same time.

Theorem 5 $\max\{\lambda^p, \lambda^d\} \geq 0.35.$

As a consequence of these two theorems, we see that it is always possible to reduce the potential function by at least $log\, 0.35 - log(1 - 0.35) = 0.78$ per iteration. We have seen in Section 4.1 that this implies convergence of the algorithm in $O(\sqrt{n})$ iterations.

4.4 Potential Reduction Method 2

The above algorithm has the disadvantage of requiring the solution of two least-squares problems per iteration. In this section we will show that a complete primal-dual algorithm can be based on the primal least-squares problem only. In linear programming this primal-dual method is called Ye's method [62], but the extension to PDPs is due to Nesterov and Nemirovsky [43] and Alizadeh [2].

We first write down the optimality conditions that characterize δx^p as the minimizer of (44):

$$\rho c_i - Tr F^{-1} F_i + \sum_{j=1}^{m} \delta x_j^p Tr \left(F_i F^{-1} F_j F^{-1} \right) = 0,$$

for $i = 1, \ldots, m$, or

$$(1/\rho) Tr F_i \left(F^{-1} - \sum_{j=1}^{m} \delta x_j^p F^{-1} F_j F^{-1} \right) = c_i.$$

In other words, the matrix $(1/\rho) \left(F^{-1} - \sum_{j=1}^{m} F^{-1} F_j F^{-1} \delta x_j^p \right)$ satisfies the equality constraints needed for dual feasibility, and the matrix

$$\delta Z^p \overset{\Delta}{=} (1/\rho) \left(F^{-1} - \sum_{j=1}^{m} \delta x_j^p F^{-1} F_j F^{-1} \right) - Z \tag{50}$$

is a dual feasible direction.

The second potential reduction algorithm takes this δZ^p as dual search direction, and performs a plane search in the plane defined by δx^p and δZ^p:

Potential Reduction Algorithm 2

given strictly feasible x and Z.

repeat

    1. Solve the least-squares problem (46) for δx^p.

    2. Find δZ^p from (50).

    3. Find $p, q \in \mathbf{R}$ that minimize $\varphi(x + p\delta x^p, Z + q\delta Z^d)$.

    4. Update: $x := x + p\delta x^p$ and $Z := Z + q\delta Z^d$.

until duality gap $\leq \epsilon$.

The effectiveness of this algorithm is a consequence of Theorem 4 and Theorem 6 below. Theorem 4 implies that

$$\varphi(x + \frac{1}{1 + \lambda^p} \delta x^p, Z) \leq \varphi(x, Z) - \lambda^p + \log(1 + \lambda_p).$$

Theorem 6 *If* $\lambda^p < 1$, *then*

$$\varphi(x, Z + \delta Z^p) \leq \varphi(x, Z) + \nu(\lambda^p - 1/2) - \lambda^p - \log(1 - \lambda^p).$$

Hence, again, there is always at least one way to reduce the potential function substantially. If λ^p is large, then a primal update only would be enough; when λ^p becomes too small, the dual update takes over. The plane search guarantees a smooth and optimal transition between both extreme cases.

4.5 Combined Algorithms

By duality, the considerations of Section 4.4 can be repeated for the dual least-squares problem (49). Solving the dual least-squares problem, we not only find a dual Newton direction δZ^d but also a primal feasible direction δx^d, and a complete primal-dual algorithm can be based on the dual least-squares problem alone.

This means that four search directions are available at each strictly feasible pair x, Z:

primal least squares	δx^p	δZ^p
dual least squares	δx^d	δZ^d

In this table, three pairs of primal and dual directions are known to reduce the potential function:

- Method 1 uses $\delta x^p, \delta Z^d$.

- Method 2 uses $\delta x^p, \delta Z^p$.

- The dual version of method 2 method uses $\delta x^d, \delta Z^d$.

These three methods all have a worst-case complexity of $O(\sqrt{n}L \log(1/\epsilon))$ iterations, and also in practice, it is not clear which one is faster (in terms of iterations). The first method has the disadvantage of requiring two least-squares problems per update, but one can think of combinations, where in each step either the primal or the dual least-squares problem (for instance, alternatingly) is solved.

4.6 Long-Step Path-Following Methods

A number of recent articles describe path-following linear programming methods with superlinear or quadratic asymptotic convergence (see Zhang and Tapia [66], Ye, Güler, Tapia and Zhang [63], Mizuno, Todd and Ye [35]). In this section we discuss a PDP variant of these methods. We use a conceptual version of an algorithm presented by Nesterov [37]. For a different approach, see Alizadeh, Haeberly and Overton [1].

The algorithm traces the central path using a predictor-corrector approach. The predictor phase makes the largest possible step without leaving a pre-determined neighborhood of the central path. Our discussion in Section 3.3 suggests using $\psi(x, Z)$ as a measure for deviation from the central path. The parameter ψ_{max} in the following algorithm is an upper bound on ψ and determines how close to the central path the iterates stay. We assume that the initial points are strictly feasible and sufficiently centered.

Long-Step Path-Following Algorithm

given strictly feasible x and Z with $\psi(x, Z) \leq \psi_{max}$.

repeat

    1. *Corrector step.*
       (Approximately) center x, Z, keeping the duality gap constant.
    2. *Predictor step.*
       Compute directions $\delta x^{prd}, \delta Z^{prd}$ (approximately) tangent to central path.
       Compute $p > 0$ such that $\psi(x + p\delta x^{prd}, Z + p\delta Z^{prd}) = \psi_{max}$.
       Take $x = x + p\delta x^{prd}, Z = Z + p\delta Z^{prd}$.

until duality gap $\leq \epsilon$.

4.6.1 Predictor Step

Assume x and Z are central with duality gap α: $F(x)Z = (\alpha/n)I$. We first derive expressions for the tangent directions to the central path at x, Z. From the first order expansion of the equality $F(x + \delta x)(Z + \delta Z) = ((\alpha - \delta\alpha)/n)I$, we find that the tangent directions δx^{prd} and δZ^{prd} satisfy

$$-I = F(x)\delta Z^{prd} + \left(\sum_{i=1}^{m} \delta x_i^{prd} F_i\right) Z, \tag{51}$$

or, using $F(x)Z = (\alpha/n)I$,

$$-I = F^{1/2}\delta Z^{prd}F^{1/2} + (\alpha/n)F^{-1/2}\left(\sum_{i=1}^{m} \delta x_i^{prd} F_i\right)F^{-1/2}, \tag{52}$$

where, as before, $F \overset{\Delta}{=} F(x)$.

This equation, together with the dual feasibility conditions $Tr F_i \delta Z^{prd} = 0$, is equivalent to the optimality conditions of the least-squares problem,

$$\delta x^{prd} = \underset{v \in \mathbf{R}^m}{\text{argmin}} \left\| (n/\alpha)I + \sum_{i=1}^{m} v_i F^{-1/2} F_i F^{-1/2} \right\|, \tag{53}$$

and the dual direction follows from $\delta Z^{prd} = -F^{-1} - (\alpha/n)F^{-1}\left(\sum_{i=1}^{m} \delta x_i^{prd} F_i\right)F^{-1}$.

Theorem 7 *If ψ_{max} is a positive constant, and $p > 0$ is computed from*

$$\psi(x + p\delta x^{prd}, Z + p\delta Z^{prd}) = \psi_{max},$$

then

$$Tr F(x + p\delta x^{prd})(Z + p\delta Z^{prd}) \le (1 - \tau/\sqrt{n}) Tr F(x)Z,$$

where τ is a positive constant less than one.

In other words, every predictor step reduces the duality gap by $1 - \tau/\sqrt{n}$. The number of predictor steps needed to reduce the duality gap by a factor ϵ is at most $\frac{\sqrt{n}}{\tau} log(1/\epsilon)$. This establishes polynomial convergence. In practice, the steps typically become larger as the algorithm approaches the solution. This results in fast (superlinear) asymptotic convergence.

4.6.2 Corrector Step

In the corrector phase one can use Newton's method, possibly combined with a plane search, to find a primal-dual central pair with a given duality gap α. As explained in Section 3.3, the number of corrector steps in every outer iteration will be bounded by an absolute constant. Therefore, the corrector steps do not contribute to the order of the worst-case complexity.

5 Conclusions and Extensions

5.1 Solving Structured Problems

A natural question is: What is the computational effort required to solve a PDP using the methods described above? The answer depends on the amount of structure in the matrix inequality $F(x)$. The three methods have the same worst-case complexity: The number of iterations to solve a PDP to a given accuracy grows as $O(\sqrt{n})$. In practice, it has been observed by many researchers that the number grows more slowly, as $\log n$ or as $n^{1/4}$, and can often be assumed to be almost constant (see Nesterov and Nemirovsky [43], or Gonzaga and Todd [25] for comments on the average behavior). Typical numbers range from 10 to 50.

The overall cost is therefore determined by the amount of work per iteration. As we have seen, the dominant part here is the solution of a least-squares problem of the form

$$\text{minimize } \left\| D - \sum_{i=1}^{m} v_i \tilde{F}_i \right\|_F , \tag{54}$$

where $\tilde{F}_i = V^T F_i V$. The matrices V and D change from iteration to iteration. Problem (54) is a least-squares problem with m variables and $n(n+1)/2$ equations. Using direct methods it can be solved in $O(m^2 n^2)$ operations.

Important savings are possible when the matrices F_i are structured. The easiest type is block-diagonal structure. Assume $F(x)$ consists of L diagonal blocks of size n_i, $i = 1, \ldots, L$. Then the number of equations in (54) is $\sum_{i=1}^{L} n_i(n_i + 1)/2$, which is often an order less than $n(n + 1)/2$. For instance, in the LP case (diagonal matrix $F(x)$), the number of variables is n, and solving the least-squares problem requires only $O(m^2 n)$ operations.

In more complicated situations, one can use iterative methods to solve (54). The LSQR algorithm of Paige and Saunders [49] appears to be very well suited. At a high level, and in exact arithmetic, it has the following properties. It solves (54) in $m + 1$ iterations. Every iteration evaluates two linear mappings

$$(v_1, \ldots, v_m) \mapsto \sum_{i=1}^{m} v_i \tilde{F}_i, \text{ and } W \mapsto (Tr\tilde{F}_1 W, \ldots, Tr\tilde{F}_m W) \tag{55}$$

on a vector v and a symmetric matrix $W = W^T$. When the matrices F_i are unstructured, these two operations take mn^2 operations. Hence, the cost of solving (54) using LSQR is $O(n^2 m^2)$, and nothing is gained over direct methods.

In most cases, however, the two operations (55) are much cheaper than mn^2 because of the special structure of the matrices F_i. A well-known example is a sparse LP. Iterative methods in interior-point algorithms for sparse LPs are addressed by Mehrotra [34], Kim and Nazareth [30], and Gill, Murray, Ponceleon and Saunders [22].

Sparsity is not the only example, however. The equations are often dense, but still highly structured in the sense that the (55) are easy to evaluate. Reference [60] discusses iterative methods for exploiting structure in PDPs arising in control theory.

Iterative methods have one important drawback. Due to round-off errors, convergence can be slow unless preconditioning or some type of reorthogonalization is used. The implementation of iterative methods is therefore very problem-dependent.

5.2 Generalized Eigenvalue Problems

As a final note, we would like to mention that great progress has recently been made in interior-point methods for generalized linear-fractional problems

$$
\begin{aligned}
&\text{minimize} \quad t \\
&\text{subject to} \quad tB(x) - A(x) \geq 0, \\
&\qquad\qquad\quad B(x) \geq 0,
\end{aligned}
$$

where $B(x)$ and $A(x)$ are affine in x. This problem is quasiconvex, and generalizes the PDP

$$
\begin{aligned}
&\text{minimize} \quad t \\
&\text{subject to} \quad tI - A(x) \geq 0.
\end{aligned}
$$

See Boyd and El Ghaoui [7], and Nesterov and Nemirovsky [42, 36] for details.

Acknowledgment

We want to thank Hervé Lebret who helped us with an early draft of this paper, and also coded and tested most of the algorithms described here. We also thank Michael Grant and Benoît Lecinq who have been involved in several software projects related to this work.

L. Vandenberghe is Postdoctoral Researcher of the Belgian National Fund for Scientific Research (NFWO). His research was supported in part by the Belgian program on interuniversity attraction poles (IUAP 17 and 50) initiated by the Belgian State, Prime Minister's Office, Science Policy Programming. The research of S. Boyd was supported in part by AFOSR (under F49620-92-J-0013), NSF (under ECS-9222391), and ARPA (under F49620-93-1-0085).

References

[1] F. Alizadeh, J. P. A. Haeberly, and M. L. Overton. "A new primal-dual interior-point method for semidefinite programming," *Proceedings of the Fifth SIAM Conference on Applied Linear Algebra, Snowbird, Utah*, June 1994.

[2] F. Alizadeh, *Combinatorial Optimization with Interior Point Methods and Semi-Definite Matrices*, Ph.D. thesis, Univ. of Minnesota, October 1991.

[3] F. Alizadeh, "Combinatorial optimization with semidefinite matrices," *Proceedings of Second Annual Integer Programming and Combinatorial Optimization conference*, Carnegie-Mellon University, 1992.

[4] F. Alizadeh, "Optimization over the positive-definite cone: interior point methods and combinatorial applications," *Advances in Optimization and Parallel Computing*, Panos Pardalos (editor), North-Holland, 1992.

[5] J. C. Allwright, "Positive semidefinite matrices: characterization via conical hulls and least-squares solution of a matrix equation," *SIAM Journal on Control and Optimization*, 26(3), 1988, 537-556.

[6] J. Allwright, "On maximizing the minimum eigenvalue of a linear combination of symmetric matrices," *SIAM J. on Matrix Analysis and Applications*, 10 (1989), 347-382.

[7] S. Boyd and L. El Ghaoui, "Method of centers for minimizing generalized eigenvalues," *Linear Algebra and Applications, special issue on Numerical Linear Algebra Methods in Control, Signals and Systems*, 188 July (1993), 63-111.

[8] S. Boyd, L. El Ghaoui, E. Feron, and V. Balakrishnan, *Linear Matrix Inequalities in System and Control Theory*, volume 15 of *Studies in Applied Mathematics*, SIAM, Philadelphia, PA, June 1994.

[9] R. Bellman and K. Fan, "On systems of linear inequalities in Hermitian matrix variables," V. L. Klee (editor), *Convexity*, 7 of *Proceedings of Symposia in Pure Mathematics*, American Mathematical Society, 1963, 1-11.

[10] A. Ben Tal and A. Nemirovskii, "Interior point polynomial time method for truss topology design," Technical Report 3/92, Faculty of Industrial Engineering and Management, Technion, Israel Institute of Technology, June 1992.

[11] A. Ben-Tal and M. P. Bendsøe, "A new method for optimal truss topology design," *SIAM Journal on Optimization*, 3 (1993), 322-358.

[12] J. Cullum, W. Donath, and P. Wolfe, "The minimization of certain nondifferentiable sums of eigenvalues of symmetric matrices," *Math. Programming Study*, 3 (1975), 35-55.

[13] B. Craven and B. Mond, "Linear programming with matrix variables," *Linear Algebra and Appl.*, 38 (1981), 73-80.

[14] D. den Hertog, *Interior Point Approach to Linear, Quadratic and Convex Programming*, Kluwer, 1993.

[15] I. Dikin, "Iterative solution of problems of linear and quadratic programming," *Soviet Math. Dokl.*, 8(3) 1967, 674-675.

[16] M. K. H. Fan, "A quadratically convergent algorithm on minimizing the largest eigenvalue of a symmetric matrix," *Linear Algebra and Appl.*, (1993) 188-189:231-253.

[17] R. Fletcher, "A nonlinear programming problem in statistics (educational testing)," *SIAM Journal on Scientific and Statistical Computing*, 2(3) 1981, 257-267.

[18] R. Fletcher, "Semidefinite matrix constraints in optimization," *SIAM J. Control and Opt.*, 23 (1985), 493-513.

[19] A. Fiacco and G. McCormick, *Nonlinear programming: sequential unconstrained minimization techniques*, Wiley, 1968. Reprinted in the *SIAM Classics in Applied Mathematics series* 1990.

[20] M. K. H. Fan and B. Nekooie, "On minimizing the largest eigenvalue of a symmetric matrix," *Proc. IEEE Conf. on Decision and Control*, December 1992, 134-139.

[21] G. Golub and C. Van Loan, *Matrix Computations*, Johns Hopkins Univ. Press, Baltimore, second edition, 1989.

[22] P. E. Gill, W. Murray, D. B. Ponceleon, and M. A. Saunders, "Preconditioners for indefinite systems arising in optimization," *SIAM J. on Matrix Analysis and Applications*, 13 (1992), 292-311.

[23] C. C. Gonzaga, "Path-following methods for linear programming," *SIAM Review*, 34(2) June 1992, 167-224.

[24] C. Goh and D. Teo, "On minimax eigenvalue problems via constrained optimization," *Journal of Optimization Theory and Applications*, 57(1) 1988, 59-68.

[25] C. C. Gonzaga and M. J. Todd, "An $O(\sqrt{n}L)$-iteration large-step primal-dual affine algorithm for linear programming," *SIAM Journal on Optimization*, 2(3) August 1992, 349-359.

[26] J. B. Hiriart-Urruty and D. Ye, "Sensitivity analysis of all eigenvalues of a symmetric matrix," Technical report, Univ. Paul Sabatier, Toulouse, 1992.

[27] F. Jarre, "An interior-point method for minimizing the maximum eigenvalue of a linear combination of matrices," *SIAM J. Control and Opt.*, 31(5) September 1993, 1360-1377.

[28] N. Karmarkar, "A new polynomial-time algorithm for linear programming," *Combinatorica*, 4(4) 1984, 373-395.

[29] M. Kojima, N. Megiddo, T. Noma, and A. Yoshise, *A Univied Approach to Interior Point Algorithms for Linear Complementarity Problems*, Lecture Notes in Computer Science, Springer-Verlag, 1991.

[30] K. Kim and J. L. Nazareth, "Implementation of a primal null-space affine scaling method and its extensions," Technical report, Department of Pure and Applied Mathematics, Washington State University, January 1992.

[31] B. Lieu and P. Huard, "La méthode des centres dans un espace topologique," *Numerische Mathematik*, 8 (1966), 56-67.

[32] I. J. Lustig, R. E. Marsten, and D. F. Shanno, "Interior point methods for linear programming: Computational state of the art," *ORSA Journal on Computing*, 6(1) 1994.

[33] L. Lovász, "On the Shannon capacity of a graph," *IEEE Transactions on Information Theory*, 25 (1979), 1-7.

[34] S. Mehrotra, "Implementations of affine scaling methods: approximate solutions of systems of linear equations using preconditioned conjugate gradient methods," *ORSA Journal on Computing*, 4(2) Spring 1992, 103-118.

[35] S. Mizuno, M. J. Todd, and Y. Ye, "On adaptive-step primal-dual interior-point algorithms for linear programming," *Mathematics of Operations Research*, 18 (1993), 964-981.

[36] A. Nemirovski, "Long-step method of analytic centers for fractional problems," Technical Report 3/94, Technion, Haifa, Israel, 1994.

[37] Y. Nesterov, "Primal-dual methods," Seminar at CORE, Université Catholique de Louvain, February 1994.

[38] Yu. Nesterov and A. Nemirovsky, "A general approach to polynomial-time algorithms design for convex programming," Technical report, Centr. Econ. & Math. Inst., USSR Acad. Sci., Moscow, USSR, 1988.

[39] Yu. Nesterov and A. Nemirovsky, *Optimization over positive semidefinite matrices: Mathematical background and user's manual*. USSR Acad. Sci. Centr. Econ. & Math. Inst., 32 Krasikova St., Moscow 117418 USSR, 1990.

[40] Yu. Nesterov and A. Nemirovsky, "Self-concordant functions and polynomial time methods in convex programming," Technical report, Centr. Econ. & Math. Inst., USSR Acad. Sci., Moscow, USSR, April 1990.

[41] Yu. Nesterov and A. Nemirovsky, "Conic formulation of a convex programming problem and duality," Technical report, Centr. Econ. & Math. Inst., USSR Academy of Sciences, Moscow USSR, 1991.

[42] Yu. Nesterov and A. Nemirovsky, "An interior point method for generalized linear-fractional programming," Submitted to *Math. Programming, Series B.*, 1991.

[43] Yu. Nesterov and A. Nemirovsky, "Interior-point polynomial methods in convex programming," *Studies in Applied Mathematics* 13, SIAM, Philadelphia, PA, 1994.

[44] M. Overton, "On minimizing the maximum eigenvalue of a symmetric matrix," *SIAM J. on Matrix Analysis and Applications*, 9(2) 1988, 256-268.

[45] M. Overton, "Large-scale optimization of eigenvalues," *SIAM J. Optimization*, (1992), 88-120.

[46] M. Overton and R. Womersley, "On the sum of the largest eigenvalues of a symmetric matrix," *SIAM J. on Matrix Analysis and Applications*, (1992), 41-45.

[47] M. Overton and R. Womersley, "Optimality conditions and duality theory for minimizing sums of the largest eigenvalues of symmetric matrices," *Mathematical Programming*, 62 (1993), 321-357.

[48] E. Panier, "On the need for special purpose algorithms for minimax eigenvalue problems," *Journal Opt. Theory Appl.*, 62(2) August 1989, 279-287.

[49] C. C. Paige and M. S. Saunders, "LSQR: An algorithm for sparse linear equations and sparse least squares," *ACM Transactions on Mathematical Software*, 8(1) March 1982, 43-71.

[50] F. Pukelsheim, *Optimal Design of Experiments*. Wiley, 1993.

[51] U. T. Ringertz, "Optimal design of nonlinear shell structures," Technical Report FFA TN 1991-18, The Aeronautical Research Institute of Sweden, 1991.

[52] R. T. Rockafellar. *Convex Analysis*. Princeton Univ. Press, Princeton, second edition, 1970.

[53] J. B. Rosen, "Pattern separation by convex programming," *Journal of Mathematical Analysis and Applications*, 10 (1965), 123-134.

[54] F. Rendl, R. Vanderbei, and H. Wolkowicz, "A primal-dual interior-point method for the max-min eigenvalue problem," Technical report, University of Waterloo, Dept. of Combinatorics and Optimization, Waterloo, Ontario, Canada, 1993.

[55] A. L. Shapiro, "Optimal block diagonal ℓ_2 scaling of matrices," *SIAM J. on Numerical Analysis*, 22(1) February 1985, 81-94.

[56] N. Z. Shor, "Cut-off method with space extension in convex programming problems," *Cybernetics*, 13(1) 1977, 94-96.

[57] G. Sonnevend, "An analytical center for polyhedrons and new classes of global algorithms for linear (smooth, convex) programming," *Lecture Notes in Control and Information Sciences*, 84, Springer-Verlag, 1986, 866-878.

[58] G. Sonnevend, "New algorithms in convex programming based on a notion of 'centre' (for systems of analytic inequalities) and on rational extrapolation," *International Series of Numerical Mathematics*, 84 (1988), 311-326.

[59] F. Uhlig, "A recurring theorem about pairs of quadratic forms and extensions: A survey," *Linear Algebra and Appl.*, 25 (1979), 219-237.

[60] L. Vandenberghe and S. Boyd, "Primal-dual potential reduction method for problems involving matrix inequalities," To be published in *Math. Programming*, 1993.

[61] G. A. Watson, "Algorithms for minimum trace factor analysis," *SIAM J. on Matrix Analysis and Applications*, **13**(4) 1992, 1039-1053.

[62] Y. Ye, "An $O(n^3 L)$ potential reduction algorithm for linear programming," *Mathematical Programming*, **50** (1991), 239-258.

[63] Y. Ye, O. Güler, R. A. Tapia, and Y. Zhang, "A quadratically convergent $O(\sqrt{n}L)$-iteration algorithm for linear programming," *Mathematical Programming*, **59** (1993), 151-162.

[64] D. B. Yudin and A. S. Nemirovski, "Informational complexity and efficient methods for solving complex extremal problems," *Matekon*, **13** (1977), 25-45.

[65] A. Yoshise, "An optimization method for convex programs–interior-point method and analytical center," *Systems, Control and Information*, Special issue on Numerical Approaches in Control Theory. In Japanese, **38**(3) March 1994, 155-160.

[66] Y. Zhang and R. A. Tapia, "A superlinearly convergent polynomial primal-dual interior-point algorithm for linear programming," *SIAM J. on Optimization*, **3** (1993), 118-133.